MANUAL OF
STROKE MODELS IN RATS

MANUAL OF STROKE MODELS IN RATS

EDITED BY

YANLIN WANG-FISCHER

CRC Press
Taylor & Francis Group
Boca Raton London New York

CRC Press is an imprint of the
Taylor & Francis Group, an **informa** business

CRC Press
Taylor & Francis Group
6000 Broken Sound Parkway NW, Suite 300
Boca Raton, FL 33487-2742

Printed in the United States of America on acid-free paper
10 9 8 7 6 5 4 3 2 1

International Standard Book Number-13: 978-0-8493-9578-9 (Softcover)

Library of Congress Cataloging-in-Publication Data

Manual of stroke models in rats / editor, Yanlin Wang-Fischer.
p. ; cm.
"A CRC title."
Includes bibliographical references and index.
ISBN 978-0-8493-9578-9 (hardcover : alk. paper)
1. Cerebrovascular disease--Animal models--Handbooks, manuals, etc. 2. Rats as laboratory animals--Handbooks, manuals, etc. 3. Mice as laboratory animals--Handbooks, manuals, etc. I. Wang-Fischer, Yanlin.
[DNLM: 1. Stroke--physiopathology. 2. Stroke--surgery. 3. Disease Models, Animal. 4. Mice. 5. Rats. WL 355 M294 2008]

RC388.5.M357 2008
616.8'1--dc22 2008014631

Visit the Taylor & Francis Web site at
http://www.taylorandfrancis.com

and the CRC Press Web site at
http://www.crcpress.com

Contents

Preface

During the last few years, exciting new insights into mechanisms and treatment of stroke have been obtained from animal experiments. Thus, animal models to induce stroke are of paramount importance as research tools. Although there are many articles in this area published in different journals, no systematic technical book is available.

This book explains in great detail the methods and techniques for accomplishing different stroke models in rats (and some techniques in mice). The contributing authors have included the most recent research information available, as well as generally recognized facts, to make the book relevant and attractive to specialists who work in stroke study.

This book, in many ways, is a very technical book. I trust and intend that this book will be helpful for researchers in this field to get started with experiments in stroke using the rat and mouse as the model animals.

Yanlin Wang-Fischer, Ph.D., M.D.

About the Editor

Yanlin Wang-Fischer, M.D., Ph.D., is a neuropharmacologist. She has worked in the pharmaceutical industry and academic environment for many years.

Dr. Wang-Fischer earned her bachelor's degree and doctor of medicine at Hunan Medical University, Changsha, China. Following clinical fellowship training in cardiovascular and metabolic diseases at Hunan Medical University, where she became an attending physician, she earned her Ph.D. in 1994. She completed her postdoctoral fellowship training at the School of Medicine at Laval University and at Boston University. Dr. Wang-Fischer joined the Howard Hughes Medical Institute at Yale University, New Haven, Connecticut, in 2000. She has contributed to several projects related to animal models and surgeries and produced an instructional video on studies of animals *in vivo* for new trainees.

In 2002, Dr. Wang-Fischer joined the team of neurological disorders at Johnson & Johnson PRD, Raritan, New Jersey. She led projects based on different animal models to discover new drugs for treatment of stroke and neurological diseases.

In 2006, Dr. Wang-Fischer joined Palatin Technologies and oversees projects on nasal and cerebral drug delivery. She has developed a novel technique to quickly assess nasal tolerance.

Dr. Wang-Fischer has published more than 40 scientific articles in prestigious medical journals. Her current interests include neurological, cardiovascular, metabolic, and inflammatory diseases, mainly concerned with physiology and pharmacology *in vivo*.

About the Editor

Contributors

Kunjan R. Dave, Ph.D.
Cerebral Vascular Disease Research Center
Department of Neurology
University of Miami Miller School of Medicine
Miami, Florida

Afshin A. Divani, Ph.D., FAHA
Department of Neurology, Neurosurgery, and Radiology
University of Minnesota
Minneapolis, Minnesota

Lee Koetzner, Ph.D.
Department of Pharmacology
Eurofins—Product Safety Labs
Dayton, New Jersey

John McCool, M.S., RLATG
Palatin Technologies
Cranbury, New Jersey

Miguel A. Perez-Pinzon, Ph.D.
Cerebral Vascular Disease Research Center
Department of Neurology
University of Miami Miller School of Medicine
Miami, Florida

Ricardo Prado, M.D.
Cerebral Vascular Disease Research Center
Department of Neurology
University of Miami Miller School of Medicine
Miami, Florida

Souvik Sen, M.D., M.S., FAHA
UNCH Stroke Center
University of North Carolina
Chapel Hill, North Carolina

Brant D. Watson, Ph.D.
Cerebral Vascular Disease Research Center
Department of Neurology
University of Miami Miller School of Medicine
Miami, Florida

1 Statistical Update of Stroke in America

Yanlin Wang-Fischer and Lee Koetzner

CONTENTS

STROKE INCIDENCE

Much of our understanding of stroke incidence comes from the Heart and Stroke Statistical Update for 2002 and 2006 by the American Heart Association,[1–3] which included a 44-year follow-up study of participants and a 20-year follow-up of their offspring by the Framingham Heart Study.[1] Stroke, the third leading killer of Americans and the leading cause of long-term disability, remains an epidemic. The American Stroke Association estimates that stroke accounts for 1 of every 15 deaths and occurs once every 45 seconds in the United States, while a death from stroke occurs every 3 minutes.[1–3] In 2002, about 600,000 people suffered a stroke; about 500,000 of these were first attacks and 100,000 were recurrent attacks.[1,2] In 2006, this number increased to 700,000, of which 500,000 were first attacks and 200,000 were recurrent attacks.[3] The 2003 costs related to stroke treatment were estimated to be $57.9 billion.[3]

INCIDENCE OF DIFFERENT TYPES OF STROKE

The most common type of stroke is atherothrombotic brain infarction, which accounts for 61% of all strokes (excluding transient ischemic attacks [TIAs]). The next most common is embolic stroke at 22%.[1]

In a 2002 study, the majority of strokes (83%) were ischemic in nature; 10% were due to intracerebral hemorrhage, and 7% were due to subarachnoid hemorrhage. In the 2006 update, these numbers had changed; 88% were ischemic, 9% were due to intracerebral hemorrhage, and 3% were due to subarachnoid hemorrhage.[3] A study report showed that among the 178 definite thrombotic brain infarctions, 38% were classified as lacunar strokes (in small blood vessels)[4]; furthermore, 7.6% of all ischemic strokes resulted in death within 30 days.[4] In addition to a high rate of death, stroke is also the leading cause of serious, long-term disability in the United States.[1–3] With the growing population and the increasing prevalence of related risk factors (like diabetes, obesity, and hypertension), the high rate of stroke-related deaths and disability is likely to go up.

Age, Sex, Race, and Ethnicity

Stroke can happen at any age, from newborn (even in a fetus) to seniors. According to the Heart and Stroke Statistical Update for 2002 and 2006 by the American Heart Association,[1–3] 28% of people who suffered a stroke in a given year were under age 65; for people over age 55, the incidence of stroke more than doubles in each successive decade. The chance of having a stroke before age 70 is 1 in 20 for both men and women.[1–3]

Stroke is more common in men than in women. In most age groups, men have a higher incidence of stroke than women.[1–3] After menopause, the incidence is a little higher in women than in men, indicating that postmenopausal hormones play an important role (refer to Chapter 5). Based on the Heart and Stroke Statistical Update,[3] the age-adjusted stroke incidence rates (per 1000 person-years) are 1.78 for white men, 4.44 for black men, 1.24 for white women, and 3.10 for black women. Blacks have a 38% greater risk of incident (first) strokes than whites.[3]

POSSIBLE TREATMENT

No specific neuroprotective treatment for stroke is available.[5] Fibrinolysis is the only approved and accepted therapy against stroke. However, thrombolytic therapy is restricted to patients who meet strict eligibility criteria,[6] including a tight 3-hour therapeutic time window.[5] This treatment is therefore eliminated for 97% of stroke patients.[6] The use of antithrombotic agents (for example, anticoagulants such as heparin, aspirin, ticlopidine, and platelet inhibitors) is based on their action in combating secondary ischemia by inhibiting the formation of new blood clots and maintaining normal cerebral blood flow; however, this has been found also to trigger increased bleeding in treated stroke patients.[7]

Scientists have not given up the search for new approaches to the detrimental consequences of stroke. Research continues to add depth to the overall understanding of the condition, including the discovery of new risk factors, new methods of screening, and most importantly, developing new drugs and improving other treatment regimens. Animal models of stroke fulfill a critical need for those purposes.

REFERENCES

1. American Heart Association, *2002 Heart and Stroke Statistical Update*, American Heart Association, Dallas, 2002.
2. American Heart Association in collaboration with the International Liaison Committee on Resuscitation, Guidelines 2000 for cardiopulmonary resuscitation and emergency cardiovascular care: An international consensus on science AHA/ILCOR guidelines: Part 7, the era of reperfusion; Section 2, acute stroke, *Circulation*, 102, I1–I384, 2000.
3. Writing Group: Thom, T., Haase, N., Rosamond, W. et al., and members of the Statistics Committee and Stroke Statistics Subcommittee: Adams, R., Friday, G., Furie, K. et al., AHA statistical update, heart disease and stroke statistics—2006 update. A report from the American Heart Association Statistics Committee and Stroke Statistics Subcommittee, *Circulation,* 113, e85, 2006.
4. Rosamond, W.D. et al., Stroke incidence and survival among middle-aged adults 9-year follow-up of the atherosclerosis risk in communities (ARIC) cohort, *Stroke*, 30, 736, 1999.
5. Alberts, M.J., tPA in acute ischemic stroke. United States experience and issues for the future, *Neurology*, 51 Suppl. 3, S53, 1998.
6. Albers, G.W. et al., Antithrombotic and thrombolytic therapy for ischemic stroke, *Chest*, 114 Suppl. 5, 683S, 1998.
7. Schellinger, P.D., Orberk, E., and Hacke, W., Antithrombotic therapy after cerebral ischemia, *Fortschr Neurol Psychiatr*, 65, 425, 1997.

2 Rationale for Using Ischemic Stroke Models

Yanlin Wang-Fischer and Lee Koetzner

CONTENTS

RATIONALE FOR ISCHEMIC STROKE MODELS

Cerebral ischemia results from a loss of blood supply to a region of the brain due to arterial blockage or hemorrhage.

Global ischemia models reduce blood flow to the entire brain, mimicking cerebral ischemia from cardiac arrest, severe hypotension, or occasionally surgical procedures that alter blood flow. In rats, global ischemia may be produced by two- or four-vessel occlusion or by cardiac arrest. The two-vessel occlusion model involves ligation of both carotid arteries and controls partial exsanguinations to decrease mean arterial pressure to 50 mm Hg. Blood is reinfused after 10 minutes and the ligatures removed.[1] Four-vessel occlusion involves permanent occlusion of the vertebral arteries with transient occlusion of the carotid arteries.[2] The cardiac arrest model attempts to mimic a clinical situation to induce ischemia through the pharmacological induction of cardiac arrest for 7 to 10 minutes, followed by injection of epinephrine and chest compression. These injuries result in neuronal necrosis in specific, vulnerable brain areas, including the cerebral cortex, hippocampus, striatum, and cerebellum. Cell death is complete 3 to 7 days after injury, and because of this delay, therapies can be tested for their ability to prevent cell death.[3]

Focal ischemia *models* are designed to induce ischemia by occlusion or blockage of major arteries supplying blood to the brain, most commonly middle cerebral artery occlusion (MCAO). Techniques include MCAO by suture ligation,[4] electric cauterization,[5] intraluminal filament occlusion,[6] blood clot injection,[7] or photochemical thrombosis.[8] The infarction from this occlusion can be found in two areas by histological staining: the core infarct area, which is in the center of the infarction, and the penumbra area, which surrounds the core area. The penumbra area is thought to be reversible. The damage occurs in two phases: (1) immediate necrotic cell death that is thought to be irreversible (core area) and (2) delayed secondary damage that can occur for days following the occlusion (penumbra). This secondary damage provides a window during which therapeutic intervention might limit the ultimate damage associated with an ischemic event. A variety of molecules have been studied for their ability to block secondary damage, including NMDA (*N*-methyl-d-aspartate) receptor antagonists, inhibitors of apoptosis, and free-radical scavengers. This model facilitates *in vivo* evaluation of the efficacy of molecules to prevent cell death and observe the resultant behavioral deficits and to optimize the timing and dosages of drug regimens before or in concert with clinical trials.[9]

WHY USE ANIMAL MODELS OF STROKE?

First, the characteristics of stroke in humans are similar to experimental models in animals. Focal vascular occlusion results in an infarct with topography that is determined by the distribution of the occluded vessel and the nature and extent of collateral channel activation. In both settings, the volume of injury is determined by the severity of ischemia (that is, the degree of cerebral blood flow [CBF] decline) and the duration of ischemia. Animals and humans have similar neural and vascular substrates (see anatomy discussion in Chapter 4) and similar biochemical and molecular mechanisms of injury.[10]

Second, the presence of an intact system of cerebral vasculature *in vivo* is essential to the study of abnormal brain perfusion. Because of the complexity of the brain and its response to injury, cerebral ischemia and its consequences cannot be evaluated with *in vitro* models alone. There is no computerized model available for this purpose. Preclinical research into the causes, pathogenesis, and therapeutic management of stroke therefore requires the use of animal models in addition to other techniques and models *in vitro*, such as tissue culture and brain slices.

Third, experimental animal models of stroke are designed to allow investigators to carefully re-create specific aspects of human stroke and study pathophysiologic and neuroprotective mechanisms as well as therapeutic responses under controlled conditions and in ways that cannot be done easily or at all in clinical patients. Thus, more rigorous histopathologic, biochemical, and physiologic measurements of stroke must be done in animals.

Finally, animal models allow investigators to study immediate and early ischemic events, events that can be difficult to examine in human patients because of the variable time delays in early recognition of a stroke and initial therapeutic intervention.

REFERENCES

1. Dietrich, W.D. et al., Thromboembolic events predispose the brain to widespread cerebral infarction after delayed transient global ischemia in rats, *Stroke*, 30, 855, 1999.
2. Poder, P. et al., An antioxidant tetrapeptide UPF1 in rats has a neuroprotective effect in transient global brain ischemia, *Neurosci Lett,* 370, 45, 2004.
3. Dietrich, W.D., Busto, R., and Bethea, J.R., Postischemic hypothermia and IL-10 treatment provide long-lasting neuroprotection of CA1 hippocampus following transient global ischemia in rats, *Exp Neurol,* 158, 444, 1999.
4. Wang, Y. et al., Diadenosine tetraphosphate protects against injuries induced by ischemia and 6-hydroxydopamine in rat brain, *J Neurosci*, 23(21), 7958, 2003.
5. Wang-Fischer, Y.L. et al., Refined technique for inducing and grading middle cerebral artery occlusion in rat stroke model. *American Association of Laboratory Animal Science 54th National Meeting,* Seattle, platform sessions speaker PS 43, 2003, October 11–16, and *AALAS Tri-branch Symposium, Biotechnology in the 21st Century and Beyond*, presenting poster 10, Philadelphia, June 8–10, 2003.
6. Belayev, L. et al., Middle cerebral artery occlusion in the rat by intraluminal suture: neurological and pathological evaluation of an improved model, *Stroke*, 27, 1616, 1996.
7. Zhang, R.L. et al., A rat model of focal embolic cerebral ischemia, *Brain Res,* 766, 83, 1997.
8. Alexis, N.E. et al., Neurobehavioral consequences of induced spreading depression following photothrombotic middle cerebral artery occlusion, *Brain Res,* 706, 273, 1996.
9. Renzi, M.J., Wang-Fischer, Y.L., and Farrell, F.X., An expanded window of opportunity for erythropoietin in stroke recovery: Separation of behavioral outcome from infarct size, Abstract No. 741.8, *Society for Neuroscience Annual Meeting*, Washington, DC, 2003.
10. Yamori, Y. et al., Pathogenetic similarity of strokes in stroke-prone spontaneously hypertensive rats and humans, *Stroke,* 7(1), 46, 1976.

3 Animal Models of Ischemic Stroke

A Historical Survey

Yanlin Wang-Fischer and Lee Koetzner

CONTENTS

HISTORY OF THE DEVELOPMENT OF STROKE MODELS

Animal experiments to produce neurological symptoms and signs of stroke have now been performed for about a half century. A number of experimental models have been developed to model this condition preclinically, including multivessel occlusion, hypotension due to hypovolemia, and middle cerebral artery occlusion (MCAO). Stroke is typically classified as either hemorrhagic or ischemic. Ischemic stroke is the most common type, both in clinical practice and in experimental studies. Different animals have been used in stroke models, including mice, rats, cats, rabbits, swine, dogs, and monkeys, but the rat is the most commonly used species.

In 1955, Hill and colleagues[1] were the first to describe injections of homologous blood clots for induction of stroke in a dog model. Since then, a wide variety of models for stroke have been proposed (see reviews by Del Zoppo, 1990[2] and Overgaard, 1994[3]). In the past 20 years, models of focal cerebral ischemia have been used most frequently. MCAO is accepted for modeling focal ischemic stroke due to its relevance to human stroke. The techniques of MCAO models include local direct occlusion of the distal middle cerebral artery (MCA) through a craniectomy,[4,5] by photochemically induced thrombus formation,[6] or by indirect occlusion of the proximal MCA from the internal carotid artery, including intraluminal suture emplacement[7] and embolic cerebral ischemia.[8]

FOCAL MIDDLE CEREBRAL ARTERY (MCA) OCCLUSION THROUGH CRANIECTOMY

Direct occlusion of an intracranial artery, usually the proximal MCA, is widely used to produce focal ischemia in large animals such as cats,[9] dogs,[10] and nonhuman primates.[11]

In 1975, Robinson and colleagues[4] adapted this technique to rats. In 1981, Tamura's group[12] provided a detailed description of this technique. Since the species are different (rats have more collateral braches to support retrograde flow),[12] a technique of permanent occlusion of the ipsilateral

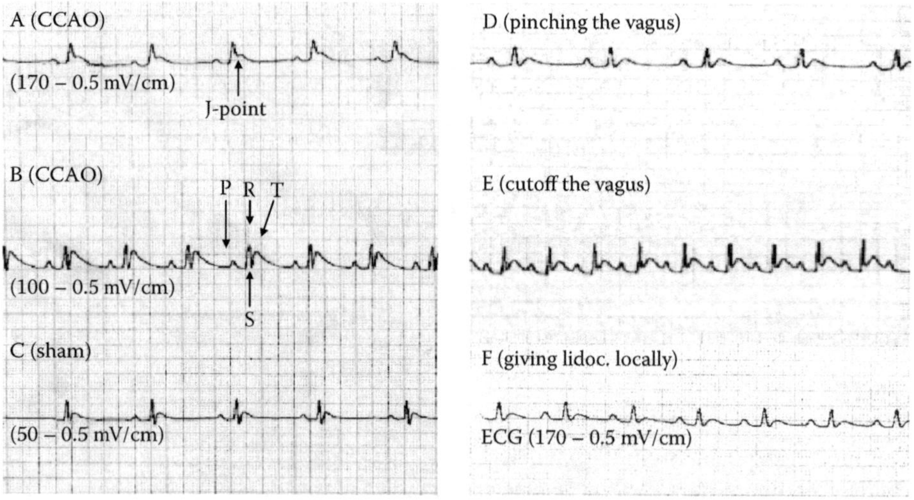

FIGURE 3.1 Electrocardiogram (ECG) showing J points elevated to 0.15 to 0.2 mV (A) and high T waves (B) in rats with common carotid artery (CCA) occlusion; brachycardia in a rat with stimulation of the vagus nerve (D); and elevated J point as well as tachycardia (450 to 500/minute) in a rat with cutoff vagus nerve (E). Locally giving lidocaine (lidoc.) blocked the ECG changes from the stimulation or damage (F). (C) A normal rat with sham surgery.

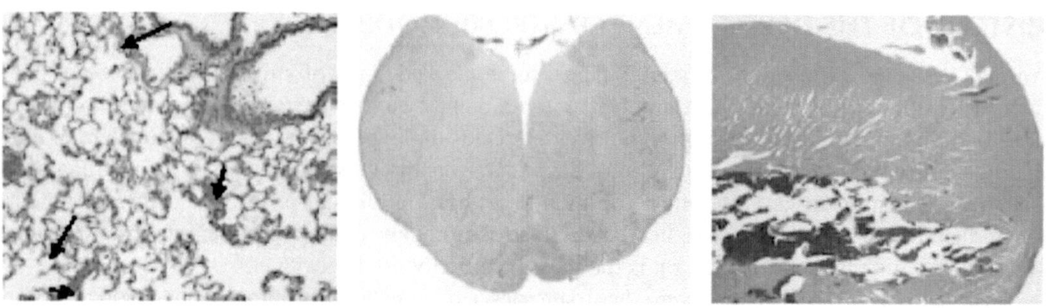

(a) (b) (c)

FIGURE 3.2 The pathological examination of the lungs (a) showed pulmonary edema (arrows, fluid in the alveoli) and congestion (arrows, blood in small arteries and capillaries, hematoxylin and eosin stain, ×20). No brainstem bleeding (b) or morphological myocardial damage (c) was observed. **(See color insert following page 146.)**

common carotid artery (CCA) with temporary clipping of the contralateral CCA was combined with focal direct MCAO.[13] This technique made this model more relevant to the clinical situation.[14]

The added procedure causes the complication of acute heart failure. In a normal situation, the heart rate is controlled by both sympathetic and parasympathetic nerves. Impulses in the sympathetic (noradrenergic) nerves to the heart increase the cardiac rate and the force of cardiac contraction. Impulses in the parasympathetic nerves (cholinergic vagal cardiac fibers) decrease the heart rate. The dissection of the soft tissues at the cervical area and occlusion of the CCAs during the surgical procedure jeopardized the balance of sympathetic and parasympathetic tone on the heart. Damage to the parasympathetic nerves during surgical procedures increases the heart rate and leads to tachycardia. The animals that experienced this complication exhibited stereotypical symptoms, including rapid heartbeat, asthma-like difficult breathing (cardiac asthma), pink bubbly discharge from the nose or mouth, wheezing, and cyanosis. The incidence of this problem was reported to be about 30%, and half of the rats having this complication died.[15] Large differences were observed between animal

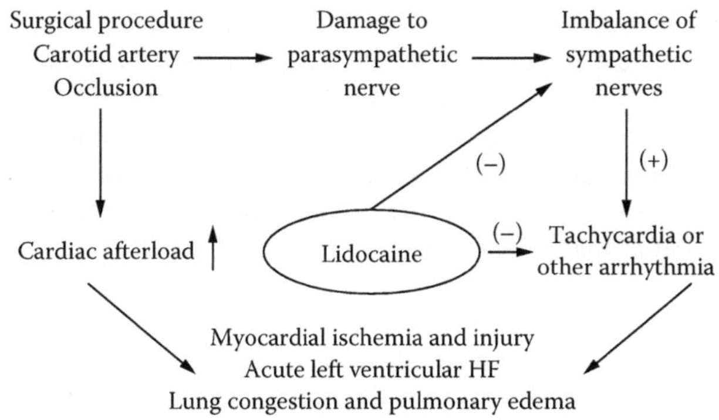

FIGURE 3.3 The mechanism and treatment for acute ventricular heart failure (HF) and respiratory distress (RD).

batches, from 0% to 57% in 11 shipments (12 to 20 rats/shipment), suggesting that the susceptibility to this complication was animal dependent and not related to the skill of the surgeon. This complication not only decreased animal survival but also increased the infarct variability.[15]

The hypothesized mechanism is supported by electrocardiogram (ECG) analysis (Figure 3.1) and pathological examination of lungs, brain, and heart (Figure 3.2). This complication could be overcome by local application of 0.25% lidocaine or bupivacaine at the cervical incision prior to CCA occlusion, leading to increased animal survival and decreased variability. The hypothesized mechanism and treatment for this complication is shown schematically in Figure 3.3.

THROMBOEMBOLIC STROKE MODEL

Thromboembolic stroke in rats was first described in 1982 by Kudo et al.[16] and in 1985 by Kaneko et al.[17] They used a suspension of clot fragments with some modifications to the composition of the resultant emboli. This model was commonly used to assess outcome of thrombolytic therapy (1994).[18] However, the size of blood clot fragments in suspension is difficult to control. Smaller pieces of blood clots may float into peripheral artery branches, causing microembolization, or a big clot may occlude the entire arterial tree. Kudo et al.[16] showed that, in addition to embolization of the MCA territory, scattered, mostly microscopic, lesions occurred in the territories of the anterior and posterior cerebral arteries (25%), even on the contralateral side (8%). The distribution of clot material is different from human stroke, for which the proximal segment is occluded, but the distal branches remain open.[16] This difference could be of importance in determining the pattern of reperfusion after thrombolytic therapy and cannot be neglected.

To overcome the problems of clot suspension, Sereghy et al. (1993)[19] used single macroclots for embolization of the rat MCA. However, to carry sufficient clot material, the single clots must be bigger than multiple clots. The larger macroclots may become trapped too far away from the MCA origin to ensure MCA occlusion. The difficulty of placing a single clot exactly at the MCA origin is the most likely explanation for the insufficient cerebral blood flow (CBF) reduction in this model to only 50% to 75% of baseline.[19]

Zhang et al. (1997)[20] described MCA occlusion with injection of a single clot 25 mm long through an internal carotid artery to the origin of its respective MCA. To perform this delicate technique, they modified a PE50 catheter with a very small end (0.3-mm diameter), which can be inserted close to the MCA origin. In this study, the infarct was limited to the MCA distribution area, and regional CBF was initially reduced to 42% of baseline, but it recovered to 66% at 2 hours after injection, indicating spontaneous reperfusion. This could cause problems for those investigating thrombolysis because it may be difficult to distinguish between spontaneous and therapeutically induced changes. This technique is very difficult to master and is likely to be operator dependent.

Busch and coworkers[21] refined this model in 1997 by injection of a small number of medium-size clots into a rat internal carotid artery. Visual inspection confirmed that the size of these clots (1.5 × 0.35 mm) was small enough to reach the MCA origin, and that the clots were trapped in the proximal MCA and anterior cerebral artery (ACA) portions. In this model, the infarct was distributed in the posterior, middle, and ACA territories, similar to a suture ischemia model. The clots were not produced under arterial pressure. These clots are looser in composition and more flexible in shape. This flexibility may be an advantage over more rigid arterial clots because it allows dense packing of clot material for complete obstruction of blood flow. They found that by washing and selection of appropriate clot segments, a fibrin content of over 70% could be reached. A high fibrin content of clots ensures durability and thereby reduces the risk of early spontaneous lysis.[19] This technique used a PE50 catheter without modification of the tip to insert to the internal carotid artery through the external carotid artery and is easier to handle the modified PE50 catheter. This approach seems the most acceptable one for an embolic stroke model, although it does not completely mimic human clinical events.

PHOTOCHEMICAL THROMBOTIC STROKE MODEL

Watson, a physicist and neurobiologist, first invented brain infarction by photochemically initiated thrombosis, by which reproducible cortical infarcts can be induced in rats.[22] In the original procedure, the photosensitive dye rose bengal is administered intravenously, and focal illumination of the (translucent) skull is performed using an optically filtered arc lamp. The light–dye interaction generates highly reactive singlet molecular oxygen,[23] which directly peroxidizes structural proteins and lipids in the endothelial cells of the underlying vasculature. Such endothelial damage stimulates intense platelet aggregation concomitant with endothelial leakiness, and the resultant thrombosis or severe edema leads to vascular occlusion or compression. Thus, the first version of this essentially noninvasive model was intended to evaluate the consequences of the cerebral infarction and not the thrombi per se. The advantage of the photochemical approach is its adaptability to any particular cortical location that one might want to infarct experimentally. In the first model (1985), an arc lamp was used for excitation of rose bengal, which induces brain cortical infarction by occluding capillaries and arterioles. The exceptional reproducibility of this model arises from the fact that the tissue density of these small vessels is the same in every rat, unlike the distribution for any arterial tree, which is extremely variable. There is a caveat to this approach, however, in that the fundamental response to the photochemical injury is endothelial leakiness (at least in the microvasculature), not platelet aggregation, so a lesion can be formed merely by microvascular compression owing to the severe edema that this model produces.

Watson and his group[24] showed in 1987 that a reproducible cortical lesion could be formed without any aggregated platelets. They did this by cooling the rat brain to 34°C, and this actually "froze" the platelets and prevented them from reacting to the severe endothelial injury. Several groups[25,26] adapted this technique and tried drugs against it, but the results were mixed because penetration of the drugs was limited owing to the edema and microvascular compression. To overcome this drawback, Watson found a method of inducing the cortical lesion without producing edema, which in retrospect was a major breakthrough. The brain is irradiated with an argon laser beam (instead of an arc lamp), which approaches the skull at 5° from horizontal and skims the surface of the parietal bone. With this beam, only the very surface of the cortex is irradiated, and all the occlusions thus appear there. The lesion appears as a uniform red color (not mottled as is the usual case) because the pial vessels have retained their distal blood, without it being squeezed out by edematous compression, and the blood stagnates in place.

Watson and colleagues then adapted the photochemical method to occlusion of the rat MCA.[27] An arc lamp beam was believed to be unsuitable for this purpose because it cannot be focused to the requisite dimension (200 μm), so a 514-nm argon ion laser was used instead. With a beam intensity of greater than 30 W/cm^2, the arterial endothelium could be photochemically damaged by the rose bengal reaction at such a rate that it caused platelets to aggregate explosively (without appreciably diluting their granular secretions) and therefore quickly to form an occlusive thrombus in the artery. It became clear that while platelets themselves are inactivated by peroxidation, they are greatly stimulated by peroxidized endothelium. Thus, in the context of photochemical damage, the functional complementarity of endothelium to platelets appears to be preserved in that thrombogenic platelets become nonthrombogenic, while nonthrombogenic endothelium becomes thrombogenic. Thrombus formation using the photochemical method in the MCA application is invasive because the overlying skull must be removed to visualize the MCA. However, the MCA can indeed be occluded by irradiating through the skull if its exact location can be determined.

As mentioned in Chapter 2, the stroke infarct occurs in two phases or two zones: (1) an infarct core that is defined by immediate and irreversible necrotic cell death and (2) a penumbra that surrounds the core and provides a window where therapeutic intervention might limit the ultimate damage. To develop a stroke model with a predefined penumbra, in 1995 Wester, Watson, and coworkers[6] invented the ring cortical model based on the photochemical technique. The exposed crania of erythrosine B-injected rats were irradiated with a 514.5-nm laser beam, configured as a 5-mm diameter and 0.5-mm-thick annulus, to yield a ring-shaped lesion caused by photochemically induced platelet occlusion of cortical vasculature. The idea was to create the penumbra by modeling its development from the outside in, instead of the usual, irreproducible inside-out development. In focal ischemia models based on cerebral arterial occlusion, the fact that perifocal tissue can be salvaged pharmacologically indicates that an amenable penumbra exists, but its precise characterization is difficult owing to individual variation. To remedy this situation, they observed that ischemia would develop reproducibly in the region interior to the ring lesion (the predesignated "zone at risk") but over a greatly extended time period compared with the original cortical method. This time dilation feature is achieved by initially circumscribing with a photothrombotically induced thin rim of ischemic tissue configured as a ring. From this initial state, ischemia proceeds toward the ring center as a concentric annulus. The enclosed tissue, while being perfused by pial vasculature that penetrates the ring, thus undergoes sequential metabolic and electrophysiological changes presumed to be involved in the development of a penumbra.

Subsequent studies in the rat model used a variant of the photothrombotic technique in a CCA to facilitate distal embolization and widespread cerebral infarction.[28,29]

FILAMENT STROKE MODEL

In 1986, Koizumi and colleagues[30] reported a novel, relatively noninvasive method of achieving reversible MCAO by the use of an intraluminal suture. Subsequently, in 1989 Longa and colleagues[7] reported a variation of this method and stated that their technique reliably produced regional infarcts. Nonetheless, brain injury (brain edema and infarct) produced by MCAO in this model varies considerably in its size and distribution; this variability in infarct volume from animal to animal necessitates the use of large numbers of animals to discern statistical significance in drug testing. To produce consistent infarcts, in 1996 Belayev et al.[31] modified this technique by coating the sutures used for intraluminal MCAO with poly-l-lysine, a polycationic polymerized amino acid, which increased adhesive forces around the suture. The polycationic poly-l-lysine molecules adsorb strongly to solid surfaces, leaving exposed cationic sites that combine with the anionic sites on the vascular endothelial surface.[32] This modification made this model more reproducible.

Intraluminal MCA occlusion can cause persistent hyperthermia in rats (lasting for 24 hours), and the hyperthermia can be correlated with ischemia in the hypothalamus and preoptic area.[33] To clarify the effects of hypothalamic artery (HTA) ischemia on body temperature and to obtain a model simulating lacunar infarction, in 1999 He et al.[34] attempted to produce small infarcts in deep structures (including the hypothalamus). They modified the method, by which a surgical suture was advanced to occlude the origin of the HTA or anterior choroidal artery (AChA) without compromise of the anterior or MCA origins. The rectal temperature and postural reflex were examined for 3 days under nonanesthetic conditions. The AChA and HTA and their link with small deep infarction were then confirmed by triphenyltetrazolium chloride (TTC), hematoxylin and eosin, and TUNEL stains (detection of DNA fragments in apoptotic cells using the terminal deoxyribonucleotidyl transferase [TdT]-mediated biotin-16-dUTP nick-end labeling) and by microsurgical dissection after colored silicone perfusion into the cerebral arteries. Their data showed that the advancement of the suture near to but not occluding the MCA origin (0.5 to 1.9 mm proximal) produced small, deep, non-neocortical strokes in 25 of 36 animals without producing MCA ischemic changes. These infarctions mainly affected the hypothalamus in 13 animals (HTA area: infarct volume 6 ± 1 mm^3) and involved both the internal capsule and hypothalamus in 12 animals (HTA plus AChA area infarct volume 48 ± 10 mm^3). Rats with HTA infarction alone exhibited persistent hyperthermia for 72 hours; the AChA plus HTA infarct group showed a transient elevation of body temperature for 24 hours. In the remaining 11 animals, the suture was inadvertently advanced across the MCA origin, producing a large infarct that affected both the neocortex and nonneocortical structures. The MCA infarct group displayed transient hyperthermia and severe postural abnormality. Their study suggested that the intraluminal suture method permits selective AChA or HTA obstruction without inducing MCA territory ischemia. This model confirms that selective HTA infarction produces significant and sustained temperature regulation abnormalities. The model also may be useful in investigating the pathophysiology of small, deep, end-vessel infarction.

REFERENCES

1. Hill, N.C. et al., Studies in cerebrovascular disease. VII. Experimental production of cerebral infarction by intracarotid injection of homologous blood clot. Preliminary report, *Mayo Clin Proc,* 30, 625, 1955.
2. Del Zoppo, G.J., Relevance of focal cerebral ischemia models. Experience with fibrinolytic agents, *Stroke,* 21 Suppl. 4, 155, 1990.
3. Overgaard, K., Thrombolytic therapy in experimental embolic stroke, *Cerebrovasc Brain Metab Rev,* 6, 257, 1994.
4. Robinson, R.G. et al., Effect of experimental cerebral infarction in rat brain on catecholamines and behavior, *Nature,* 255, 332, 1975.
5. Tamura, A. et al., Focal cerebral ischemia in the rat: Description of technique and early neuropathological consequences following middle cerebral artery occlusion, *J Cereb Blood Flow Metab,* 1, 53, 1981.
6. Wester, P. et al., A photothrombotic "ring" model of rat stroke-in-evolution displaying putative penumbral inversion, *Stroke*, 26, 444, 1995.
7. Longa, E.Z. et al., Reversible middle cerebral artery occlusion without craniectomy in rats, *Stroke,* 20(1), 84, 1989.
8. Zhang, Z. et al., A new rat model of thrombotic focal cerebral ischemia, *J Cereb Blood Flow Metab,* 17, 123, 1997.
9. Sundt, T.M. and Waltz, A.G., Experimental cerebral infarction: Retro-orbital, extradural approach for occluding the middle cerebral artery, *Mayo Clin Proc,* 41, 159, 1966.
10. Suzuki, J. et al., Production of various models of cerebral infarction in the dog by means of occlusion of intracranial trunk arteries, *Stroke,* 11, 337, 1980.
11. Hudgins, W.R. and Garcia, J.H., Transorbital approach to the middle cerebral artery of the squirrel monkey: a technique for experimental cerebral infarction applicable to ultrastructural studies, *Stroke,* 1, 107, 1970.
12. Tamura, A. et al., Focal cerebral ischaemia in the rat: description of technique and early neuropathological consequences following middle cerebral artery occlusion, *J Cereb Blood Flow Metab,* 1, 53, 1981.
13. Brint, S. et al., Focal brain ischemia in the rat: methods for reproducible neocortical infarction using tandem occlusion of the distal middle cerebral and ipsilateral common carotid arteries, *J Cereb Blood Flow Metab*, 8(4), 474, 1988.
14. Brines, M.L. et al., Erythropoietin crosses the blood–brain barrier to protect against experimental brain injury, *Proc Natl Acad Sci USA*, 97(19), 10526, 2000.

15. Wang-Fischer, Y.L. et al., Refined technique for inducing and grading middle cerebral artery occlusion in rat stroke model. *American Association of Laboratory Animal Science 54th National Meeting*, Seattle, platform sessions speaker PS 43 2003, October 11–16 and *AALAS Tribranch Symposium, Biotechnology in the 21st Century and Beyond*, presenting poster 10, Philadelphia, June 8–10, 2003.
16. Kudo, M. et al., An animal model of cerebral infarction, *Stroke,* 13, 505, 1982.
17. Kaneko, D., Nakamura, N., and Ogawa, T., Cerebral infarction in rats using homologous blood emboli: development of a new experimental model, *Stroke,* 16, 76, 1985.
18. Overgaard, K., Thrombolytic therapy in experimental embolic stroke, *Cerebrovasc Brain Metab Rev,* 6, 257, 1994.
19. Sereghy, T., Overgaard, K., and Boysen, G., Neuroprotection by excitatory amino acid antagonist augments the benefit of thrombolysis in embolic stroke in rats, *Stroke,* 24, 1702, 1993.
20. Zhang, Z. et al., A new rat model of thrombotic focal cerebral ischemia, *J Cereb Blood Flow Meta,* 17, 123, 1997.
21. Busch, E., Krüger, K., and Hossmann, K.A., Improved model of thromboembolic stroke and rt-PA induced reperfusion in the rat, *Brain Res,* 778(1), 16, 1997.
22. Watson, B.D. et al., Induction of reproducible brain infarction by photochemically initiated thrombosis, *Ann Neurol,* 17(5), 497, 1985.
23. Foote, C.S., Photosensitized oxidation of singlet oxygen: consequences in biological systems. In: Pryor, W.A., ed., *Free Radicals in Biological Systems*, Academic Press, New York, 1976, pp. 85–133.
24. Dietrich, W.D. et al., Photochemically induced cerebral infarction, *Acta Neuropathol (Berl),* 72, 315, 1987.
25. Schroeter, M., Jander, S., and Stoll, G., Non-invasive induction of focal cerebral ischemia in mice by photothrombosis of cortical microvessels: Characterization of inflammatory responses. *J Neurosci Meth,* 117, 43, 2002.
26. Arii, K. et al., The effect of ozagrel sodium on photochemical thrombosis in rat: Therapeutic window and combined therapy with heparin sodium, *Life Sci,* 71, 2983, 2002.
27. Dietrich, W.D. et al., Middle cerebral artery thrombosis: Acute blood-brain barrier consequences, *J Neuropath Exp Neurol,* 47(4), 443, 1988.
28. Futrell, N. et al., A new model of embolic stroke produced by photochemical injury to the carotid artery in the rat, *Ann Neurol,* 23(3), 251, 1988.
29. Dietrich, W.D. et al., Thromboembolic events predispose the brain to widespread cerebral infarction after delayed transient global ischemia in rats, *Stroke*, 30, 855, 1999.
30. Koizumi, J. et al., Experimental studies of ischemic brain edema, I: A new experimental model of cerebral embolism in rats in which recirculation can be introduced in the ischemic area, *Jpn J Stroke*, 8, 1, 1986.
31. Belayev, L. et al., Middle cerebral artery occlusion in the rat by intraluminal suture, *Stroke,* 27, 1616, 1996.
32. Mazia, D., Schatten, G., and Sale, W., Adhesion of cells to surfaces coated with polylysine, *J Cell Biol,* 66, 198, 1975.
33. Zhao, Q. et al., Hyperthermia complicates middle cerebral artery occlusion induced by an intraluminal filament, *Brain Res,* 649, 253, 1994.
34. He, Z. et al., Experimental model of small deep infarcts involving the hypothalamus in rats, changes in body temperature and postural reflex, *Stroke*, 30, 2743, 1999.

4 Anatomy and Cerebral Circulation of the Rat

Yanlin Wang-Fischer, Ricardo Prado, and Lee Koetzner

CONTENTS

ANATOMY OF ARTERIES IN THE CERVICAL AND BRAIN AREAS IN RATS

The anatomy of the arterial and nerve supply in rats is essentially similar to that of humans (Figure 4.1).[1–3] Knowledge of this anatomy is a prerequisite for successfully implementing the stroke models described in this book.

AORTA

The ascending aorta appears as a short segment in the pericardium, from which arise the right and left coronary arteries, which leave the aorta close to its origin and run downward to supply the heart wall.

The arch section of the aorta crosses the ventral trachea, turning dorsally to the left of the trachea, and continues downward as the descending aorta. From it, the innominate, left common carotid, and left subclavian arteries arise. The right subclavian and right common carotid arteries comprise the continuation of the innominate (Figure 4.1).[3,4]

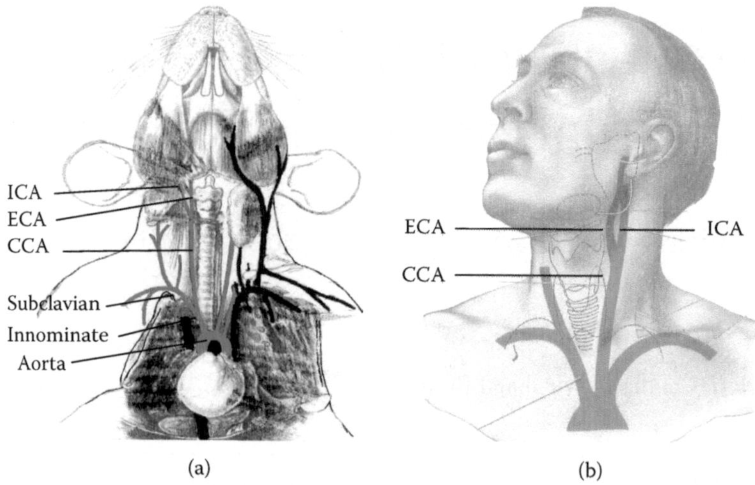

FIGURE 4.1 Vessels at the cervical area. (a) The vessels in the rat and (b) the vessels in the human. The anatomy and artery supplies in rats are essentially similar to humans. CCA, common carotid artery; ECA, external carotid artery; ICA, internal carotid artery. (Adapted from References 3 and 4. With permission.)

Subclavian Artery

The subclavian artery (Figure 4.1 and Figure 4.2)[1,3,4] is quite short, extending only from the innominate artery to the border of the first rib. It yields the following branches: the costocervical trunk and the internal mammary, vertebral, and cervical trunk arteries (for detailed illustration of these branches, see Reference 1; here, only the vertebral is illustrated in Figure 4.2).[1]

Vertebral Artery

The vertebral artery arises from the anterior surface of the subclavian artery, crosses the roots of the brachial plexus (Figure 4.2), and then runs upward under the carotid tubercle of the sixth cervical vertebra. Through the foramen magnum, it enters the skull and joins the vertebral artery

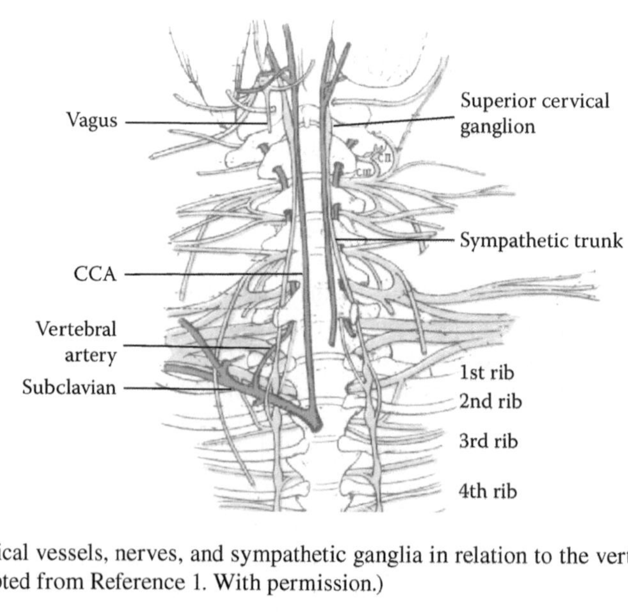

FIGURE 4.2 Cervical vessels, nerves, and sympathetic ganglia in relation to the vertebrae. CCA, common carotid artery. (Adapted from Reference 1. With permission.)

of the contralateral side to form the basilar artery (BA) at the level of pons.[1] The vertebral artery occasionally arises from the cervical trunk. The basilar artery is a component of the circle of Willis (discussed separately in this chapter).

Common Carotid Arteries

The right and left common carotid arteries supply the head and neck. The anatomy of the arterial supply in rats is essentially similar to that of humans (Figure 4.1a,b). The internal and external carotid arteries are derived from their respective common carotid artery. The right common carotid artery is a continuation of the innominate artery. The left common carotid artery arises from the arch of the aorta (Figure 4.1).[1–4]

External Carotid Artery

The external carotid artery (Figure 4.3a)[1] begins at the level of the posterior corner (greater horn) of the hyoid bone, about 2 to 4 mm below the corner in adult rats, as a direct continuation of the common carotid. Before reaching the angle of the jaw, it branches into five major arteries: the occipital, superior thyroid, ascending pharyngeal, lingual, and ascending palatine (for detailed illustrations of these branches, see Reference 1). Note that the hyoid bone is shaped like a horseshoe and is suspended from the tips of the styloid processes of the temporal bones by the stylohyoid ligaments. It consists of five segments: a body; two greater corners, also called greater horns; and two lesser corners.

Internal Carotid Artery

The internal carotid artery (Figure 4.3)[1] arises from the common carotid artery and follows the same general direction through the neck as the external carotid artery but at a deeper level along the base of the skull. On reaching the tympanic bulla, it branches into the pterygopalatine artery (PPA) (Figure 4.3b).[1] This artery represents what is in the human a portion of the internal maxillary branch of the external carotid artery,[1,3] and it is located behind the internal carotid artery. The internal carotid artery continues along the medial surface of the bulla for a short distance and enters the carotid canal between the bulla and the basal plate of the occipital bone to reach the base of the skull.

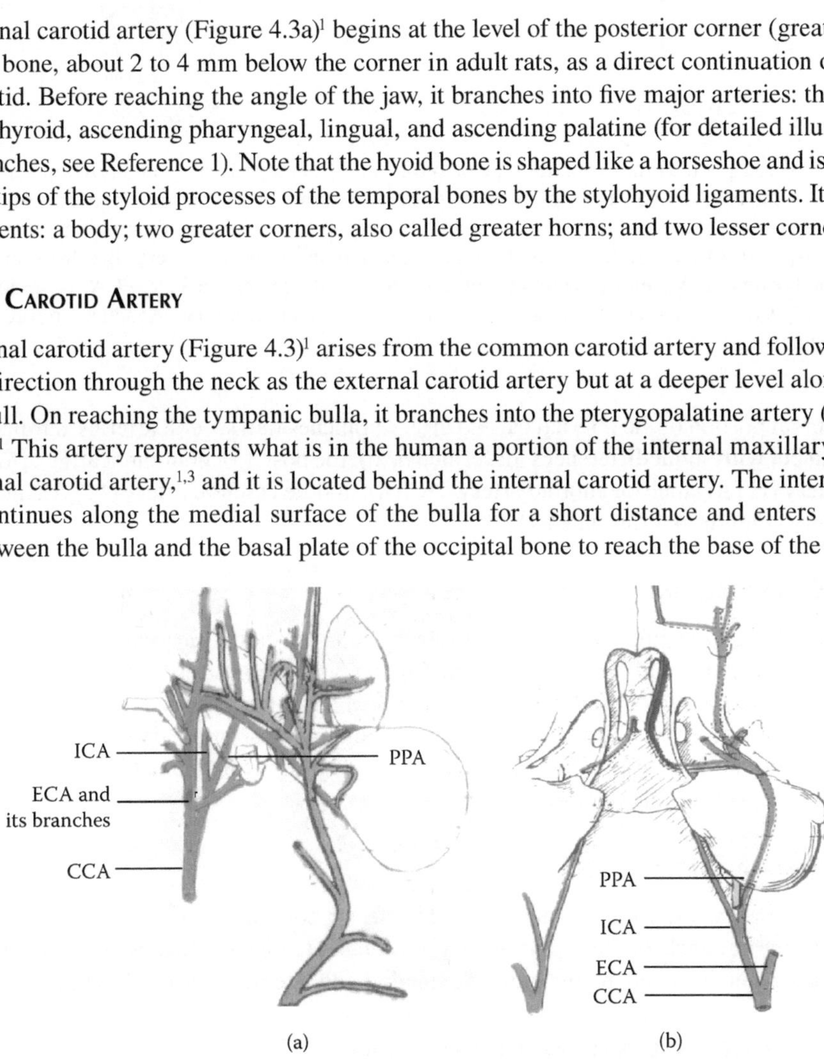

FIGURE 4.3 External and internal carotid arteries and their branches in the rat. (a) The branches of the external and internal carotid arteries in the rat at the neck area. (b) The pterygopalatine artery; this artery is in the back of internal carotid artery. CCA, common carotid artery; ECA, external carotid artery; ICA, internal carotid artery; PPA, pterygopalatine artery. (Adapted from Reference 1. With permission.)

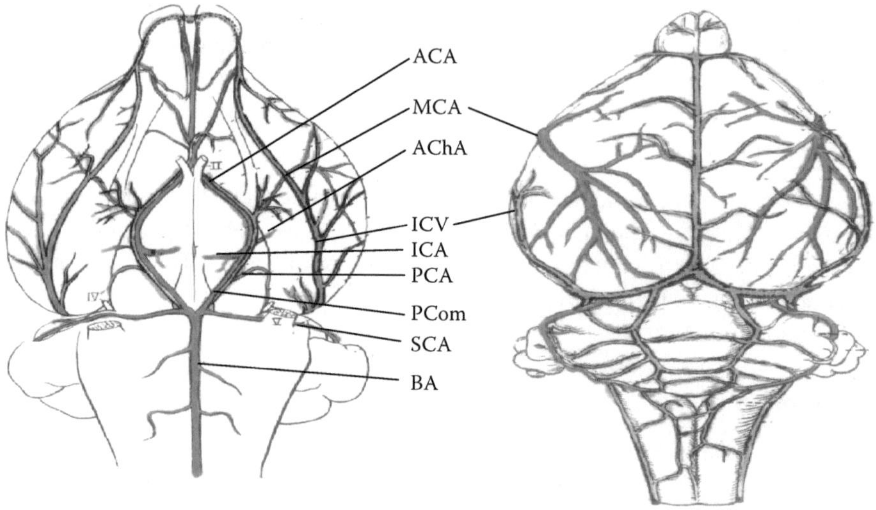

FIGURE 4.4 Vessels of the brain. (a) The circle of Willis, middle cerebral artery, and cerebral veins on basal aspect in the rat; the carotid and basilar parts of the posterior communicating artery are approximately equal in caliber. (b) Top view of the rat brain. Middle cerebral artery gives branches to the cerebral cortex and supports the hemisphere. ACA, anterior cerebral artery; AChA, anterior choroid artery; BA, basal artery; ICA, internal carotid artery; ICV, inferior cerebral vein; MCA, middle cerebral artery; PCA, posterior cerebral artery; P Com, posterior communication artery; SCA, superior cerebral artery. (Adapted from Reference 1. With permission.)

The cerebral portion of the internal carotid artery branches into several arteries, similar to those in the human but with slight differences (described next): the posterior communicating artery, hypothalamic artery (HTA), anterior choroid artery (AChA), middle cerebral artery (MCA), and anterior cerebral artery (Figures 4.4 through 4.7).[1,3]

Posterior Communicating Artery

The posterior communicating artery arises from the internal carotid artery on the basal surface of the brain lateral to the stalk of the infundibulum and curves posteriorly around the hypophysis, lying in a groove of the hypophysis, to reach the median basilar artery formed by the union of the vertebral arteries from either side. The posterior communicating arteries of the two sides thus form the posterior portion of the arterial circle of Willis of the brain, while the anterior cerebral arteries form the anterior portion (Figure 4.4a).[1]

Posterior Cerebral Artery

The posterior cerebral artery is a branch of the posterior communicating artery (Figure 4.4a). It curves around the peduncle of the cerebrum and runs upward along the fold of the tentorium to supply the surface of the hemisphere, including the medial and lateral surfaces of the occipital lobe.

Hypothalamic Artery and Anterior Choroid Artery

In rats, two small arteries arise from the distal internal carotid artery proximal to the MCA bifurcation: the HTA and the AChA (Figure 4.5).[5]

The HTA arises from the internal carotid artery just before the beginning of the posterior communicating artery; the distance between the HTA and the MCA is not always clear.

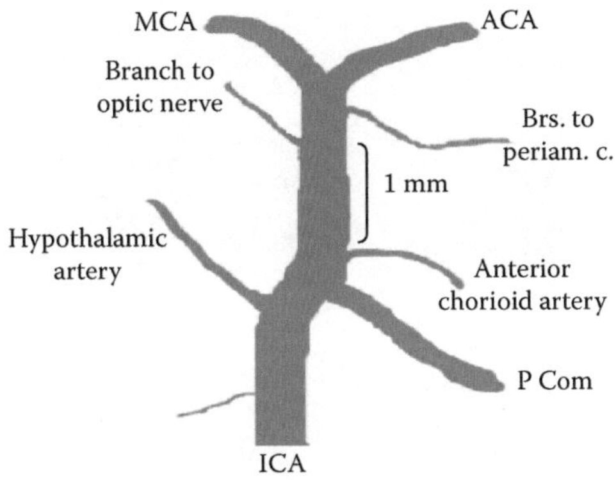

FIGURE 4.5 The relationships of the hypothalamic artery, anterior choroid artery, posterior communicating artery (P Com), and middle cerebral artery (MCA). The distance between the anterior choroid artery and the MCA is a little more than 1 mm in adult rats. The hypothalamic artery arises just before the origin of P Com. The anterior choroid artery arises just after the origin of P Com. ACA, anterior cerebral artery; Brs. to periam. c., branches to periamygdaloid cortex; ICA, internal carotid artery.

The AChA arises from the internal carotid artery just after the beginning of the posterior communicating artery; it is a little more than 1 mm from the MCA in an adult female rat (250 to 300 g).[5] The choroid artery runs for a short distance in the groove between the temporal lobe and the cerebral peduncle, curves dorsally around the latter, and pushes into the lateral ventricle along the choroid fissure, carrying the thin membranous medial wall of ventricle with it in such a way that the plexus formed by the artery is entirely surrounded by this membrane and the cerebral nerves actually penetrate into the cavity of the ventricle. In addition to the choroid plexi of the lateral ventricles, the choroid artery enters into the formation of the tela choroidea of the third ventricle in a similar fashion.

Understanding the anatomy of these two small vessels is very important for the suture stroke model since in the suture stroke model the intraluminal suture advancing toward the MCA bifurcation probably causes the obstruction of these two small arteries concomitant with MCA trunk obstruction. Studies in humans or dogs indicated that AChA occlusion does not cause hypothalamic damage in these species.[6–8] In rats, when using an intraluminal suture to permanently block these two arteries, the body temperature consistently increased to 38.5°C to 40.5°C.[5] The body temperature returned to normal at 24 hours after reperfusion (see Chapter 13 regarding the suture stroke model).

Anterior Cerebral Artery

The anterior cerebral artery (Figure 4.4)[1] is one of the terminal branches of the internal carotid artery. It begins at the base of the brain lateral to the optic chiasm and crosses the olfactory tract, joining the corresponding artery of the opposite side to form a zygos vessel (Figure 4.4), then curves upward over the genu of the corpus callosum and extends dorsally along the entire body of the corpus callosum.

Middle Cerebral Artery

The MCA (Figures 4.4, 4.6, and 4.7)[1,3] is the larger of the two terminal branches of the internal carotid artery. It starts at the base of the brain lateral to the infundibulum, running upward over the lateral surface of the olfactory tract, and then branches into the cerebral cortex to support the hemisphere (Figure 4.4b). The diameter of MCA in adult rats (250 to 320 g) is about 200 ± 40 μm in our unpublished data. Detailed size for each vessel is described in Chapter 13. The blood supply of the rat thalamus and basal ganglia is also similar to that in human.[3,9–11]

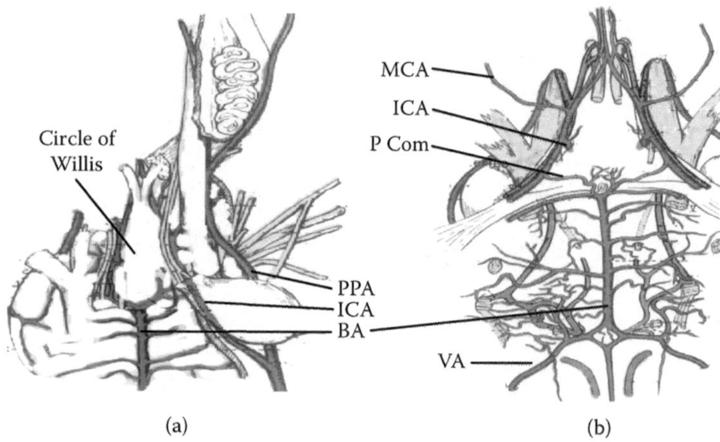

FIGURE 4.6 Variations of the circle of Willis in rats in a basal view of the brain. (a) The posterior cerebral artery gets its main supply from the basilar artery. (b) The internal carotid artery is the major source of supply for the posterior cerebral artery. BA, basal artery; ICA, internal carotid artery; P Com, posterior communication artery; PPA, pterygopalatine artery; VA, vertebral artery. (Adapted with modification from Reference 1. With permission.)

Variations of MCA anatomy have been found in the human.[12,13] In 1962, Crompton[14] described the accessory middle cerebral artery, which included duplication of the MCA, and the anomalous vessel originating from the A1 portion of the anterior cerebral artery, which coursed parallel to the MCA. In 1973, Teal et al.[15] proposed using the term *middle cerebral artery duplication* to characterize the two vessels originating from the distal end of the internal carotid artery and the term *accessory middle cerebral artery* to describe the anomalous vessel originating from the anterior cerebral artery. The accessory middle cerebral artery is a variation of MCA branching, and its incidence has been reported to be 0.3% to 4.0%.[14,16] The accessory middle cerebral artery originates from either the proximal or distal horizontal portion of the anterior cerebral artery coursing parallel to the horizontal portion of the MCA and reaches the anterior frontal lobe (Figure 4.7c).[3]

An association between the duplicated middle cerebral artery or accessory middle cerebral artery and cerebral aneurysms has been well documented.[12–17] The terminal portions of the bilateral internal carotid arteries and their vicinities were markedly stenotic and so-called moyamoya vessels developed at the base of the brain. Moyamoya disease is characterized by angiographic features of stenoocclusive changes of the terminal portions of the bilateral intracranial internal carotid arteries as well as dilated perforating arteries at the base of the brain known as "moyamoya" vessels. The clinical manifestation of moyamoya disease is typically brain ischemia in the pediatric population and brain hemorrhage in adults.

Similar variations of MCA anatomy are also found in rats. Fox et al.,[18] in a study consisting of 263 Sprague-Dawley rats, found that the majority (82.9%) of MCAs conformed to the typical bifurcating vessel commonly referred to in the literature. However, the remaining 17.1% were far more variable, with major, atypical branching. This variation leads to the variation of infarct size in focal stroke models.

Niiro et al.[19] studied the branching patterns of the proximal MCA and their influence on infarct size following MCA occlusion in 106 Sprague-Dawley rats. They classified the branching patterns of the posterior extending surface branches as follows: type 1 (57.5%), one prominent proximal surface branch; type 2 (30.2%), no prominent branch but two or more small surface branches; and type 3 (12.3%), no surface branches with a visible junction with the MCA but surface branches probably arising from the internal carotid artery at the origin of the MCA, which is similar to the duplicate middle cerebral artery described in a human anomalous MCA. When the proximal surface branches

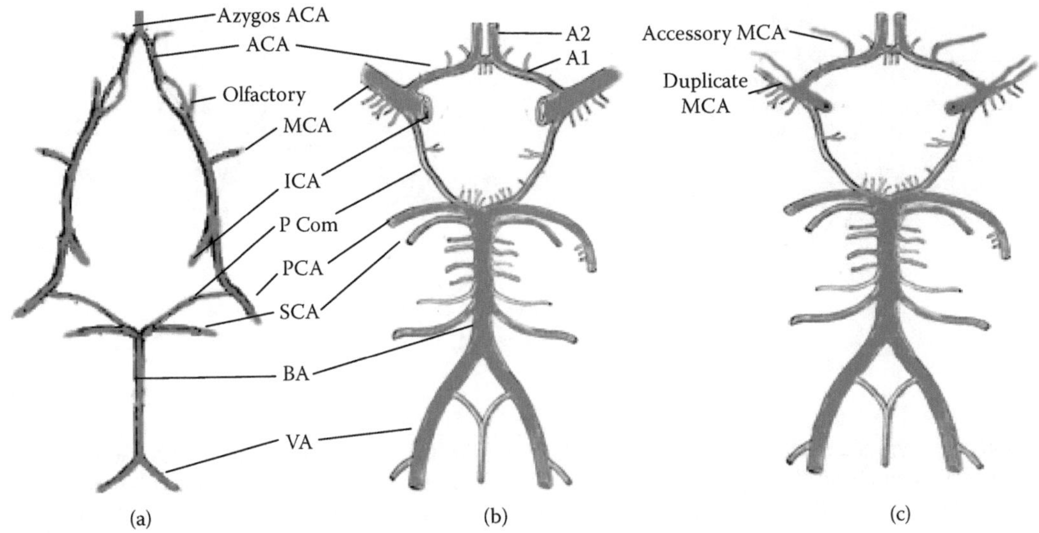

FIGURE 4.7 Comparison of cerebral blood vessel variations between rats and humans. (a) The circle of Willis in rats with a buttonhole-like structure on the anterior cerebral artery (ACA); (b) the circle in most normal people; and (c) the variations of the middle cerebral artery that usually accompany stenotic vessels or aneurysm in humans. (Figure 7b adapted from Reference 3. With permission.)

(especially those of type 1) remained uncoagulated, the infarct size in both the cortex and striatum was significantly reduced compared with that when coagulation was proximal to the most proximal surface branch. The reduction of cortical and striatal infarct sizes was particularly prominent in the posterior portion. To increase the infarct size and reduce variability, it is suggested that the MCA trunk should be coagulated from the origin to the rhinal fissure, including the surface branches.

Herz et al.[20] analyzed the morphology of the major cerebral arteries in Wistar and Fischer 344 rats and showed a significantly higher number of side branches in the proximal MCA segment in Wistar rats than in Fischer 344 rats, which resulted in a smaller cortical infarction in Wistar compared to Fischer rats in a focal stroke model. The variation of vessel branches leads to the variation of infarct size in different stroke models.

Circle of Willis and Its Variations

The circle of Willis, formed by the anterior cerebral and posterior communicating arteries, appears to be similar in humans and rats, and for that reason this term has been granted widespread acceptance for use in the rat.[11] It has been variously interpreted by different people[1–4] both in rats (Figure 4.4a,b) and in humans (Figure 4.7b).

Variations in circle of Willis anatomy have been reported in humans[21] and rats.[1,11] In 1928, Adachi from Kaiserlich-Japanischen University at zu Kyoto showed 83 individual human cases displaying many types of variation, which are similar to those in rats. Figures 4.4a and 4.6a,b show these variations in rats.[1] The figures are from random samples. All three samples show a marked difference in the branching and relative size of the various vessels. This fact would lead one to suppose that we may find as extensive variation in the rat as found in the human. Figure 4.4a depicts a case for which the carotid and basilar parts of the posterior communicating artery are approximately equal in caliber. Figure 4.6a is a case for which the posterior cerebral artery acquires its main supply from the basilar artery. Figure 4.6b is a case for which the internal carotid artery is the major source of supply for the posterior cerebral artery.

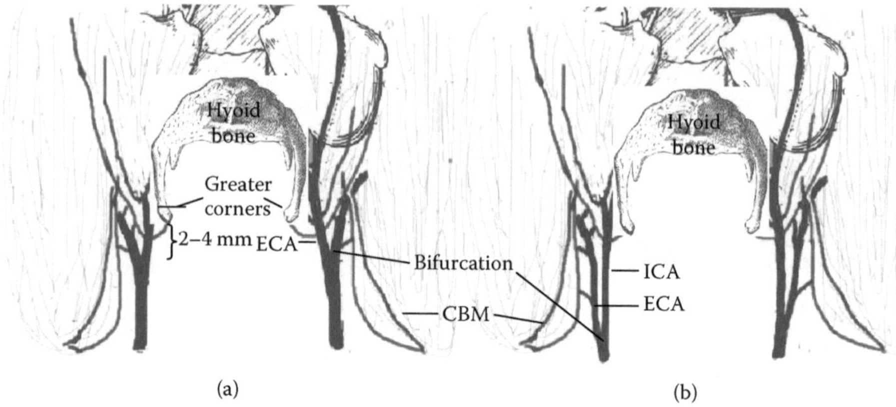

FIGURE 4.8 Variation in carotid bifurcation anatomy. (a) The most common location of bifurcation in normal rats, 2 to 4 mm beneath the greater corner of the hyoid bone. (b) The variation of the lower bifurcation in the rat. CBM, caudal belly of digastric muscle; ECA, external carotid artery; ICA, internal carotid artery.

Variations on Carotid Bifurcation

The variation also can be seen in other vessels; for example, the location of the carotid bifurcation into the external and internal carotid arteries can be higher or lower in humans[22,23] and in rats. Figure 4.8 demonstrates the variations of carotid bifurcations in Sprague-Dawley rats. Figure 4.8a is the most common location of the bifurcation. It is about 2 to 4 mm below the greater horns of hyoid bone in adult rats (250 to 320 g) in our unpublished data. Figure 4.8b shows the variation of lower bifurcation. The incidence of this variation was 1.5% to 2% in our study (5/340 Sprague-Dawley adult rats). Rats with lower bifurcation had no infarction or very small infarction after a routine 2-hour filament insertion. It is thus understandable why rat stroke models can have a large range of infarct size in a given experimental group even if the stroke procedure is uniformly administered.

Comparison of the Major Arteries between Rat and Human

Lee[11] compared the anatomy of the major cerebral arteries between humans and rats. The comparison showed many similarities, including anomalies in their general organization, the structure of these vessels at the light and electron microscopic levels, and their morphological changes associated with cerebral vascular diseases. The general organization of the major cerebral arteries shows the main differences between humans and rats discussed next.

In rats, the internal carotid arteries have become an integral part of the circle of Willis. In the anterior cerebral arteries, a common variation in humans is the underdevelopment of one of the two arteries, whereas in rats, buttonhole-like structures are common in one or both arteries (Figure 4.7a). The anterior communicating artery present in humans is absent in rats. The olfactory artery is prominent in rats but absent in humans. The posterior communicating artery in humans is the most variable component of the circle of Willis, being asymmetric in its origin, diameters, and branches. Similarly, the posterior cerebral arteries in rats often exhibit asymmetrical origins from the basilar artery. In humans, most aneurysms occur in the anterior half of the circle of Willis; the MCA is most often the one to become occluded, and the vertebral arteries are also common sites for thrombosis. The various channels that constitute collateral circulation in humans provide a margin of safety, so that in case of cerebral occlusion due to thrombosis, atherosclerosis, or vasospasm related to hemorrhage, blood supply to the affected area can be maintained through these collaterals. Collateral circulation is also present in rats. However, in rats, information on the presence of various types of aneurysms and their location and frequency in normal and experimental models of

TABLE 4.1
Summary of Variations in Cerebral Vasculature: Rats and Humans

	Rats	Humans
ACA	Buttonhole-like structures	Underdeveloped
A comm	Absent	Present
Olfactory artery	Prominent	Absent
P Com	Relatively stable; no aneurysm was reported	Most variable, abnormalities; common site of aneurysm
PCA	Similar	Similar
Duplicated or accessory MCA	12.3% (SD rats)	0.3% to 4.0%
Collaterals between ACA and MCA	4 to 5 times more than human (Wistar rat)	Less than rats
Circle of Willis	Uncompleted or completed	Completed
Lower bifurcation	1.5% to 2.0%	13% to 15%23

Notes: ACA, anterior cerebral artery; A comm, anterior communicating artery; MCA, middle cerebral artery; PCA, posterior cerebral artery; P Com, posterior communicating artery; SD, Sprague-Dawley.

hypertension and stroke is still lacking. The major differences between rats and humans are summarized in Table 4.1.[11,16,17,22]

NERVES IN THE CERVICAL AREA

Carotid Sinus and Carotid Body

The distributions of nerves and nerve-related sensors in rats are similar to those in the human.[3] The carotid sinus is a slight dilation in the carotid artery at its bifurcation into the external and internal carotid arteries: It contains baroreceptors (pressure sensors), which detect vasodilation and thus a decrease in blood pressure. The carotid body is near the bifurcation of the internal carotid artery. It has a chemoreceptor to modulate respiratory and cardiovascular function in response to fluctuations in arterial pH, carbon dioxide, and oxygen concentrations. Its blood supply is from the external carotid artery. The carotid sinus and body are very vascular and abundantly supplied with nerves. The sensory innervations are from the branches of glossopharyngeal and vagus trunks.

Vagus Nerve

In the neck, the vagus nerve supplies cardiac branches and the laryngeal recurrent nerve. On the right side, the recurrent nerve makes a loop around the beginning of the subclavian artery. On the left side, the vagus enters the thorax between the left carotid and subclavian arteries, behind the left innominate vein. It crosses the left side of the arch of the aorta and descends behind the root of the left lung, forming there the posterior pulmonary plexus. In the cervical area, the vagus nerve and common carotid artery are enveloped in the same sheath and run very closely together.

Sympathetic Nerve

The cervical part of the sympathetic trunk runs beside the capsule[1] (Figure 4.2). The ganglia in this area are consolidated into three pairs: the superior, middle, and inferior cervical ganglia. The superior cervical ganglion lies at the level of bifurcation of the common carotid into external and internal carotid arteries and is in close proximity to the latter vessel, carotid body, and carotid sinus. It is accompanied by the sympathetic trunk through the carotid canal. The sympathetic trunks lie dorsal to the common carotid artery and the vagus nerve. At the first rib level, it displays the middle cervical ganglion. The inferior cervical ganglion is at the level between the second and third ribs. The cardiac branch is from the inferior cervical ganglion.

The cervical ganglions, sympathetic trunk, and vagus nerve can be mechanically traumatized during the procedures for separating the common carotid artery or external or internal carotid arteries. Such stimulation of these nerves can cause acute cardiac arrhythmia, myocardial ischemia, and even left ventricular heart failure. Local application of bupivaine or lidocaine can block the response of nerves or sensors to such stimulation and decrease the incidence of heart failure.[24]

REFERENCES

1. Greene, E.C., *Anatomy of the Rat*, Hafner, New York, 1963, pp. 85–177.
2. Popesko, P., Rajtova, V., and Horak, J.A., *Colour Atlas of Anatomy of Small Laboratory Animals*, Wolfe, 1990, 2:33.
3. Gray, H., *Gray's Anatomy, the Anatomical Basis of Medicine and Surgery,* 38th ed., Bannister, L.H., Berry, M.M., and Williams, P.L. eds., Churchill Livingstone (Elsevier Science), New York, 1995, p. 1515.
4. Walker, W.F., Jr. and Homberge, D.G., *Anatomy and Dissection of the Rat,* 3rd ed., Freeman, New York, 1997.
5. He, Z. et al., Experimental model of small deep infarcts involving the hypothalamus in rats, changes in body temperature and postural reflex, *Stroke*, 30, 2743, 1999.
6. Hupperts, R.M.M. et al., Infarcts in the anterior choroidal artery territory: Anatomical distribution, clinical syndromes, presumed pathogenesis and early outcome, *Brain,* 117, 825, 1994.
7. Ghika, J.A., Bogousslavsky, J., and Regli, F., Deep perforators from the carotid system, *Arch Neurol,* 47, 1097, 1990.
8. Bertan, V. and Wilson, C.B., Anatomy of the anterior choroidal artery in the dog, *Arch Neurol,* 14, 526, 1966.
9. Rieke, K.G., Bowers, E.D., and Penn, P., Vascular supply pattern to rat caudoputamen and blobus pallidus, scanning electron microscopic study of vascular endocast of stroke-prone vessels, *Stroke,* 12, 840, 1981.
10. Menzies, S.A., Hoff, J.T., and Betz, A.L., Middle cerebral artery occlusion in rats: A neurological and pathological evaluation of a reproducible model, *Neurosurgery,* 31(1), 100, discussion 106, 1992.
11. Lee, R.M., Morphology of cerebral arteries, *Pharmacol Ther,* 66(1), 149, 1995.
12. Komiyama, M. et al., Middle cerebral artery variations: Duplicated and accessory arteries, *Am J Neuroradiol*, 19(1), 45, 1998.
13. Komiyama, M. and Yasui, T., Accessory middle cerebral artery and moyamoya disease, *J Neurol Neurosurg Psychiatry*, 71(1), 129, 2001.
14. Crompton, M.R., The pathology of ruptured middle-cerebral aneurysms with special reference to the differences between the sexes, *Lancet,* 2, 421, 1962.
15. Teal, J.S. et al., Anomalies of the middle cerebral artery, accessory artery, duplication, and early bifurcation, *AJR Am J Roentgenol,* 118, 567, 1973.
16. Jain, K.K., Some observations on the anatomy of the middle cerebral artery, *Can J Surg,* 7, 134, 1964.
17. Kudo, T., Spontaneous occlusion of the circle of Willis: A disease apparently confined to Japanese, *Neurology,* 18, 485, 1968.
18. Fox, G. et al., Anatomic variation of the middle cerebral artery in the Sprague-Dawley rat, *Stroke,* 24(12), 2087, discussion 2092, 1993.
19. Niiro, M. et al., Proximal branching patterns of middle cerebral artery (MCA) in rats and their influence on the infarct size produced by MCA occlusion, *J Neurosci Methods*, 64(1), 19, 1996.
20. Herz, R.C. et al., Middle cerebral artery occlusion in Wistar and Fischer-344 rats: Functional and morphological assessment of the model, *J Cereb Blood Flow Metab*, 16(2), 296, 1996.
21. Hoksbergen, A.W.J. et al., Collateral variations in circle of Willis in atherosclerotic population assessed by means of transcranial color-coded duplex ultrasonography, *Stroke,* 31, 1656, 2000.
22. Gailloud, P., Murphy, K.J., and Rigamonti, D., Bilateral thoracic bifurcation of the common carotid artery associated with Klippel-Feil anomaly, *Am J Neuroradiol,* 21, 941, 2000.

23. Schulz, U.G.R. and Rothwell, P.M., Major variation in carotid bifurcation anatomy: a possible risk factor for plaque development? *Stroke,* 32, 2522, 2001.
24. Wang-Fischer, Y.L. et al., Refined technique for inducing and grading middle cerebral artery occlusion in rat stroke model. *American Association of Laboratory Animal Science 54th National Meeting,* Seattle, platform sessions speaker PS 43, October 11–16, 2003, and *AALAS Tri-Branch Symposium, Biotechnology in the 21st Century and Beyond*, presenting poster 10, Philadelphia, June 8–10, 2003.

5 Which Animal to Choose?

Yanlin Wang-Fischer, Brant D. Watson, and Lee Koetzner

CONTENTS

WHY USE THE RAT TO MODEL STROKE?

A variety of animals have been used for stroke models, including large animals (cats, swans, rabbits, pigs, dogs, and subhuman primates) and small animal species (for example, rats, gerbils, and mice). In this book, we focus on rats owing to the following benefits:

1. Rats in particular are useful for stroke research because their cerebrovascular anatomy and physiology closely resemble that of humans.[1–3] For example, the distribution of the hypothalamic artery in rats is similar to that of humans, such that during stroke modeling by means of filament insertion, which results in blockage of the entire internal carotid artery tree, including the hypothalamic artery, hyperthermia is produced in rats apparently to the same degree as occurs in human stroke patients. The same procedure did not cause hyperthermia in most mice in our study (our unpublished data).
2. The rat has a small brain volume that is nonetheless well suited to different analytical procedures and thus can allow more extensive and comprehensive evaluation of the entire brain without excessive cost, time, and labor.
3. The possibility of genomic modulation is particularly efficacious for future understanding of the complexities of stroke; for example, the spontaneously hypertensive rat (SHR) is a good genetic disease model given that hypertension is an important risk factor for human stroke.
4. Rats are reasonably inexpensive in terms of purchase and maintenance costs compared with those for larger animals.
5. Investigators have begun to look at functional outcomes in experimental stroke by evaluating a battery of behavioral, cognitive, and sensorimotor tests. A number of neurosensory and motor behavioral outcomes have been well described and standardized for rats, for example, the foot fault test,[4] adhesive tape test, and rod-walking test.[5]
6. The public tends to have fewer animal welfare concerns regarding the use of rodents in stroke research compared with the use of dogs and nonhuman primates.

7. Commercially available inbred rats are relatively homogeneous genetically, allowing researchers to minimize confounding effects arising from a heterogeneous background and to achieve good reducibility (small standard deviation).

EFFECTS OF ANIMAL STRAIN AND VENDOR ON INFARCT VOLUME

It is well recognized that genetic factors play an important role in the outcome of focal cerebral ischemia. Rat strain and vendor can affect infarct volume and the complications from the stroke modeling. Oliff et al.[6,7] and Sauter and Rudin[8] used different strains from different vendors to study transient focal cerebral ischemia.

Cortical and subcortical infarct volumes were compared between the middle cerebral artery (MCA) suture occlusion model and the distal MCA cauterization model, which combined MCA occlusion and common carotid artery occlusion (CCAO) for 1 hour. Wistar rats and Simonsen Sprague-Dawley (SD) rats developed smaller infarcts compared to SD rats from Taconic and Charles River Laboratories.[7] Their data are summarized in Table 5.1 (means ± standard error were converted into means ± standard deviation).

Duverger and MacKenzie[9] showed in 1988 that in genetically hypertensive strains of rats (SHRs), distal MCA occlusion resulted in larger infarct volume and lower variability of infarct size than normotensive strains. This is summarized in Table 5.2.

Oliff et al.[10] in 1996 studied the neuroprotective efficacy of MK-801 (a glutamate receptor antagonist) in focal cerebral ischemia. MK-801 (0.12 mg/kg i.v. bolus followed by 0.108 mg/kg/hour infusion or 0.60 mg/kg i.v. bolus followed by 0.540 mg/kg/hour infusion) or saline were administered just after intraluminal MCA occlusion. Administration of 0.54 mg/kg/hour MK-801 provided strain-/line-dependent neuroprotection in the following rank order: Simonsen Laboratories SD rats > Simonsen Laboratories Wistar rats > Taconic Laboratories SD rats. After 0.108 mg/kg/hour MK-801 treatment, the Simonsen Laboratories Wistar rat was the only strain/line that was significantly

TABLE 5.1
Effects of Animal Strains and Vendors on Total Infarct Volumes (hemisphere volume, mm^3)

Surgery	Simonsen Labs Wistar	Simonsen Labs SD	Taconic Labs SD	Charles River Labs SD
MCAO (suture)	172 ± 72 n = 19	73 ± 22 n = 12	396 ± 69 n = 15	424 ± 90 n = 17
MCAO/CCAO	266 ± 156 n = 18	222 ± 53 n = 11	389 ± 58 n = 15	422 ± 76 n = 16

Source: From Reference 7.
Notes: Data were summarized for the whole hemisphere (cortical plus subcortical infarct) and converted to mean ± standard deviation. CCAO, common carotid artery occlusion; MCAO, middle cerebral artery occlusion; SD, Sprague-Dawley rats.

TABLE 5.2
Effect of Genetic Factors on Infarct Size (mm^3) in Distal Stroke Model

Surgery	Wistar Kyoto	Wistar	SD	Fisher 344	SHR
MCAO (permanent), mean ± standard deviation	68 ± 40 n = 25	121 ± 70 n = 10	134 ± 88 n = 30	157 ± 54 n = 20	198 ± 60 n = 16

Source: From Reference 9. In this study, they only occluded the middle cerebral artery without occlusion of common carotid arteries.
Notes: MCAO, middle cerebral artery occlusion; SD, Sprague-Dawley rats.

TABLE 5.3
The Effect of Different Producers on Acute Heart Failure

Producer Places	Incidence of Heart Failure	Total Average of Incidence
Kinston K97	9% (5/55)	
Kinston K93	41% (7/17)	30%
Raleigh	19% (6/32)	

protected. These results indicate that the neuroprotective effect of an experimental drug may be influenced by rat strain and vendor differences.

The effect is not only on infarct size, but also on the complication from the stroke modeling. In our unpublished data, the incidence of acute heart failure from the stroke model by occlusion of the distal MCA and two common carotid arteries was related to the breeder and the colony. The SD rats were purchased from the same company, Charles River; they have, however, three colonies that produce rats. The incidences for acute heart failure in the distal stroke model varied between different producers. Table 5.3 shows the difference. This blinded study was done by the same experienced surgeon.

The concept of a strain difference on stroke was also found in humans. Based on the National Heart, Lung, and Blood Institute's ARIC (Atherosclerosis Risk in Communities) study, the age-adjusted stroke incidence rates (per 1000 person-years) are 1.78 for white men and 4.44 for black men. Blacks have a 38% greater risk of incident (first) strokes than whites.[11]

ANIMAL SEX AFFECTS THE INFARCT SIZE

In 1998 a study by Alkayed et al.,[12] endogenous estrogen was found to improve stroke outcome. In this study, male, female, and ovariectomized female Wistar rats underwent 2 hours of MCA suture occlusion. The infarct size was measured in cortical and subcortical tissues at 24 hours postocclusion. The female rats (bars in the middle) displayed smaller infarct sizes compared to the males. However, ovariectomized females developed the same infarct size as male rats (Figure 5.1). Another conclusion from this study was that female rats maintained higher striatal, but not cortical, cerebral blood flow (CBF) than males at the end of vascular occlusion. Use of males was justified as a means of reducing experimental variability caused by female hormone cycling and was based on

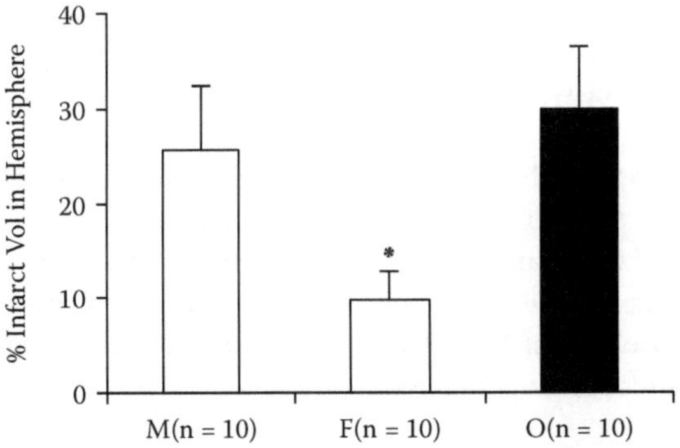

FIGURE 5.1 Sex effects on infarct volume in Wistar rats. The middle cerebral artery (MCA) was occluded by an intraluminal filament for 2 hours in age-matched male (M), female (F), and ovariectomized female (O) rats. Infarct volumes were measured in brain sections by triphenyltetrazolium chloride (TTC) staining after 24 hours postischemia and expressed as a percentage of the ipsilateral cortex. *Significantly different from M and O groups ($p < .05$). (Data summarized from Reference 12.)

the assumption that mechanisms of cell injury or treatment effects observed in males would also apply to females.

The effect of sex on stroke was also found in human beings.[11] Stroke is more common in men than in women. In most age groups, more men than women will have a stroke in a given year. However, after menopause the incidence of stroke is equal or a little higher in women than in men.[11]

ANIMAL AGE AFFECTS INFARCT DEVELOPMENT AND RECOVERY

As we know, stroke is a major cause of death and disability in elderly people.[11] Elderly patients tend to have a worse outcome than younger patients because they usually have hypertension as well as other conditions (for example, diabetes) that heavily affect outcome. Some patients have poor collateral circulation and a smaller penumbra or no penumbra. However, almost all *in vivo* experimental studies, including the evaluation of neuroprotective drugs, have been performed on young adult animals.[13]

Aging was associated with a significant increase in infarct size (40% of the hemisphere volume in aged Wistar male rats at 28 to 36 months compared with 30% of that in adult male rats aged 11 to 17 months).[14]

OTHER FACTORS THAT CAN AFFECT BRAIN DAMAGE

CRANIAL AND BODY TEMPERATURE

It is well known that higher temperatures increase brain damage, and lower temperatures protect against damage. A regimen of low temperature in the head area has been used in clinical practice for treatment of acute neurological diseases for many years.

In the classic 1987 paper by Busto and coworkers,[15] the effect of cranial and body temperature was investigated in a model of global cerebral ischemia. When animals were kept normothermic (36°C) during ischemia, severe neuronal damage in the hippocampus resulted. When the temperature was decreased to 33°C or 30°C during ischemia, the hippocampal neurons were spared.

HYPERGLYCEMIC ISCHEMIA INCREASES BRAIN DAMAGE

Prado et al.[16] investigated the effect of hyperglycemia on rat stroke models. They found that hyperglycemia increases infarct size in collaterally perfused but not end-arterial vascular territories.

Kagansky et al.[17] reviewed the effects of hyperglycemia on stroke in human and animal studies. They concluded that most human studies have shown that hyperglycemia on admission for acute stroke in patients with or without diabetes is associated with a worse clinical outcome than in patients without hyperglycemia. This association was more consistent in the nonlacunar type of stroke. Animal studies supported these findings by showing in both global and focal postischemic models that hyperglycemia exaggerates the following damaging processes: intracellular acidosis, accumulation of extracellular glutamate, brain edema formation, blood–brain barrier disruption, and tendency for hemorrhagic transformation. Insulin treatment of hyperglycemic animals was found to have a beneficial effect in focal and global brain ischemia, which may be mediated by glucose reduction or by direct neuroprotection.[17]

ANIMAL NUTRITION AND FOOD EFFECT ON BRAIN DAMAGE

Sometimes the animal-producing companies improve their standards for the care of animals without telling their customers; for example, they add vitamin E to the animal food, or they add more soybeans to the food, which contain more natural estrogen. The improvement of food can reduce the infarct volume due to its neuroprotective function.[18]

REFERENCES

1. Greene, E.C., *Anatomy of the Rat*, Hafner, New York, 1963, pp. 85–177.
2. Gray, H., Bannister, L.H., Berry, M.M., and Williams, P.L., *Gray's Anatomy, the Anatomical Basis of Medicine and Surgery*, 38th ed., Churchill Livingstone (Elsevier Science), New York, 1995, p. 1515.
3. Lee, R.M., Morphology of cerebral arteries, *Pharmacol Ther*, 66(1), 149, 1995.
4. Hernandez, T.D. and Schallert, T., Seizures and recovery from experimental brain damage, *Exp Neurol*, 102, 318, 1988.
5. Tucker, J.C., McDaniel, W.F., and Smith, S.R., A behavioral study of bilateral middle cerebral artery hemorrhagic ischemia in rats, *Neuroreport*, 3(8), 725, 1992.
6. Oliff, H.S. et al., Infarct volume varies with rat strain and vendor in focal cerebral ischemia induced by transcranial middle cerebral artery occlusion, *Brain Res*, 20, 699(2), 329, 1995.
7. Oliff, H.S. et al., The role of strain/vendor differences on the outcome of focal ischemia induced by intraluminal middle cerebral artery occlusion in the rat, *Brain Res*, 675(1–2), 20, 1995.
8. Sauter, A. and Rudin, M., Strain-dependent drug effects in rat middle cerebral artery occlusion model of stroke, *J Pharmacol Exp Ther*, 274(2), 1008, 1995.
9. Duverger, D. and MacKenzie, E.T., The quantification of cerebral infarction following focal ischemia in the rat: influence of strain, arterial pressure, blood glucose concentration, and age, *J Cereb Blood Flow Metab*, 8(4), 449, 1988.
10. Oliff, H.S. et al., The neuroprotective efficacy of MK-801 in focal cerebral ischemia varies with rat strain and vendor, *Brain Res*, 26, 731(1–2), 208, 1996.
11. American Heart Association, *2002 Heart and Stroke Statistical Update*, American Heart Association, Dallas, 2002.
12. Alkayed, N.J. et al., Gender-linked brain injury in experimental stroke, *Stroke*, 29(1), 159, discussion 166, 1998.
13. Gladstone, D.J., Black, S.E., and Hakim, A.M., Toward wisdom from failure lessons from neuroprotective stroke trials and new therapeutic directions, *Stroke*, 33, 2123, 2002.
14. Davis, M. et al., Experimental stroke and neuroprotection in the aging rat brain, *Stroke*, 26(6), 1072, 1995.
15. Busto, R. et al., Small differences in intra-ischemic brain temperature critically determine the extent of ischemic neuronal injury, *J Cereb Blood Flow Metab*, 7(6), 729, 1987.
16. Prado, R. et al., Hyperglycemia increases infarct size in collaterally perfused but not end-arterial vascular territories, *J Cereb Blood Flow Metab*, 8(2), 186, 1988.
17. Kagansky, N., Levy, S., and Knobler, H., The role of hyperglycemia in acute stroke, *Arch Neurol*, 58(8), 1209, 2001.
18. Noguchi, T. et al., Effects of vitamin E and sesamin on hypertension and cerebral thrombogenesis in stroke-prone spontaneously hypertensive rats, *Clin Exp Pharmacol Physiol*, 31 Suppl 2, S24, 2004.

6 Which Model to Use?

Yanlin Wang-Fischer and Lee Koetzner

CONTENTS

INTRODUCTION

The choice of an animal model is very important to everyone who works in stroke research. The primary criteria for choice should be how closely the model approximates the clinical condition and what the research question is.

It is obvious that there can never be one model that mimics all types of strokes because the pathogenesis of stroke is multifaceted with overlapping and interacting mechanisms.

Stroke models include ischemia and hemorrhage. Ischemic stroke is responsible for 85% of clinical cases; ischemic stroke models are very well studied (refer to Chapter 3).

In 1989, Myron Ginsberg and Raul Busto[1] published a review of rodent models of cerebral ischemia. They categorized ischemic stroke models as global or focal.

GLOBAL STROKE MODELS

The most popular models of global cerebral ischemia are four-vessel occlusion and two-vessel occlusion. Four-vessel occlusion is induced by permanent bilateral vertebral artery occlusions, followed by transient bilateral carotid artery occlusion, while two-vessel occlusion is induced by occlusion of both carotid arteries with hypotension. The models mimic human "cardiac arrest" with survival. It is a reversible, high-grade circulatory interruption to widespread areas of forebrain. These models result in death of neurons within "selectively vulnerable" brain areas, such as pyramidal neurons of the hippocampus, small neurons of the caudoputamen (striatum and basal ganglia), and neocortical neurons of layers 3, 5, and 6. The global stroke model has been accepted for study of the neuroprotective action or mechanism of new compounds, for example, SP600125, a new inhibitor of JNK (c-Jun-N-terminal kinase)[2] and estrogen.[3]

FOCAL STROKE MODELS

Focal stroke models include those for multifocal cerebral ischemia and focal hemisphere cerebral ischemia. These models mimic clinical situations of stroke with atherosclerosis, thrombus, and embolus. Emboli commonly come from the heart, where different diseases (for example, bacterial endocarditis, atrial fibrillation, ball thrombus, mitral valve stenosis, mural thrombi, and myocardial infarction) can cause thrombus formation.

Multifocal stroke has multiple sites of ischemia resulting from embolization of autologous or heterologous blood clots, microspheres, or photochemically induced thrombi at common carotid arteries. This model mimics stroke caused by multiple parietal (valvular) thrombi from heart diseases or infective thrombus from systemic infectious diseases or leukemia. Embolic models resemble clinical vascular dementia. Multifocal ischemia models have limited applications but can provide a unique opportunity to investigate potential antiplatelet and thrombolytic regimens. Although this model is easy to develop, its main disadvantages are inconsistency in the location of the infarctions, difficulty in detecting where injuries are relative to infarction areas, and possible microvascular injury with photochemical reactions.

Focal hemisphere cerebral ischemia results in a single site of ischemia in the caudoputamen or cortex from occlusion of the middle cerebral artery. This model mimics the most common site of human ischemic stroke. The degree and distribution of blood flow and infarction depend on duration of occlusion, site of occlusion along the middle cerebral artery, and amount of collateral blood flow into the middle cerebral artery territory.

Middle cerebral artery occlusion (MCAO) can be proximal or distal, permanent or temporary, alone or combined with carotid artery occlusions (ipsi-, contra-, or bilateral; temporary or permanent), on a normotensive or hypertensive rat strain. Occlusion methods include electrocoagulation, surgical cut, microclip, intraluminal suture, embolus (blood clot), photochemical-initiated MCAO (laser light plus intravenous administration of rose bengal), and pharmacological occlusion (vasoconstriction by local endothelin 1). Many of these models have been used extensively because of their relevance to human thromboembolic stroke.

One widely used technique of MCAO involves cauterization of the middle cerebral artery via a craniectomy. This technique is invasive and does not permit reperfusion.

In the rat, mechanical clipping of the middle cerebral artery and photothrombotic occlusion of the vessel are in common use, but these techniques also involve craniectomy.

The suture model and embolic clot model involve cervical vessel surgery without craniectomy, but the techniques are more difficult than those used in photochemical occlusion.

WHY DO WE NEED SO MANY DIFFERENT STROKE MODELS?

Although many animal stroke models have been developed and characterized, no single model can fully mimic human stroke because of the heterogeneity of human clinical disease. Some limitations of these models include variations in the size and distribution of infarction from interanimal variations in collateral flow and substantial mortality in acute and chronic survival studies.

Different models produce different infarct volumes; for example, proximal MCAO produces a big infarction, which involves cortical and subcortical areas. Distal MCAO produces only cortical infarction (see Figure 6.1).

Each drug needs to use its own model. For example, if we want to study an antithrombotic agent, we need a model that produces thrombi; if we have an antiedema drug, we need a model that produces edema, and so on.

CLASSIFICATION OF STROKE MODELS AND ANIMAL SPECIES USED IN DIFFERENT MODELS

In 2004, Murphy et al.[4] and Graham et al.[5] provided a detailed description of classifications of ischemic stroke and species used in different models. We have updated the classifications and included both ischemic stroke and brain hemorrhage in Table 6.1.

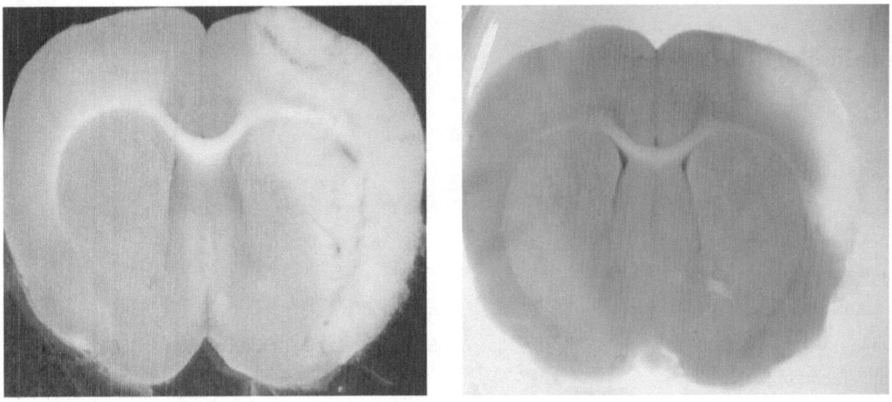

FIGURE 6.1 Proximal middle cerebral artery (MCA) occlusion produces a big infarction that involves cortical and subcortical areas (A). Distal occlusion produces a smaller infarction that only involves the cortex (B). **(See color insert following page 146.)**

This book introduces detailed techniques for MCAO models that are well established and accepted in neurological studies, including:

1. Distal MCAO model by craniectomy alone or with tandem occlusion of common carotid arteries (CCAs).
2. Intraluminal suture proximal MCAO models, including procedures with whole internal carotid artery and MCA occlusion and procedures with only MCA occlusion.
3. Thromboembolic stroke model.
4. Photochemical MCAO model and "ring" stroke model.

We also briefly introduce the techniques for induced subarachnoid hemorrhage and four-vessel occlusion models.

COMPARISON OF DIFFERENT MODELS

Table 6.2 compares the focal ischemia models based on a search of publications from 1975 to 2004 and our own experiences in using these models for many years.

TABLE 6.1
Classification of Stroke Models

Stroke Models	Animal Species
Ischemic Stroke	
Focal cerebral ischemia	
MCAO	Nonhuman primate, dog, cat, rabbit, guinea pig, rat, mouse
MCAO + ipsi- or bilateral common carotid artery occlusion (CCAO)	Dog, rat
Photochemical-initiated "ring" model[6]	Rat, mouse
Spontaneous brain infarction	Spontaneously hypertensive rat (SHR)
Multifocal cerebral ischemia	
Autologous or heterologous blood clot embolization	Rat, dog, swine[7]
Microsphere embolization	Rabbit, rat
Photochemical-initiated multifocal embolization	Rat
Complete global cerebral ischemia	
Decapitation	Rat
Aortic and vena caval occlusion	Dog
Neck tourniquet or cuff inflation	Nonhuman primate, dog, cat, rat
Cephalic artery occlusion (neck, thorax)	Nonhuman primate, cat
Cardiac arrest ± cardiopulmonary resuscitation	Nonhuman primate, dog, sheep, pig, rat, mouse
Bilateral CCAO	Gerbil
Incomplete global cerebral ischemia	
Hemorrhage/hypotension	Cat, pig
Hypoxia-ischemia	Dog, cat, sheep, pig, rat, mouse
Intracranial hypertension ± unilateral CCAO	Rat
Two-vessel occlusion ± hypotension	Rat, mouse
Four-vessel occlusion	Rat
Unilateral CCAO	Gerbil
Brain Hemorrhage Stroke	
Subarachnoid hemorrhage[8–11]	Rat, dog, rabbit, swine, nonhuman primates
Intracerebral hemorrhage[12–14]	Rabbit, rat, mice, pig, nonhuman primates

Notes: CCAO, common carotid artery occlusion; MCAO, middle cerebral artery occlusion.

TABLE 6.2
Comparison of Different Models

Models	Distal MCAO	Filament Model	Photochemical Embolic Model	Blood Clot Embolic Model
Year invented	1966	1986	1985	1982
Inventors	T.M. Sundt[15]	J. Koizumi[16]	B.D. Watson[17]	M. Kudo[18]
Journal of first publication	*Mayo Clin Proc*	*Jpn J Stroke*	*Ann Neurol*	*Stroke*
No. of laboratories/ countries used the model	32/10	160/27	83/17	116/20
No. of publications used the model	41	199	203	294
Mechanism	MCA cautery or ligation	Insert filament to block MCA	Photochemically induced clot	Inject blood clot to block MCA
Craniectomy	Yes	No	Yes	No
Permanent	Yes (cauterization)	Yes (suture remaining)	Yes	Yes
Transient	Yes (ligation)	Yes (suture removed)		
Time for surgery	15 to 20 minutes	15 to 20 minutes	10 minutes	25 minutes
Ease of preparing	A little difficult	Difficult	Easy	Very difficult
Reproducibility	+++	++	++++	+
Infarct area	Cortex	Cortex and subcortex	Cortex	Cortex and subcortex
Infarct size	++	++++	+	+++
Degree of edema	++	+++	++++	+++

Source: Publications searched with Medline from 1975 to 2004 by Kerry Kushinka. Additional observations from Drs. Wang-Fischer and Watson's unpublished data.

Notes: MCA, middle cerebral artery; MCAO, middle cerebral artery occlusion. + – ++++ indicates the arbitrary range or degree of damage.

REFERENCES

1. Ginsberg, M.D. and Busto, R., Rodent models of cerebral ischemia, *Stroke*, 20(12), 1627, 1989.
2. Guan, Q.H. et al., The neuroprotective action of SP600125, a new inhibitor of JNK, on transient brain ischemia/reperfusion-induced neuronal death in rat hippocampal CA1 via nuclear and non-nuclear pathways, *Brain Res*, 21, 1035(1), 51, 2005.
3. Merchenthaler, I., Dellovade, T.L., and Shughrue, P.J., Neuroprotection by estrogen in animal models of global and focal ischemia, *Ann N Y Acad Sci*, 1007, 89, 2003.
4. Murphy, S.J., McCullough, L.D., and Smith, J.M., Stroke in the female: Role of biological sex and estrogen, *ILAR J*, 45(2), 147, 2004.
5. Graham, S.M., McCullough, L.D., and Murphy, S.J., Animal models of ischemic stroke: balancing experimental aims and animal care, *Comp Med*, 54(5), 486, 2004.
6. Weste, P. et al., A photothrombotic "ring" model of rat stroke-in-evolution displaying putative penumbral inversion, *Stroke*, 26(3), 444, 1995.
7. Ringer, A.J., Guterman, L.R., and Hopkins, L.N., Site-specific thromboembolism: A novel animal model for stroke, *AJNR Am J Neuroradiol*, 25(2), 329, 2004.
8. Endo, H. et al., Akt/GSK3beta survival signaling is involved in acute brain injury after subarachnoid hemorrhage in rats, *Stroke,* 37(8), 2140, Epub 2006.
9. Hacein-Bey, L. et al., Reversal of delayed vasospasm by TS-011 in the dual hemorrhage dog model of subarachnoid hemorrhage, *AJNR Am J Neuroradiol,* 27(6), 1350, 2006.

10. Hemphill, J.C. III et al., Brain tissue oxygen monitoring in intracerebral hemorrhage, *Neurocrit Care,* 3(3), 260, 2005.
11. Clatterbuck, R.E. et al., Controlled release of a nitric oxide donor for the prevention of delayed cerebral vasospasm following experimental subarachnoid hemorrhage in nonhuman primates, *J Neurosurg,* 103(4), 745, 2005.
12. Thai, Q.A. et al., Lysis of intracerebral hematoma with stereotactically implanted tissue plasminogen activator polymers in a rabbit model, *J Neurosurg,* 105(3), 424, 2006.
13. Gong, Y. et al., Systemic zinc protoporphyrin administration reduces intracerebral hemorrhage-induced brain injury, *Acta Neurochir Suppl*, 96, 232, 2006.
14. Roertgen, K.E. et al., A beta-associated cerebral angiopathy and senile plaques with neurofibrillary tangles and cerebral hemorrhage in an aged wolverine (Gulo gulo), *Neurobiol Aging,* 17(2), 243, 1996.
15. Sundt, T.M., Jr. and Waltz, A.G., Experimental cerebral infarction: Retro-orbital extradural approach for occluding the middle cerebral artery, *Mayo Clin Proc* 41, 159–168, 1966.
16. Koizumi, J., Yoshida, Y., Nakazawa, T., and Ooneda, G., Experimental studies of ischemic brain edema. I: A new experimental model of cerebral embolism in rats in which recirculation can be introduced in the ischemic area. *Jpn J Stroke* 8, 1–8, 1986.
17. Watson, B.D., Dietrich, W.D., Busto, R., Wachtel, M.S., and Ginsberg, M.D., Induction of reproducible brain infarction by photochemically initiated thrombosis, *Ann Neurol* 17, 497–504, 1985.
18. Kudo, M., Aoyama, A., Ichimori, S., and Fukunaga, N., An animal model of cerebral infarction. Homologous blood clot emboli in rats. *Stroke* 13(4), 505–508, 1982.

7 Failure Is the Mother of Success

Why Neuroprotective Therapies That Work in Laboratories Fail in Clinical Trials

Lee Koetzner and Yanlin Wang-Fischer

CONTENTS

THE CHALLENGE

Neuroprotective drugs, while promising in experimental models, have nearly all failed in clinical trials.[1–11] The failure has been so profound as to lead some to question whether stroke pharmacotherapy is "a fantasy invented by basic scientists."[1] Indeed, when Kidwell et al.[4] reviewed articles for clinical trials of ischemic stroke therapies through 1999 (178 clinical studies), they found that about 2% of these trials met strict criteria defining them as having a positive outcome. These included rt-PA (recombinant tissue plasminogen activator),[5] prolyse,[6] low molecular weight heparin,[7] and ancrod.[8] The successful translation of these trials from preclinical research to clinical trials has demonstrated that stroke is a treatable disorder in the hyperacute stage.

There is a common saying in Chinese: "Failure is the mother of success." To serve as the midwife for the translation of a stroke therapeutic from preclinical program to successful trial, one must understand previous failures. These can happen in any of three points: at the preclinical stage, at the clinical stage, or in the translation from one stage to the next. The myriad factors that are important to preclinical models are discussed at length throughout this volume; for a review of fundamental experimental design issues that can compromise model studies, consult Chapter 24. This chapter focuses on the path to clinical data and the aspects of trial design that may predispose to failure.

THE CHALLENGE IN TRANSLATION OF PRECLINICAL RESEARCH TO CLINICAL TRIALS

WHEN IS THE TREATMENT GIVEN?

Preclinical studies use very short time windows, whereas clinical trials use longer time windows. Most neuroprotective studies in animal models have shown positive effects with drug administration either before the ischemic insult or very soon after the onset of ischemia.[4,12,13] For example, in our study of dextrorphan (30 mg/kg i.p.) in rats, drug administration just after middle cerebral artery occlusion decreased the infarct volume, whereas giving dextrorphan 24 hours after stroke showed no effect on infarct volume. In contrast, in clinical trials, the time windows for entry into acute neuroprotection trials have been longer and highly variable.[4] Trials at short time intervals should not be impossible to run: Successful trials of thrombolytics have stayed within a short post-stroke interval.[14]

HOW ARE THE DOSES CHOSEN?

The strategies used for selection of drug doses differ between preclinical and clinical trials.[2] Specifically, preclinical drug treatments are usually dosed to effect: The dose is increased until an effect is observed, and side effects are only assessed after a demonstration of efficacy. In contrast, regulatory concerns drive clinical trials to start with low doses and only increase doses in the absence of untoward effects. Potential side effects are usually monitored very closely, in accordance with clinical trial plans set in place before patients are dosed. As a result, preclinical studies offer few data on treatment tolerability, and clinical studies may never approach active doses.[1,2,14]

HOW AND WHEN ARE THE RESULTS MEASURED?

Studies in animals have relied on infarct size, measured early, to judge therapeutic efficacy, whereas clinical trials have relied on behavioral outcomes, measured some time after stroke.[1,15] These discrepancies suggest that measurement of infarct size alone in animal study may be misleading as an indicator of therapeutic efficacy in clinical trials.[16,17] Therefore, assessment of therapeutic efficacy in animal studies should require, in addition to infarct size, functional measures of motor, sensory, and cognitive deficits,[18] such as rotorod performance, foot fault test, and sticky tape test. Ideally, the time course of these treatment-induced changes should be established.[19,20]

THE CHALLENGE IN CLINICAL TRIALS

IS THE STUDY POWERED CORRECTLY?

A poorly designed clinical trial will never give useful results. Some stroke trials have lacked the statistical power to show treatment effects.[1] To detect the efficacy of neuroprotective compounds in traditional trial designs, large trials are necessary (thousands of patients, according to some experts) to prevent type 2 statistical error.[16] Alternative designs may prove more economical; the use of "adaptive randomization" may reduce sample size requirements.[19,20]

ARE THE PATIENTS SIMILAR?

Individual patients present different stroke syndromes; this creates another challenge in clinical trial design. Even among patients with occlusive strokes, the site(s) of occlusion, type of occlusion, severity of symptoms, and general health and comorbid diseases will all vary.[1] Good experimental design will ensure that stroke severity is balanced across treatment groups. However, the inclusion of a large number of severely affected patients could introduce a floor effect (that is, disability so great it overwhelms possible drug effects). Even more troubling is the possibility that different stroke pathologies might be differentially responsive to therapeutic interventions. In this case, patients with a nonresponsive variety of stroke could lead to an underestimate of therapeutic activity (if the patients were randomly distributed among treatment groups) or a complete failure to observe activity (if the patients were predominant in one group).[4]

Is the Penumbra a Factor?

The penumbra poses a final challenge in the design of clinical trials. As some scientists[14–16] have emphasized, the target of current neuroprotective therapy is the penumbra, ischemic tissue that surrounds the infarct core and is functionally impaired but with damage that is potentially reversible.[17,18] If reversible ischemic tissue is not present at the time of treatment, then neuroprotective therapy cannot be expected to work. However, the extent of the penumbra is difficult to establish in trials and appears highly variable.[19] In the future, clinical trials may need to set up stricter entry criteria to target patients with a sufficient penumbra volume.

The historical record does not offer many encouraging examples of successful stroke medication development.[22–34] However, paying close attention the preclinical and clinical stages and the translation between stages offers the opportunity for these failures to give birth to a successful trial.

REFERENCES

1. Gladstone, D.J., Black, S.E., and Hakim, A.M., Toward wisdom from failure lessons from neuroprotective stroke trials and new therapeutic directions, *Stroke*, 33, 2123, 2002.
2. Danton, G.H. and Dietrich, W.D., The search for neuroprotective strategies in stroke, *Am J Neuroradiol*, 25, 181, 2004.
3. Markgraf, C.G., Kraydieh, S., Prado, R., Watson, B.D., Dietrich, W.D., and Ginsberg, M.D., Comparative histopathologic consequences of photothrombotic occlusion of the distal middle cerebral artery in Sprague-Dawley and Wistar rats, *Stroke*, 24, 286, discussion 292, 1993.
4. Kidwell, C.S., Liebeskind, D.S., Starkman, S., and Saver, J.L., Trends in acute ischemic stroke trials through the 20th century, *Stroke,* 32, 1349, 2001.
5. National Institute of Neurological Disorders and Stroke rt-PA Stroke Study Group, Tissue plasminogen activator for acute ischemic stroke, *N Engl J Med,* 333, 1581, 1995.
6. Furlan, A. et al., Intra-arterial prourokinase for acute ischemic stroke: The PROACT II study: A randomized controlled trial: Prolyse in acute cerebral thromboembolism, *JAMA,* 282, 2003, 1999.
7. Kay, R. et al., Low-molecular-weight heparin for the treatment of acute ischemic stroke, *N Engl J Med,* 333, 1588, 1995.
8. Sherman, D.G. et al., Intravenous ancrod for treatment of acute ischemic stroke: The STAT study: a randomized controlled trial: Stroke treatment with ancrod trial, *JAMA,* 283, 2395, 2000.
9. Martinez-Vila, E. and Sieira, P.I., Current status and perspectives of neuroprotection in ischemic stroke treatment, *Cerebrovasc Dis,* 11 (Suppl 1), 60, 2001.
10. Fisher, M. and Schaebitz, W., An overview of acute stroke therapy: Past, present, and future, *Arch Intern Med,* 160, 3196, 2000.
11. Fisher, M., Neuroprotection of acute ischemic stroke: Where are we? *Neuroscientist,* 5, 392, 1999.
12. Jonas, S. et al., Does effect of a neuroprotective agent on volume of experimental animal cerebral infarct predict effect of the agent on clinical outcome in human stroke? *Ann N Y Acad Sci,* 825, 281, 1997.
13. Grotta, J.C., Acute stroke therapy at the millennium: Consummating the marriage between the laboratory and bedside: The Feinberg lecture, *Stroke,* 30, 1722, 1999.
14. Fisher, M., Characterizing the target of acute stroke therapy, *Stroke,* 28, 866, 1997.
15. Baron, J.C., Perfusion thresholds in human cerebral ischemia: Historical perspective and therapeutic implications, *Cerebrovasc Dis,* 11 (Suppl 1), 2, 2001.
16. Lees, K.R., Advances in neuroprotection trials, *Eur Neurol,* 45, 6, 2001.
17. Hakim, A.M., The cerebral ischemic penumbra, *Can J Neurol Sci,* 14, 557, 1987.
18. Hakim, A.M., Ischemic penumbra: The therapeutic window, *Neurology,* 51, S44, 1998.

19. Heiss, W.D., Thiel, A., Grond, M., and Graf, R., Which targets are relevant for therapy of acute ischemic stroke? *Stroke,* 30, 1486, 1999.
20. Duncan, P.W., Jorgensen, H.S., and Wade, D.T., Outcome measures in acute stroke trials: a systematic review and some recommendations to improve practice, *Stroke,* 31, 1429, 2000.
21. Sacco, R.L. et al., The Glycine Antagonist in Neuroprotection Americas investigators: glycine antagonist in neuroprotection for patients with acute stroke: GAIN Americas: A randomized controlled trial, *JAMA,* 285, 1719, 2001.
22. Hunter, A.J., Mackay, K.B., and Rogers, D.C., To what extent have functional studies of ischaemia in animals been useful in the assessment of potential neuroprotective agents? *Trends Pharmacol Sci,* 19, 59, 1998.
23. Corbett, D. and Nurse, S., The problem of assessing effective neuroprotection in experimental cerebral ischemia, *Prog Neurobiol,* 54, 531, 1998.
24. Alonso de Lecinana, M., Diez-Tejedor, E., Carceller, F., and Roda, J.M., Cerebral ischemia: from animal studies to clinical practice: Should the methods be reviewed? *Cerebrovasc Dis,* 11 (Suppl 1), 20, 2001.
25. Demchuk, A.M. and Buchan, A.M., Predictors of stroke outcome, *Neurol Clin,* 18, 455, 2000.
26. Jorgensen, H.S. et al., Potentially reversible factors during the very acute phase of stroke and their impact on the prognosis: is there a large therapeutic potential to be explored? *Cerebrovasc Dis,* 11, 207, 2001.
27. Counsell, C. and Dennis, M., Systematic review of prognostic models in patients with acute stroke, *Cerebrovasc Dis,* 12, 159, 2001.
28. Boysen, G. and Christensen, H., Early stroke: A dynamic process, *Stroke,* 32, 2423, 2001.
29. Kagansky, N., Levy, S., and Knobler, H., The role of hyperglycemia in acute stroke, *Arch Neurol,* 58, 1209, 2001.
30. Demchuk, A.M. et al., Serum glucose level and diabetes predict tissue plasminogen activator-related intracerebral hemorrhage in acute ischemic stroke, *Stroke,* 30, 34, 1999.
31. Capes, S.E. et al., Stress hyperglycemia and prognosis of stroke in nondiabetic and diabetic patients: A systematic overview, *Stroke,* 32, 2426, 2001.
32. Sandercock, P. et al., Mega trials versus small trials in stroke. In: Fisher, M. and Bogousslavsky, J., eds., *Current Review of Cerebrovascular Disease*, Current Medicine, Philadelphia, 2001, p. 241.
33. Malakoff, D., Bayes offers a "new" way to make sense of numbers, *Science,* 286, 1460, 1999.
34. Stroke Therapy Academic Industry Roundtable, Recommendations for standards regarding preclinical neuroprotective and restorative drug development, *Stroke*, 30, 2752, 1999.

8 Anesthesia of Laboratory Rats

Yanlin Wang-Fischer and Lee Koetzner

CONTENTS

OBJECTIVES

The objectives of anesthesia on laboratory rats are to provide humane restraint, a reasonable degree of muscle relaxation so that surgery can be done easily, and most importantly, a sufficient degree of analgesia or loss of consciousness to prevent the animal from experiencing pain and to decrease the variability of experimental outcome due to distress.

PREANESTHESIA CONSIDERATIONS

It is important to use expert care when handling experimental animals, with consideration of ways to minimize their pain and stress during surgical procedures. It is essential to ensure that the animals are healthy before anesthesia inasmuch as anesthesia disturbs normal physiology and can induce severe distress. Animals with chronic disease have a high degree of risk during anesthesia. Biological factors other than the health of experimental animals must also be taken into account, such as variations in body weight and age, differences between strains, sex, nutrition, acclimatization, and endocrine status (female rats), all of which may influence outcomes such as efficacy of the test drug. Further, animals that are frequently handled and gently stroked are much easier to anesthetize.

The reasons for imposing fasting at 12 to 24 hours before anesthesia depend on the study purpose. If the study results could be influenced by mild hyperglycemia, the animals should be fasted overnight. However, in some species such as dogs, cats, pigs, and the like, preanesthesia fasting is necessary to reduce vomiting and aspiration of vomit; fasting is not necessary in rats because they do not vomit.

SCIENTIFIC VALIDITY OF RAT MODELS AFTER ANESTHESIA

Researchers induce disease models in which procedures are carefully defined and for which the outcomes have the smallest attainable degree of variability. If a rat is allowed to recover following surgery, then it should be maintained with sufficient care to return to physiological normality, or to a defined state of abnormality, as soon as possible. These scientific goals are easily interfered with by poor anesthesia practice. Pain, fear, and distress can become uncontrolled variables and will interfere with proper conduct of a research protocol. Similarly, if an animal does not eat or drink well for 24 to 48 hours but is nonetheless scheduled for experiment, the animal will be at risk of developing severe hypothermia, dehydration, respiratory acidosis, and hypoxia—a clearly undesirable situation if the model is to yield stable results.

RESPONSE TO ANESTHETIC AND SURGICAL STRESS

It is important to consider the effects on the experimental protocol of the interaction of anesthesia with the surgical procedures. The response in rats to anesthetic and surgical stress is similar to that in humans and is based on mobilization of reserves of substrates, such as glucose and amino acids, to facilitate survival of important organs. This response has clear evolutionary advantages but is sometimes inappropriate in humans and rats.[1]

The major aspects of this response include elevation in plasma concentrations of different hormones,[1–6] including catecholamines, corticosteroids,[3,4] growth hormone, vasopressin, renin, aldosterone, thyroid hormones, and prolactin[5] and decreases in plasma concentration of follicle-stimulating hormone (FSH), luteinizing hormone (LH) and testosterone.[1] The effects of surgical and anesthetic procedures on plasma insulin and glucagon are more complicated. Initially, insulin concentrations decrease at 2 hours and increase at 4 hours postoperation,[2] but glucagon concentrations increase initially and decrease later. Grundy et al. (2001)[3] and DeKeyser et al. (2000)[4] studied the hypothalamus-pituitary-adrenal (HPA) axis response to experimental traumatic injury. Their data indicated that anesthesia associated with surgery or sham surgery induced a generalized activation of the HPA axis. These hormonal responses to tissue trauma produced an increase in glycogenolysis and lipolysis, resulting in hyperglycemia. The duration of hyperglycemia varies but after major surgery may persist for 4 to 6 hours. More prolonged changes in protein metabolism occur as well, leading to negative nitrogen balance that lasts for several days.[7] Even relatively minor surgical procedures can produce prolonged effects. For example, blood vessel cannulation in rats evinced an elevation of corticosterone for several days.[8]

Sometimes, investigators are reluctant to refine their anesthetic methods because a new method may affect the postoperative response of the animal model and thus complicate data interpretation. In some instances, there are scientific reasons to avoid certain agents, but more often, the effects of anesthesia are overshadowed by those induced by surgical stress. Of similar concern, administration of analgesics could alter the fundamental nature of an animal model—even to the extent of superimposing neuroprotective effects on models of stroke.

ASSESSMENT OF DEPTH OF ANESTHESIA

Anesthesia is applied to eliminate the perception of pain; therefore, it is essential, by definition, that painful stimuli elicit no response. Indicators to assess the depth of anesthesia in rats include the following:

1. Reflex tests such as pinching the tail or the interdigital skin (pedal reflex) or abdominal skin with a hemostat are common. Other common reflexes for assessment of central nervous system (CNS) function include eye closure response (the palpebral reflex, which is relatively insensitive to anesthesia), head shake following pinna stimulation, and jaw closure.
2. Changes in skin color in the rat are best seen at the ear, nose, and feet (albino rat) and in the mucosa of the mouth. If the animals become cyanotic or extremely pale, a low oxygen level in the blood is indicated, mostly caused by poor circulation.
3. Rate and depth of respiration (particularly useful with inhalation anesthesia) are assessed.
4. The color and arterial beat of the common carotid artery supply useful information during the surgical procedure.

To classify depth of anesthesia into different stages or planes in rats, we adopted Waynforth and Flecknell's method.[9]

1. Light anesthetic stage: The animal loses its righting reflex and is immobilized but responds markedly to painful stimuli.
2. Light surgical anesthetic stage: A deeper level of anesthesia is attained, allowing minor superficial surgical procedures such as a skin biopsy to be undertaken.
3. Medium surgical anesthetic stage: This stage is used to perform some procedures such as laparotomy without the animal moving in response to surgical stimuli.
4. Deep surgical anesthesia: Deep surgical anesthesia is used to perform some major procedures or when operating on particularly sensitive structures, for example, drilling a hole on cranium.

ASSESSMENT OF ANESTHESIA METHODS

To conduct a study successfully, it is necessary to assess carefully the available methods of anesthesia and to minimize the interaction between the anesthetic regimen and the particular animal model. Achieving this goal is difficult, but a careful assessment of the available alternative anesthetics and their particular physiological and pharmacological effects can at least minimize this interference. It must be appreciated that publication of anesthesia methods in detail does not ensure that the investigators have carried out this type of assessment. Simply adopting a method of anesthesia that is published for an application to a particular animal model will not necessarily ensure that this technique will be appropriate in a new study using the same model. For example, ketamine plus xylazine can be used for surgical procedures of vessel cannulation but cannot be used for stroke study since ketamine has neuroprotective effects, making it difficult to explain the results.

The response of rats to anesthetics varies considerably among different strains.[10] A dose rate that anesthetizes one strain of rat may be ineffective in another; for example, in our experience, 30 mg/kg i.p. sodium pentobarbital anesthetizes Sprague-Dawley rats very well but fails to do so in Zucker diabetic fat rats. Strain-sensitive variations in anesthesia response have not been well characterized in rats. The response to anesthesia of male and female rats is also different in our study (unpublished data).

Owing to these variations in response and unknown side effects of anesthesia methods, a pilot study will help the larger study. In pilot studies, only rats are anesthetized before beginning a formal study to assess the depth of anesthesia and its aftermath over several days and to ensure that recovery has been completed without untoward side effects. In some instances, the method of anesthesia may interfere with the purpose of the study. For example, we found that use of 100% CO_2 or a mixture of 30% O_2 plus 70% CO_2 to anesthetize rats could destroy the blood–brain barrier, which could interfere with any study of CNS drugs (Figure 8.1).

Studies have shown that some anesthetics, such as ketamine, may evince neuroprotective effects in rats (but this has not been shown in humans). Zhang et al. (2004)[11] studied the effects of ketamine-midazolam anesthesia on focal cerebral ischemic injury in rats. They found that, compared with ketamine alone, anesthesia with the combination of ketamine and midazolam may provide neuroprotection from ischemic cerebral injury following middle cerebral artery occlusion (MCAO) in rats, as evidenced by reduction in infarct size and reduced numbers of apoptotic cells. Brunson et al. (2001)[12] studied the effect of the noncompetitive NMDA (*N*-methyl-d-aspartate) antagonists MK-801 and ketamine on the spastic Han-Wistar mutant, used as a rat model of excitotoxicity. They found that ketamine and MK-801 have some neuroprotective effects on these mutant rats. Spandou et al. (1999)[13] studied the effect of ketamine on hypoxic-ischemic brain damage in newborn rats and found that ketamine offered partial protection. On the other hand, in a clinical study of 106 patients undergoing cardiopulmonary bypass surgery, in 2004 Nagels et al.[14] compared the effect

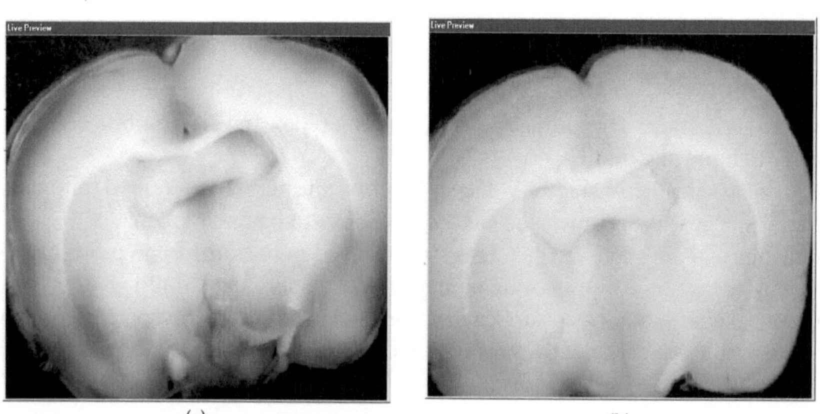

(a) (b)

FIGURE 8.1 Results from studies with rats perfused (35 minutes) with 2% triphenyltetrazolium chloride (TTC) showed that CO_2 anesthesia damaged the blood–brain barrier (BBB). Sprague-Dawley adult rats sacrificed under a mixture gases of 70% CO_2 and 30% O_2 (a) or pentobarbital (b). The red color in the brain tissue indicates the BBB leakages. **(See color insert following page 146.)**

on neurocognitive outcome of S(+)-ketamine to remifentanil, both alone and in combination with propofol, but could not demonstrate any relative neuroprotective advantage due to S(+)-ketamine.

To reiterate, an anesthetic regimen must be chosen carefully to implement the purpose of a proposed study without confounding it. After establishing that the anesthetic regimen is safe and effective and that its side effects do not interfere fundamentally with the purpose of the study, it can be used with larger numbers of animals.

SELECTING AN ANESTHESIA TECHNIQUE

When facing a long list of anesthetic drugs, it is very difficult to justify an arbitrary choice of one or two of these. The chosen anesthetic procedure should cause minimum interference with the conduct and outcome of the experiment and the interpretation of results. Another important factor is the nature and the duration of the surgical procedure to be performed.

Anesthesia can be induced by either inhalable or injectable anesthetics. Here, it is advisable to use inhalation anesthetics for stroke studies because most injectable anesthetics are neuroprotective, whereas inhalation agents are largely expelled via the lungs with only a very small portion of the drugs metabolized by the liver. An increase in cerebral spinal fluid pressure, however, may arise through the vasodilator action of most inhalational agents, but this increase can be prevented or reversed by hyperventilating the rat before or during anesthesia (see the sections on isoflurane and halothane).

This book gives more details about inhalational anesthetics than about injectable agents. A vast range of injectable anesthetics is available, for which detailed, in-depth information is provided in the book by Kohn et al. (1997).[15]

INHALATIONAL ANESTHETICS

OPEN METHOD: COTTON WOOL SWAB IN A GLASS JAR

Open methods for laboratory rat anesthesia are widely used because they are relatively simple and inexpensive. A cotton wool swab is soaked with a volatile anesthetic and placed in a glass jar before introducing an animal (Figure 8.2). However, this method is inefficient and wasteful and may release high concentrations of unused anesthetic gas into the environment, which can be hazardous to humans. It is also hard to maintain a stable and profound degree of surgical anesthesia. This method is not recommended in experimental stroke surgery. An anesthesia machine eliminates many of these drawbacks.

FIGURE 8.2 A glass jar with a metal filter inside; a cotton wool swab with volatile anesthetic solution can be placed beneath the metal filter. A rat can be anesthetized in the jar.

INHALATION THROUGH AN ANESTHESIA APPARATUS

With the correct apparatus, administration of inhalational anesthetics and control of their effects on brain vasculature become much easier. It is thus highly advisable to use an anesthesia machine in conjunction with an appropriate stroke model.

Next, the needed apparatuses are described.

FIGURE 8.3 Downdraft table for inhalational anesthetics. The table is specially installed and has a perforated tabletop that draws air evenly through the small openings. The tabletop has no large openings to create areas of high-velocity airflow and noise or for tools to drop through. The height of the table is adjustable. A gas tube system for oxygen or mixed gases is installed into the table.

Downdraft Table or Fume Hood

A special downdraft surgical table is needed to scavenge the waste anesthetic gases. Figure 8.3 shows a very efficient open downdraft table used by many laboratories in the pharmaceutical and biotech industry. Since gases (100% O_2, air, or a mixture of 70% N_2O [nitrous oxide] and 30% O_2) are needed for vaporizing inhaled anesthetics, this table is fitted with gas outlets. A general surgical room may contain several of these tables for doing surgeries simultaneously. (The table in Figure 8.3 was installed by TBJ, Chambersburg, Pennsylvania.)

An alternative to the open downdraft table is the downdraft booth by Dualdraw, which offers an efficient and economical way for companies or academic laboratories to protect their workers from harmful gases. Figure 8.4 is a representative sample of their products.

FIGURE 8.4 The downdraft booth manufactured by Dualdraw. With the downdraft booth, working on parts high above the table is not a problem. The downdraft table has a perforated table and back wall that evenly draw air contaminants like waste anesthetic gases through 2500 openings away from the operators' breathing zones. (Courtesy of Dualdraw, Commerce City, Colorado.)

Gas Filters or Chemical Scavengers

Gas filtering means that the gas is percolated through an absorbent material, such as activated charcoal, which binds (adsorbs) the anesthetic molecules while allowing the oxygen and CO_2 to pass through. Activated charcoal, when used as a filter medium, has oversight requirements as well. Once the charcoal is saturated, a new infusion of anesthetic gases will *not* be adsorbed. Activated charcoal can adsorb roughly 25% of its own weight. Small charcoal canisters should have their saturation levels monitored by weight on a daily basis. Protocols should include weighing the filter canister prior to use, recording the initial weight on the canister, and then reweighing the canister. Once saturated, the filter should be discarded. Figure 8.5 shows a charcoal filter.

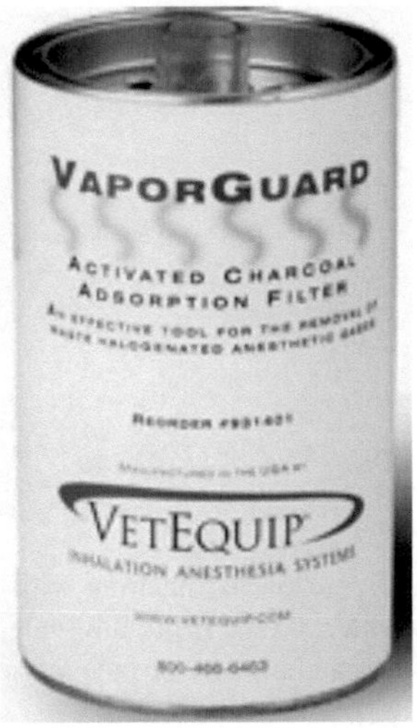

FIGURE 8.5 VaporGuard activated charcoal absorption filter from VetEquip (Item No. 931401). (Courtesy of VetEquip, Pleasanton, California.)

Compared to installing a downdraft table, this kind of filter is convenient to use and cost efficient; however, if you perform a large number of surgical procedures, you may need to change the charcoal every day or sometimes every couple of hours. This will increase cost.

Induction Chamber

An induction chamber is necessary to anesthetize rats. Figure 8.6 shows a simple plastic chamber purchased through VWR.

Anesthesia Machine and Its Accessories

It is recommended that every research laboratory engaged in rat stroke studies invest in an anesthesia machine and its accessories. This system has the following benefits:

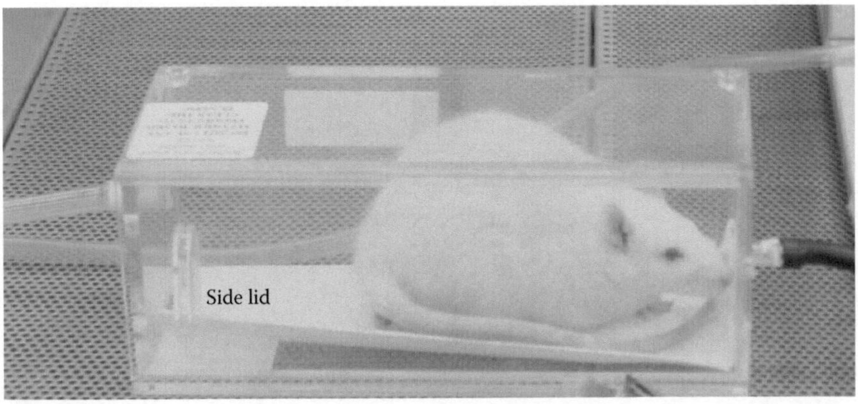

FIGURE 8.6 A transparent plastic induction chamber. The chamber has two outlets: One is connected to the anesthetic machine with anesthetic gas coming into the chamber; another one is connected to a tube for extracting the waste gas, which is drawn out by the downdraft table.

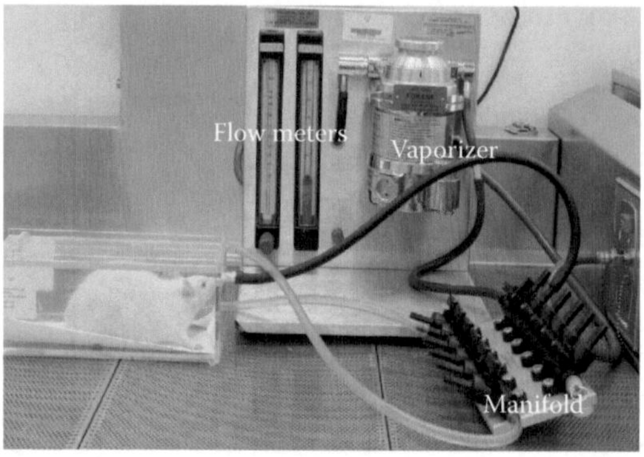

FIGURE 8.7 Anesthetic machine with the manifold connected to an induction chamber. It has precision flow meters with a 0 to 1 L/minute scale for O_2 and a calibrated vaporizer for isoflurane. The manifold can connect to several chambers or face masks for working on several animals simultaneously.

1. Simplicity of use: Instructions are straightforward and easy to follow.
2. The system supplies a specific rate of gas flow to each animal, which can be adjusted to accommodate changes in the number of circuits in use or length of tubing.
3. One vaporizer and one to six individual connections (see Figure 8.7) can supply as many circuits or chambers as needed, in any combination, which allows anesthesia and surgical procedures to be performed on multiple rats in parallel.

Many different anesthesia machines are available commercially to fit specific needs. Figure 8.7 shows the machine used in our laboratory. (The system was installed by System Specialties, 1800 Mearns Road, Building 3T, Warminster, Pennsylvania 18974, 215-443-9293.) Figure 8.8 is a machine from Colonial Medical Supply (504 Wells Road, Franconia, New Hampshire 03580, 888-446-8427). This delivery system is accurate, dependable, safe for the operator and animal as well, and cost-effective. The system uses a precision-machined flow meter with a scale of either 0 to 1000 mL per minute or 0 to 6 L per minute. The flow meter controls the amount of carrier gas to the vaporizer, which then flows to your choice of delivery devices.

A manifold (Figure 8.7 and Figure 8.9) can be employed to direct the gas to multiple chambers or a breathing face mask (Figure 8.10 and Figure 8.11) as well as to collect the waste gas and direct it to a single collection system.

FIGURE 8.8 Anesthetic machine from Colonial Medical Supply shows the flow meter and a vaporizer for isoflurane, which maintains accurate flow to the chamber or to the animal mask. (Courtesy of Colonial Medical Supply [cms@colmedsupply.com].)

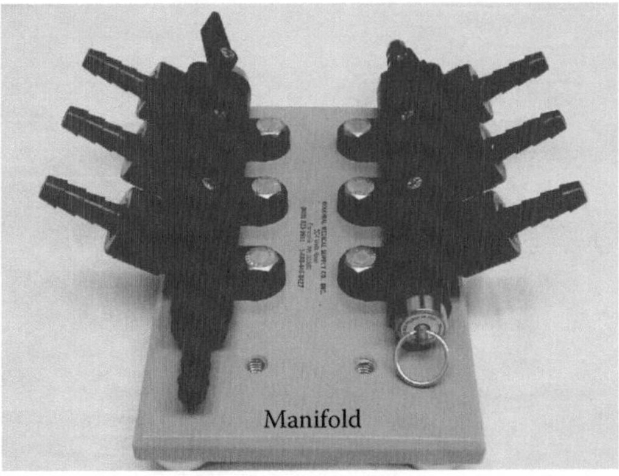

FIGURE 8.9 Manifold connects to the anesthetic machine, directs the gas to multiple stations, collects the waste gas, and directs it to a collection system.

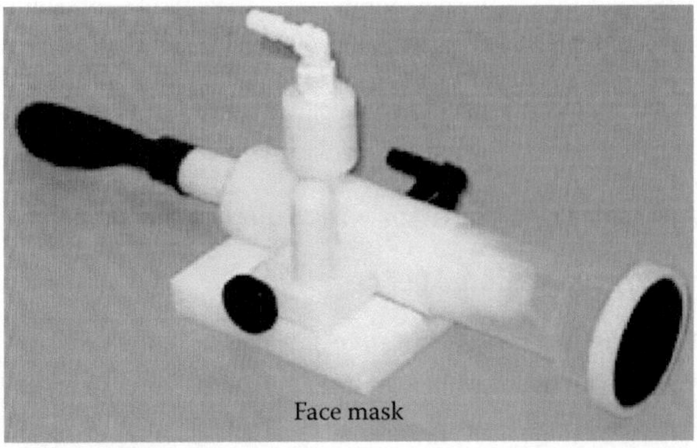

FIGURE 8.10 A face mask for anesthetic maintenance. A balloon is attached in the back portion. It flaps when the rat breathes. The black diaphragm covers the tube with a cross cut in the center of diaphragm to fit the rat's nose.

Pretesting the Waste Gas

Anesthetic waste gases can include any gas from the anesthesia machine that passes through the subject animal without being inhaled as well as any metabolic gases the animal exhales. Trace gas can occur due to leaking equipment, the method of filling vaporizers, spillage, and the like. It is assumed that prolonged and repetitive exposure to anesthetic gases can be toxic to personnel in the procedure area. The recommended exposure limits for halogenated anesthetic agents is 2 parts per million (ppm) for 1 hour.[16] NIOSH (National Institute of Occupational Safety and Health) made this recommendation in 1978, when modern anesthesia agents such as isoflurane, sevoflurane, and desflurane were unknown; while comments have been solicited for a revision, a draft has not been published. Similarly, although there is a recommended exposure limit of 25 ppm for nitrous oxide, final standards have not been published.[17] Additional regulations may exist in some jurisdictions (for example, in California).

Because gas is constantly being delivered to the animal, there is always some excess that either must be evacuated from the area or filtered to remove the entire amount of exhaled anesthetic agent. Testing must be done to ensure that the concentration of waste gas in the work area is lower than the recommended or exposure limits. Laboratories are available to test the waste samples. For this purpose, our laboratory employs Broadspire (95 Oakwood Road, Lake Zurich, Illinois 60047, 888-576-7522).

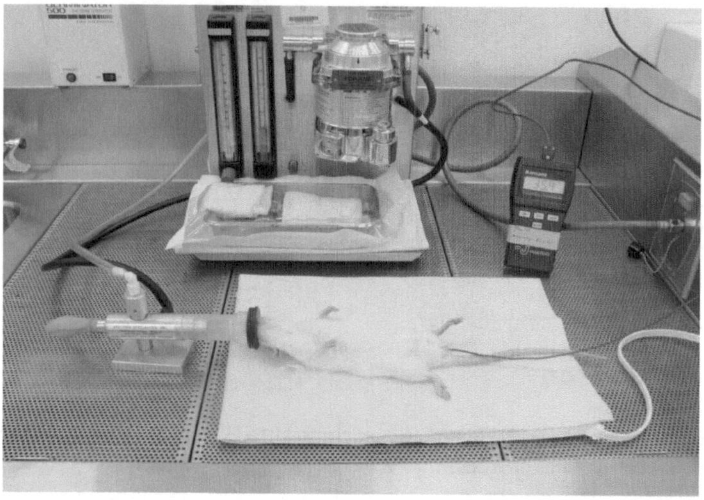

FIGURE 8.11 An anesthetic machine set on a downdraft table. Anesthesia is maintained by placing the animal's nose into the face mask. The machine has precision flow meters for O_2 and a calibrated vaporizer for isoflurane. The body temperature is maintained at 36°C to 37°C by placing the rat on a heating pad and monitoring through a rectal thermometer. The special downdraft tables connect to a central exhaust tube system and draft the waste gases into the exhaust system to be sent out of the building. (Courtesy of Kent Scientific Corporation.)

Induction and Maintenance of Anesthesia with Inhalation Agents

Induction of Anesthesia by Inhalation Agents

Induction of anesthesia is straightforward, but many investigators are somewhat daunted at first by the apparatus. It is useful to provide a step-by-step guide for new investigators; the guide can be attached to or displayed beside the apparatus. The guide shown in Table 8.1 is used in our laboratory and can be modified for use with different apparatuses.

Problems of Overdose and Its Resolution

If a rat starts to gasp and make violent respiratory movements or its respirations become very shallow, overanesthesia is indicated. Remove the rat from the mask or chamber until it begins to breathe normally and then continue but with a reduced concentration of isoflurane. If the rat stops breathing, remove it from the chamber or mask, lay it on its back, and gently squeeze (45 to 60 times/minute) its chest between the thumb and forefinger to help its breathing. When the depth and frequency of breathing recover to normal, the animal may be reanesthetized. But, if the breathing was interrupted for too long (over 3 minutes) the animal should be discarded because hypoxia would likely interfere with its response to the intended stroke model.

The use of an anesthetic chamber made of inert transparent material such as plastic is probably the easiest and least stressful means of inducing anesthesia with a volatile anesthetic (Figure 8.7). A layer of paper towels should be placed on the floor of the chamber to absorb any urine and to ease cleaning. All of the commonly used volatile anesthetics (halothane, methoxyflurane, enflurane, and isoflurane) produce a smooth induction of anesthesia. Induction is slower with methoxyflurane, but this can be advantageous for an inexperienced investigator. A flow of 0.8 to 1 L/minute of oxygen is suitable for use in a small induction chamber (for example, 25 × 10 × 10 cm) (see Figure 8.6 and Figure 8.7). When agents are delivered with nitrous oxide, the second gas effect will speed induction.[18]

The recommended concentrations of volatile agents for induction are summarized in Table 8.2. When the rat has lost its righting reflex and remains immobile in the chamber, it can be removed. Induction takes about 2 or 3 minutes. After induction, a deep anesthesia persists for 60 to 90 seconds;

TABLE 8.1
Instructions for Using an Anesthetic Machine for Rodent (Using Isoflurane)

a. Turn on the general oxygen cylinder using the switch on the oxygen regulator. The regulator will show you the pressure and how much oxygen is in the cylinder.

b. Check that the tubing from the anesthetic machine is connected to the induction chamber or manifold or a rat nose mask.

c. Check the level of liquid anesthetic in the vaporizer. If necessary, refill the liquid anesthetic.

d. Turn on the oxygen on the anesthetic machine. Adjust the gas flow rate to about 800 mL/minute or about 1 L/minute O_2. The total gas flow from the anesthetic machine should at least be 2.5 to 3 times the minute volume of the rat (about 220 mL/minute) (i.e., 550 to 800 mL/min).

e. Pressing down the vaporizer switch, turn to the appropriate settings at 3 (3%); you can use 4% to 5% to speed up the induction time, but these concentrations can result in overdose and death.

f. Turn on the switch on the manifold so the anesthetic gas can get into the chamber or the rat face mask.

g. Place a rat into the induction chamber.

h. When the rat has lost its righting reflex and is sleeping quietly, turn off the switch on the manifold, move the rat out of the chamber for shaving the fur, and close the lid of the induction chamber.

i. Place the rat's nose in the face mask after shaving.

j. Reduce the vaporizer setting to 2% to 2.5% for isoflurane; rats vary slightly in their response to the anesthetic.

k. Repeat the above steps for multiple-animal surgeries.

l. After completion of the surgery, place the rat in an intensive care unit box. It should recover in 5 to 10 minutes and can be returned to a clean cage.

m. At the end of procedure, turn off the vaporizer, flow meter, and oxygen cylinder.

n. Clean the induction chamber and intensive care unit box.

during this period, a brief procedure can be carried out without using a face mask (for example, shaving the fur, orbital bleeding, or intranasal drug administration). Full recovery of consciousness will take a few minutes, depending on the duration of anesthesia.

Gas (O_2 or mixture gases of 70% N_2O and 30% O_2) is delivered at a flow rate of 550 to 800 mL/minute when used with isoflurane. Note that some varieties of vaporizer, even newer models, can deliver different agent concentrations at different flow rates[19]; checking delivery with an anesthetic gas monitor is helpful.

Maintenance of Anesthesia

Following induction, anesthesia should be maintained by placing the animal into a small face mask connected to the anesthesia machine set at a gas concentration of 1.55% to 2.5% (see Figure 8.11). It is important to remember at this stage to adjust the anesthetic gas to the recommended concentration.

TABLE 8.2
Concentrations of Volatile Anesthetics for the Rat

Anesthetics	Induction (%)	Maintenance (%)
Isoflurane	3 to 4	1.5 to 2.5
Enflurane	3 to 4	0.5 to 2
Halothane	3 to 4	1 to 2
Ether	10 to 20	4 to 5
Methoxyflurane	4	0.4 to 1

Note: Agents are inhaled with mixture gases of 30% O_2 and 70% N_2O.

Overdose of anesthesia is the most common cause of animal death during experiments. The important requirement for successful anesthesia is careful observation of the rat during the induction and remembering to reduce the anesthesia level during the surgery. This may be difficult when the same person performs the role of anesthesiologist and surgeon. If combining the duties of anesthesia and surgery is unavoidable, then careful planning and attention to detail are required.

The recovery time from anesthesia is dependent on the duration of anesthesia. If anesthesia has been maintained for 30 minutes, then the recovery time is usually within 5 to 10 minutes, but after several hours of anesthesia, full recovery can take 15 to 20 minutes or even longer. This will vary considerably according to the depth of anesthesia, which is dependent on the attention given to varying the concentration of anesthetic delivered to the rat. Following the completion of major surgery, a further reduction in depth of anesthesia and thus in recovery time can be made during surgical wound closure. The capability to rapidly vary the depth of anesthesia is one of the major advantages of using inhalational agents.

Available Inhalational Agents

Halogenated methylethyl ethers such as isoflurane and enflurane are good choices as inhalational anesthetics for stroke studies because they are highly volatile and thus largely expelled via the lungs and induce few neuroprotective effects (our unpublished data).

Isoflurane

The 2005 data sheet prepared by Abbott Laboratories contains a detailed description of this agent and its properties.

Presentation

Isoflurane is a clear, colorless, volatile, nonflammable liquid intended for general inhalation anesthesia. It emits a very slightly pungent odor but is a nonirritant to rats. Isoflurane is nonflammable and nonexplosive. It is commercially available from different animal or human health care companies in amounts of 100 to 250 mL/bottle. Figure 8.12 shows a 100-mL bottle purchased from J.A. Webster (800-225-7911).

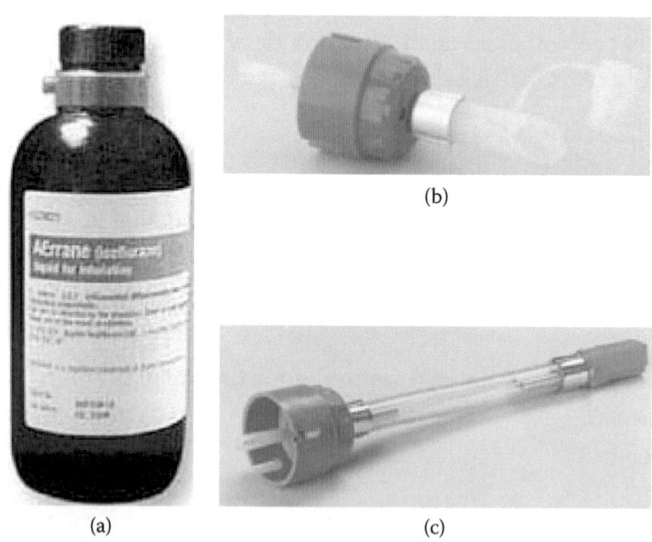

FIGURE 8.12 Isoflurane purchased from J.A. Webster (a). The antispill adapter allows the filling of a funnel-fill vaporizer without excessive spillage or pollution. It is fully reusable and can be left on the bottle and recapped. (b) For isoflurane; (c) for the filling of pin-indexed vaporizers.

Actions

Isoflurane induces and maintains general anesthesia by depression of the central nervous system, resulting in loss of consciousness. Administration of isoflurane produces rapid induction and subsequent recovery from anesthesia.

Pharmacokinetics

The pharmacokinetics of isoflurane is inhaled gaseous or volatile anesthetics. The blood concentrations of isoflurane are related to alveolar concentrations through established partition coefficients, and its distribution to tissues is also determined by solubility coefficients which are relatively constant under a wide variety of circumstances. Isoflurane shows very low solubility in blood and body tissues, much lower than for enflurane and halothane, thus its partial pressure (concentration) in alveolar gas or arterial blood rises to 50% of the inspired partial pressure (concentration) within 4 to 8 minutes of the start of its inhalation, and to 60% within 15 minutes. This rate of rise is slightly faster than that obtained with enflurane (a structural isomer of isoflurane) and considerably faster (40%) than the more soluble halothane. Throughout maintenance of anesthesia, the lungs eliminate a high proportion of the inspired isoflurane. When administration is stopped and the inspired concentration becomes zero, the bulk of the remaining isoflurane is eliminated unchanged from the lungs. In keeping with its low solubility, recovery from isoflurane anesthesia in rats is quick. Biotransformation of isoflurane is significantly less than that of enflurane or halothane. Animals and humans biotransform a small fraction of administered isoflurane. In man about 0.2% administered is evident as recoverable metabolites (fluoride and organic fluorine), with approximately 50% of these excreted in the urine, the principal metabolite being trifluoracetic acid. It has virtually few hepatic metabolisms. Enzyme induction associated with pre-existing drug therapy would not appear to be an important factor in the metabolism of isoflurane in rats and humans, mainly because the overall rate of metabolism of isoflurane is so low.

Dosage and Administration

Induction: 3% to 5% isoflurane with 800 mL/minute of gas flowing into an induction chamber.
Maintenance: 1.5% to 2.5% isoflurane with 800 mL/ minute of gas delivered via a face mask.

Recovery

The concentration of isoflurane can be reduced to 0.5% during surgical wound closing and then to 0% at the end of this procedure.

Warnings and Precautions

1. Because levels of anesthesia can be altered easily and quickly with isoflurane, only a vaporizer should be used that produces a predictable concentration with a good degree of accuracy (see Figures 8.7, 8.8, and 8.11; refer to "The Anesthesia Machine and Its Accessories" section earlier in this chapter). The open method (a cotton wool swab in a glass jar) cannot be used for isoflurane.
2. There are some side effects of isoflurane: cardiovascular depression, including hypotension and arrhythmias (mainly with high exposures); and respiratory system depression, especially following overdose. A minor side effect is an increase in the white blood cell count (even in the absence of surgical stress).
3. Isoflurane causes an increase in cerebral blood flow at deeper levels of anesthesia (1.5%), which may increase cerebral spinal fluid pressure. If appropriate, this can be prevented or reversed by hyperventilating the rat before or during anesthesia.
4. Isoflurane is a powerful systemic and coronary arterial dilator. Keeping the rat horizontal can suppress its effect on systemic arterial pressure.
5. Vapor from this and other inhalational gases should be efficiently extracted from the area of use (this is highly recommended).

6. As with all halogenated anesthetics, repeating an anesthesia procedure within a short period of time should be approached with caution since the risk of hepatotoxicity is not fully understood. There is insufficient experience with repeated episodes of anesthesia to make a definite recommendation in this regard. In our experience of using isoflurane, repetition three times in 2 hours (2 to 3 minutes exposure to isoflurane each time) did not show hepatotoxicity in rats (our unpublished data).
7. Although peak inorganic fluoride concentrations resulting from the breakdown of isoflurane are generally much lower than those considered to be nephrotoxic, a study of using isoflurane (3% isoflurane for 15 minutes of exposure) on rats did not show kidney damage (our unpublished data).
8. Isoflurane has been reported to interact with dry carbon dioxide adsorbents during closed-circuit anesthesia to form carbon monoxide. Inhalation of carbon monoxide may lead to formation of significant levels of carboxyhemoglobin in exposed animals.

In many respects, isoflurane could be considered the most suitable volatile anesthetic for use in rats, but its high cost relative to halothane, coupled with the additional cost of a calibrated vaporizer, have restricted its use in research animal units.

Halothane (Fluothane)

Presentation

Halothane (Fluothane) is a fluorinated hydrocarbon and has similar characteristics and actions to isoflurane. Halothane is a clear, colorless, volatile, nonexplosive, and nonflammable liquid with a pleasant odor. The information here is provided by AstraZeneca (303 Manukau Road, Epsom, P.O. Box 1301, Auckland). (Courtesy of AstraZeneca.) Halothane has been out of market in the United States (since 2006) due to its side effects (for example, liver damage).

Actions

"When inhaled, halothane is absorbed through the alveoli into the bloodstream and thence to its principal site of action, the brain. Here, halothane causes a progressive depression of the central nervous system, beginning with the higher centers (cerebral cortex) and spreading to the vital centers in the medulla. This depression is reversible. However, its mode of action, as with all anesthetic agents, is unknown.

"Halothane may cause bronchodilation. Bronchial relaxation is usually dose-related and may be due to blockage of pathways causing bronchoconstriction, or depression of bronchial muscular tone. Halothane causes a reversible, dose-related decline in renal blood flow, glomerular filtration rate and urinary flow. The detailed mechanism for blockage of the pathway is not clear.

"The rubber used in some anesthesia circuits (tubes connected to manifold, induction chamber, face mask and vaporizer) may absorb halothane."

Pharmacokinetics

> Halothane has a relatively low solubility in blood and therefore blood concentrations equilibrate rapidly with the alveoli. The triexponential decline in halothane blood concentrations following the end of administration is thought to represent distribution into three compartments: the vessel-rich group (brain/heart/liver), the musculature, and the adipose tissue. Approximately 80% of the inhaled halothane is eliminated unchanged by the lungs. The remaining 20% is metabolized in the liver by oxidative and, under hypoxic conditions, reductive pathways. The main metabolites are trifluoroacetic acid, bromide and chloride salts (via the oxidative pathway) and fluoride salts (via the reductive pathway). Their concentrations peak at 24 hours postoperatively and are eliminated by renal excretion the following week.
>
> During recovery from anesthesia, halothane is exhaled from the lungs, but significant quantities of this anesthetic are metabolized in the liver and a postanesthetic increase in liver microsomal enzymes

occurs. Liver enzyme induction is likely to be significant only after prolonged periods of anesthesia (>30 minutes), but detectable effects on drug metabolism both during and postanesthesia have been reported.

Dosage and Administration

Induction in a chamber: 3% to 4% halothane at 1 L/minute in a mixture of 30% oxygen and 70% room air or in a mixture of 50% N_2O and 50% oxygen.

Maintenance by a face mask: 1% to 2% halothane at 800 mL/minute in a mixture of 30% oxygen and 70% room air or in a mixture of 50% N_2O and 50% oxygen.

Warnings and Precautions

1. Owing to its potent anesthetic effect and ease of vaporization, halothane should be used only in a calibrated vaporizer so that the concentration delivered to a rat can be carefully controlled. Halothane should never be used in open system anesthesia machine.
2. Repeated exposure to halothane in a short period is not recommended.
3. The side effects of halothane are more numerous than those of isoflurane. Halothane causes liver and kidney damage in a dose-dependent manner. A syndrome called *halothane hepatitis* occurs in 1 in 10,000 halothane-induced anesthesia patients. Halothane has induced hepatic lesions and necrosis of the liver in rats and has reproductive effects on both humans and animals (refer to the AstraZeneca data sheet).
4. Cardiac arrhythmias are reportedly very common during anesthesia with halothane.
5. Halothane is a potent cerebral vasodilator. Increases in cerebral blood flow or intracranial pressure may be observed during anesthesia with halothane. These may be more marked in the presence of intracranial space-occupying lesions. The use of moderate hyperventilation during neurosurgery is recommended to counteract the rise in cerebrospinal fluid pressure that may occur with halothane.
6. In healthy rats, the above effects are not usually significant at normal maintenance concentration.

Enflurane (Ethrane)

We adopted the information for enflurane (Ethrane) from the Abbott Laboratories data sheet. (Courtesy of Abbott Laboratories.)

Presentation

Enflurane is a clear, colorless, stable liquid with a mild, sweet odor and contains no chemical stabilizers. It is a nonflammable liquid administered by vaporization in a general inhalation anesthesia medicine.

Actions

Similarly to isoflurane, enflurane induces anesthesia rapidly and recovery is also rapid. Progressive increases in depth of anesthesia produce corresponding increases in hypotension. The heart rate remains relatively constant without significant bradycardia. Electrocardiographic monitoring indicates that the cardiac rhythm remains stable.

Pharmacokinetics

The anesthetic response in rats to enflurane is similar to that of isoflurane, inducing rapid induction of and recovery from anesthesia in rats, and is thus safe and effective. Hepatic biotransformation of enflurane results in low peak levels of serum fluoride averaging 15 m*M*/L. These levels are well below the 50 m*M*/L threshold level which can produce minimal renal damage in normal subjects. Enflurane does not depress lymphocytic immune response.

Dosage and Administration

Induction in a chamber: 3% to 4% enflurane in oxygen or in a mixture of 2:1 N_2O/O_2 atmosphere.
Maintenance via a face mask: 2% in oxygen or in a mix of 2:1 N_2O/O_2 atmosphere.

Warnings and Precautions

1. Because levels of anesthesia may be altered easily and rapidly, only vaporizers producing predictable concentrations should be used (see Figure 8.7 and Figure 8.8) (refer to The Anesthesia Machine and Its Accessories section earlier in this chapter).
2. Hypotension and respiratory exchange can serve as a guide to depth of anesthesia. Deep levels of anesthesia may produce marked hypotension and respiratory depression.

Ether

Ether is flammable; when volatilized, it produces an irritant vapor that forms explosive mixtures with air or oxygen. Because of the proven dangers of ether explosions and fires, many research laboratories have abandoned the use of ether due to safety hazards. Aside from the safety hazard, ether is an irritant and is unpleasant to inhale. It has remained a popular anesthetic in some research laboratories because it can be used directly in a simple chamber and is inexpensive.

Carbon Dioxide (CO_2)

A need for an acceptable replacement for ether or injectable anesthetics in rats has led to the use of carbon dioxide.[15] The sedation or anesthetic role of carbon dioxide has been used most frequently for blood collection through retroorbital sinus or cardio puncture. The advantages of using carbon dioxide are the relatively rapid induction and recovery, ready availability, low cost (no expensive vaporizer), and safety for operating personnel.

Carbon dioxide is a colorless gas at normal air pressure without irritant odor. It is recommended for induction using a mix of 70% carbon dioxide and 30% oxygen or 50:50 ratios with oxygen in an induction chamber. It induces anesthesia in about 2 minutes, and the anesthesia persists for 30 to 90 seconds. Full recovery of consciousness will take 4 to 5 minutes after removing the carbon dioxide. For euthanization of rats, 100% carbon dioxide is recommended.

Carbon dioxide is an asphyxiant and a powerful cerebral vasodilator. Fenwick and Blackshaw (1989)[20] showed that the concentration of oxyhemoglobin is decreased during CO_2 anesthesia. In our study, anesthesia induced by a mix of CO_2 and O_2 caused blood–brain barrier damage in a 5- to 6-minute inducing period on normal rats (see Figure 8.1).

It is recommended that carbon dioxide anesthesia cannot be used for stroke surgery or studies related to the blood–brain barrier.

Table 8.2 summarizes the dose and administration of the inhalation agents. Table 8.3 compares the different inhalational anesthetics.

TABLE 8.3
Comparison of Different Inhalational Anesthetics

	Halothane	Isoflurane	Enflurane	Ether	CO_2
Induce and recover	Rapid	More rapid	More rapid	Slow	Middle
Administration	Vaporizer	Vaporizer	Vaporizer	Jar	Chamber
Flammable	Non	Non	Non	Yes	Non
Irritating odor	Slight	No	No	Irritant	No
Liver and kidney	Liver	Few	Few	Liver	Few damage
Cerebral blood	Increase	Increase	Increase	No data	Increase flow
Cost	+++	++++	+++	+	+

INJECTABLE ANESTHETICS

While the use of inhalational anesthetics is recommended for stroke study, injectable agents have some advantages in neurological studies and have been used by many laboratories. Injectable agents offer the following advantages: (1) The administration of injectable agents is simple and relatively easy to master. (2) No special equipment is needed; a simple needle and syringe can complete the anesthesia work. (3) Inexpensive equipment and drugs lower the cost of study. (4) Many anesthetic regimens have been developed and in fact are still utilized in some studies. For these reasons, we summarize the common injectable agents in this section. Some agents have neuroprotective effects, which may interfere with stroke studies. We also include the major side effects and recommended dose of each agent for your specific consideration. Investigators can select the best regimen for each study.

ROUTE OF ADMINISTRATION FOR RAT ANESTHESIA

Most injectable anesthetics are administered to rats by intraperitoneal routes as a single dose, although intramuscular, intravenous, and subcutaneous injections are preferable in certain situations. The rates of absorption and anesthetic effect vary considerably depending on the route of drug delivery. When choosing intraperitoneal or intramuscular administration, drugs with a wide safety margin are preferable. Compared to other routes of delivery, the intravenous route usually produces the most rapid and predictable dose/time response of anesthetic action. Each route has some advantages and disadvantages.

Sex differences in anesthetic action are also common in injectable agents, depending on the strains and type of drug; for example, female rats are more susceptible to pentobarbital than male rats.[21]

Intraperitoneal administration is probably the most popular parenteral method of drug delivery to rats, because (1) this method enables large volumes of anesthetic to be administered due to the large volume of the peritoneal cavity; (2) it appears to cause a minimum of pain and distress to rats; (3) the skills for intraperitoneal injection are easy to master; (4) peripheral veins for intravenous injection are harder to access in rats; and (5) many published regimens for anesthesia of rats have been formulated for intraperitoneal delivery. It is possible, however, inadvertently to administer anesthetic into the lumen of the intestine, the bladder, or the subcutaneous fat. This method usually cannot be adapted to humans.

Intramuscular administration provides more reliable absorption and drug delivery. A single small volume of anesthetic is usually administered into the quadriceps muscle mass in rats. Unfortunately, a few anesthetics have been shown to produce muscle damage when administered by this route.[22]

Intravenous administration provides quick delivery and anesthesia. This route enables the adjustment of the dose according to the individual animal response, so it is easy to avoid an overdose or underdose. Intravenous injection has obvious advantages in controlling depth of anesthesia and enabling a wider range of anesthetics to be used. The perceived difficulties are restraining the animal and the technical problem of intravenous administration. In a rat, the lateral tail vein is mostly used for intravenous injection, particularly after this vessel has been vasodilated using warm water, alcohol, a heating lamp, or a heating restrainer. Continuous intravenous administration can be carried out by tail vein cannulation or jugular vein cannulation; for a detailed method, see Chapter 23.

INJECTABLE AGENTS AVAILABLE

A wide range of injectable anesthetics is available for use in rats. The duration and quality of anesthesia produced by different agents vary considerably. Although the degree of analgesia produced by some agents is inadequate for major stroke surgery, addition of a low concentration of volatile anesthetic to the regimen can provide improved levels of analgesia.

Short-Duration (Up to 10 Minutes) Anesthesia

Propofol (Diprivan, Rapinovet)

Propofol (Diprivan, Rapinovet) is a novel hypnotic agent that is chemically classified as an alkyl phenol. The advantages of propofol are its rapid onset after intravenous administration, short duration of action after bolus injection, lack of accumulation after repetitive injections, and rapid recovery after stopping intravenous administration. Studies have shown neuroprotective actions; sevoflurane and propofol are neuroprotective possibly due to attenuation of central or peripheral catecholamines.[23]

Propofol (10 mg/kg i.v.) produces about 5 minutes of surgical anesthesia in the rat. When administered by continuous intravenous infusion, more prolonged periods of anesthesia can be produced. Cardiovascular and respiratory depression are generally mild in this dose range. Due to the rapid redistribution and metabolism of propofol, intramuscular or intraperitoneal routes cannot be used to administer propofol effectively.

The recommended doses are 10 to 25 mg/kg i.v. for repetitive bolus administration or induction using 7.5 to 10 mg/kg i.v. followed by 44 to 55 mg/kg/hour for up to 1 to 2 hours of continuous intravenous infusion.[15]

Alphaxolone/Alphadolone (Saffan, Althesin)

The mixture of steroids alphaxalone (0.9% weight/volume) and alphadolone acetate (0.3% weight/volume) (Cremophor) is prepared in a soluble agent, which promotes histamine release in some species; in rats, it does not appear to be a problem. Cross et al. (1991)[24] studied the neuroprotective effect of alphaxolone/alphadolone on transient forebrain ischemia in the gerbil. They found that Saffan (alphaxalone, 45 mg/kg^{-1} plus alphadolone 15 mg/kg^{-1} i.p.) had no neuroprotective effect when given 1 hour after the ischemic episode, while pentobarbitone (30 mg/kg^{-1} i.p.) had a modest protective effect.

Alphaxalone/alphadolone at 10 to 15 mg/kg i.v. or 25 to 30 mg/kg i.p. produces about 5 minutes of surgical anesthesia with a good degree of muscle relaxation. Anesthesia can be prolonged for up to 8 hours by administering 3 to 4 mg/kg by intravenous bolus every 15 to 20 minutes or 0.25 to 0.45 mg/kg by continuous intravenous infusion.

Thiopental

Thiopentone is a short-acting barbiturate. It produces about 10-minute surgical anesthesia when administered intravenously or intraperitoneally at 30 mg/kg of a 2.5% solution. With a 100-mg/kg dose, the duration of surgical anesthesia is up to 4 hours. These dosages have been shown to cause dose-dependent hypothermia, hypercarbial acidosis, and hypoventilation, but the recovery is rapid. There are some contradictious results from studies about its neuroprotection. Chen et al. (2003)[25] studied the neuroprotective effects of propofol, midazolam, and thiopental sodium. Both propofol and midazolam attenuated neurological deficits and reduced infarct and edema volumes. Propofol showed better neurological protection than midazolam, while thiopental sodium did not exhibit any protective effect. Fischer et al. (1998)[26] showed thiopental has a neuroprotective effect by attenuating the expression of vascular endothelial growth factor in hypoxic cultures and decreasing brain edema formation.

Recommended doses are 20 to 40 mg/kg of a 1.25% to 2.5% solution given intraperitoneally or intravenously.

Methohexital (Brevital)

Methohexital (Brevital) is a methylated oxybarbiturate. The advantages of this drug include its short duration of action and rapid recovery, similar to inhalant agents. This drug produces moderate cardiovascular and respiratory depression. Recovery is rapid but is associated with excitement. The neuroprotective effect is similar to other barbiturates.

This short-acting barbiturate produces 5 minutes of surgical anesthesia when administered intravenously (7 to 15 mg/kg). Anesthesia can be prolonged by administration of up to two additional doses without unduly prolonging the recovery period. Administration of 40 mg/kg i.p. produces 15 to 20 minutes of profound restraint but insufficient anesthesia for surgical manipulations in rats.

Medium-Duration (Up to 1 Hour) Anesthesia

Pentobarbital (Nembutal)

The oxybarbiturate pentobarbital (Nembutal) has been one of the most widely used injectable anesthetics for laboratory rats. The advantages of this drug are generalized availability, modest cost, widely available database encompassing decades of use, nonirritant nature, and ease of administration to rats. A single intraperitoneal injection (40 mg/kg) can produce light surgical anesthesia. This drug has a narrow safety margin and produces severe cardiovascular and respiratory depression. The onset time is about 5 to 10 minutes after intraperitoneal injection. Recovery time is prolonged, and no specific antagonist is available. It has neuroprotective effects like other barbiturates.

The recommended doses are 30 to 40 mg/kg i.v. and 30 to 60 mg/kg i.p. in Sprague-Dawley rats.

Ketamine (Vetalar, Ketaset)

Ketamine (hydrochloride) (Vetalar, Ketaset) is a nonbarbiturate, rapid-acting agent with pharmacological action that is characterized by profound analgesia, normal pharyngeal-laryngeal reflexes, mild cardiac stimulation, and respiratory depression. The anesthetic state produced does not fit into the conventional classification of stages of anesthesia, but instead ketamine produces a state of unresponsiveness with amnesia and appears to selectively interrupt association pathways to the brain, which has been called "dissociative" anesthesia.

This agent is the most common dissociative anesthetic used in rats. Ketamine alone causes muscle rigidity, which complicates surgeries. To induce a sufficient depth of surgical anesthesia, it is recommended that ketamine be combined with other compounds. Such combinations will produce surgical anesthesia without muscle rigidity. Ketamine stimulates the sympathetic nervous system and causes an increase of heart rate and blood pressure. It is well documented that ketamine is neuroprotective.[22]

The recommended doses for ketamine in rats range between 10 and 175 mg/kg i.m. or i.p. and are reported to produce effects from sedation to surgical anesthesia.

Combination of Ketamine and Xylazine

Ketamine plus xylazine is the most widely used combination. The addition of xylazine, an α_2-adrenergic agonist with sedative and analgesic properties, results in planes of surgical anesthesia. Xylazine produces polyuria owing to inhibition of antidiuretic hormone and transient hyperglycemia.[15] The addition of xylazine's muscle relaxant effects allows the use of lower doses of ketamine; this in turn allows the reversal of anesthesia with α-antagonists such as yohimbine.[27] Although no data are available for xylazine, other α_2-agonists have been reported to have neuroprotective effects (see the section on ketamine plus medetomidine).

The recommended doses for the combination of ketamine and xylazine are 40 to 60 mg/kg ketamine and 3 to 5 mg/kg xylazine given intramuscularly or intraperitoneally.

Combination of Ketamine and Acetylpromazine

The combination of ketamine and acetylpromazine, a phenothiazine tranquilizer, produces light surgical anesthesia in most rat strains. The blood pressure is reduced, primarily because acetylpromazine induces peripheral vasodilation and moderate respiratory depression.

Recommended intraperitoneal doses are 40 to 75 mg/kg ketamine and 0.75 to 2.5 mg/kg acetylpromazine.[9,15]

Combination of Ketamine and Medetomidine

Medetomidine is a potent and selective α_2-adrenoreceptor agonist with both sedative and analgesic effects and has been reported to have fewer side effects than xylazine. When used in combination with ketamine, it produces moderate surgical anesthesia. Like xylazine, medetomidine produces diuresis and transient hyperglycemia.

Recovery is prolonged but can be reduced considerably by the administration of atipamazole.[9] Dexmedetomidine, an isomer of medetomidine, has been reported to have the α_2-antagonist neuroprotective actions.[28,29]

Recommended doses in Sprague-Dawley rats are 60 to 75 mg/kg i.m. of ketamine plus 0.25 to 0.5 mg/kg s.c. of medetomidine. It produces moderate surgical anesthesia of up to 20 to 30 minutes, with prolonged immobility of up to 300 minutes postinjection.

Combination of Ketamine and Diazepam

Ketamine is also commonly combined with diazepam for use in rat surgery. Diazepam is a benzodiazepine tranquilizer that is believed to facilitate the synaptic actions of γ-aminobutyric acid (GABA), the predominant inhibitory neurotransmitter of the CNS. Diazepam does not act at the same site as GABA but at an allosterically linked site called the benzodiazepine receptor. It is through this site that the anticonvulsant, sedative, skeletal muscle relaxant, and amnestic properties of diazepam are mediated.

Recommended doses in rats are 40 to 80 mg/kg ketamine i.m. plus 2.5 to 10 mg/kg diazepam i.p.[15] This will produce surgical-depth anesthesia for 45 to 60 minutes.

Combination of Tiletamine and Zolezapam (Telazol)

Tiletamine, a dissociative anesthetic similar to ketamine, when used alone does not produce even light anesthesia. It is marketed commercially in combination with the benzodiazepine tranquilizer zolezapam (Telazol). This combination produces a light-to-medium plane of anesthesia. With this combination, the corneal, pedal, and swallowing reflexes remain intact, such that these commonly used anesthetic parameters are not reliable for judging the depth of Telazol anesthesia. The neuroprotective effect is similar to that of ketamine.

The recommended doses of 20 to 40 mg/kg i.p. in rats will produce an anesthetic duration of 30 to 60 minutes.[15]

Combination of Tiletamine and Xylazine

Tiletamine combined with xylazine produces 130 to 200 minutes of surgical anesthesia but with marked cardiovascular depression and mild respiratory inhibition.

Recommended intraperitoneal doses are 20 to 40 mg/kg tiletamine plus 5 to 10 mg/kg xylazine.[15]

Neuroleptanalgesics

Several commercial neuroleptanalgesic preparations are available that combine a potent opioid analgesic with a phenothiazine or butyrophenone tranquilizer.

Fentanyl-Fluanisone (Hypnorm)

Fentanyl is a short-acting narcotic analgesic. Fluanisone is a butyrophenone tranquilizer used as an anti-anxiety agent. Their combination contains 0.3 mg fentanyl citrate and 10 mg fluanisone per milliliter.

When used as the sole anesthetic agent, anesthesia is produced sufficient to facilitate surgery but may be accompanied by poor muscle relaxation and pronounced respiratory depression. These disadvantages can be overcome by adding a benzodiazepine such as midazolam to the regimen. Of further interest, Johansen et al. (1994)[30] found that Hypnorm induced hyperglycemia in rats, whereas pentobarbital did not change blood glucose. No neuroprotective effects of Hypnorm were discovered during a literature search.

The recommended doses of Hypnorm are 0.2 to 1.5 mL/kg, with optimal doses 0.2 to 0.4 mL/kg i.p. or i.m.[15] or 0.3 to 0.65 mL/kg Hypnorm plus either midazolam or diazepam at 2.5 mg/kg i.p. or i.m. The commercial preparation of Hypnorm and midazolam called Hypnovel (Roche) can be given as a single injection intraperitoneally if prediluted with water. A mixture of two parts water, one part Hypnorm, and one part midazolam (5 mg/mL) is recommended for injection at 2.7 to 4 mL/kg i.p. This mixture can last for 2 months.

Fentanyl-Droperidol (Innovar-Vet)

Innovar-Vet contains 0.4 mg fentanyl citrate and 20 mg droperidol per milliliter. It has been used for rodents for many years and produces profound anesthesia. The disadvantages of this drug are the muscle rigidity and the irritant nature, which may cause tissue necrosis and self-trauma to the digits following intramuscular injection. A literature search found no evidence for neuroprotection.

The recommended doses for Innovar-Vet are 0.13 to 0.4 mL/kg i.m. or 0.2 to 0.6 mL/kg i.p. Administration of these doses produces mild sedation to a surgical plane of anesthesia. Redosing with a quarter to a half of the initial dose can prolong anesthesia.[9,15]

Miscellaneous Anesthetics

Chloral Hydrate

Chloral hydrate is a hypnotic agent and was the first depressant developed for the specific purpose of inducing sleep. It has been a popular agent for stroke studies in rodents, particularly for studies of CNS function, as it may depress such function less than other injectable agents.[31,32] Literature searches did not find evidence for neuroprotective effects.

Adverse effects include severe respiratory, cardiovascular, and thermoregulatory depression accompanied by acidosis, hypercarbia, hypoxia, and hypothermia. An incident of fatal paralytic ileus was reported, which might be related to the concentration of the solution used.[33] Use of a dilute solution (<5% concentration) can prevent the development of this injury in rats.[33]

The recommended doses are 300 to 400 mg/kg i.p. or 400 to 600 mg/kg s.c. with 3.6% to 5% concentration, resulting in 45 to 60 minutes of light anesthesia. Addition of 40 mg/kg i.p. after the first dose can produce prolonged anesthesia for about 2 hours.[9,15]

Urethane (Ethyl Carbamate)

Urethane (ethyl carbamate) has been popularized for producing long-lasting anesthesia in the rat with minimal cardiovascular and respiratory system depression and good muscle relaxation. Unfortunately, it is a carcinogen[34] and should be used only when no suitable alternative is available. If urethane is used, it should be prepared and administered under controlled conditions. Urethane is a high-grade irritant and can cause peritonitis and so should not be used for recovery procedures. Only one study of the neuroprotective effects of urethane has been reported.[35] Yokoyama et al. (1997) studied the influence of urethane on bladder hyperactivity induced by middle cerebral artery occlusion in the rat. They found that urethane and MK-801 similarly inhibited the development of stroke-induced bladder hyperactivity and hypothesized that the most likely mechanism was blockade of glutamatergic transmission in the brain.

The recommended dose in rats is 0.5 to 1.5 g/kg i.p. This produces a prolonged period (up to 24 hours) of deep surgical anesthesia.

A summary of recommended doses for injectable agents is in Table 8.4.

LOCAL ANESTHESIA

Compared to inhalational and injectable anesthesia, local anesthesia is less frequently used in research rats. The advantages of its use are

1. Local anesthesia decreases postsurgical relief of pain in rats.
2. Most local anesthetics are not neuroprotectants and thus do not interfere with studies of new compounds on neurological function.

However, a local agent cannot provide general anesthesia during surgery. Applications for local anesthetics in stroke studies include

TABLE 8.4
Injectable Anesthetics and Techniques Used in Rats

Drugs	Dosage	Route of Injection	Duration of Anesthesia
Propofol	10 to 25 mg/kg	i.v.	5 minutes
	7.5 to 10 mg/kg	i.v. followed	1 to 2 hours
	44 to 55 mg/kg/hour	By continuous i.v. infusion	
Saffan	10 to 15 mg/kg	i.v.	5 minutes
	25 to 30 mg/kg	i.p.	5 minutes
	3 to 4 mg/kg	i.v./15 to 20 minutes	8 hours
	0.25 to 0.45 mg/kg	Continuous i.v. infusion	8 hours
Thiopental (1.25% to 2.5%)	20 to 40 mg/kg	i.v.	5 to 10 minutes
	40 mg/kg	i.p.	5 to 10 minutes
Methohexital (1%)	7 to 15 mg/kg	i.v.	5 to 10 minutes
	40 mg/kg	i.p.	15 to 20 minutes
Pentobarbital (Nembutal)	30 to 40 mg/kg	i.v.	20 to 60 minutes
	30 to 60 mg/kg	i.p.	20 to 60 minutes
	Induce with 50 mg/kg	i.p.	2 to 3 hours
	then 500 μg/kg/minute	Continuous i.v. infusion	
Ketamine	80 to 100 mg/kg	i.m.	20 to 30 minutes
	100 mg/kg	i.p.	Sedation only
	50 mg/kg	i.v.	
Ketamine +	40 to 100 mg/kg	i.m. or i.p.	20 to 60 minutes
Xylazine	3 to 15 mg/kg		
Ketamine +	40 to 75 mg/kg	i.p.	20 to 30 minutes
Acetylpromazine	0.75 to 2.5 mg/kg	i.p.	
Ketamine +	60 to 75 mg/kg	i.m. or i.p.	20 to 30 minutes
Medetomidine	0.25 to 0.5 mg/kg	s.c.	
Ketamine +	40 to 80 mg/kg	i.m. or i.p.	45 to 60 minutes
Diazepam	2.5 to 10 mg/kg	i.m. or i.p.	
Tiletamine +	20 to 40 mg/kg	i.p.	30 to 60 minutes
Zolezapam	Commercially prepared		
Tiletamine +	20 to 40 mg/kg	i.p.	2 to 3 hours
Xylazine	5 to 10 mg/kg	i.p.	
Fentanyl + fluanisone (Hypnorm)	0.2 to 0.4 mL/kg	i.p. or i.m.	20 to 90 minutes
1 part Hypnorm + 1 part midazolam + 2 parts water (Hypnovel)	2.7 to 4 mL/kg	i.p.	60 to 90 minutes
Fentanyl + droperidole (Innovar-Vet)	0.13 to 0.4 mL/kg	i.m.	20 to 30 minutes
	0.2 to 0.6 mL/kg	i.p.	Analgesia, muscle rigidity
	Repeat one-fourth to one-half of initial dose		Up to 1 hour

TABLE 8.4 (Continued)
Injectable Anesthetics and Techniques Used in Rats

Drugs	Dosage	Route of Injection	Duration of Anesthesia
Chloral hydrate (3% to 5%)	300 to 400 mg/kg	i.p.	45 to 60 minutes
	400 to 600 mg/kg	s.c.	Up to 2 hours
	Add 40 mg/kg i.p. after the first dose		
Urethane (50%)	0.5 to 1.5 g/kg	i.p.	Up to 24 hours

Notes: i.m., intramuscular; i.p., intraperitoneal; i.v., intravenous; s.c., subcutaneous.

1. A locally applied agent can provide analgesia during surgery in combination with other anesthetic regimens at lower doses.
2. Postoperative administration of long-acting forms of local anesthetics relieves postsurgical pain.
3. Administration of local anesthetics to the neck area blocks the reflex of the nerve-carotid sinus, which can be stimulated during surgery, and reduces the mortality rate resulting from such stimulation.[36]
4. Epidural and/or spinal administration is possible.

Available Anesthetic Agents

There are many drugs available commercially, including short- and long-duration forms. Here, we list the local anesthetics most commonly used in rat neurological studies.

Lidocaine HCl Solution

Lidocaine (2%, 50 mL/bottle or 5 to 10 mL/sample, Henry Schein) is a short-acting local anesthetic. When applied to the site of a skin wound, it relieves pain by inhibiting the ionic fluxes required for the initiation and conduction of impulses, thereby effecting local anesthetic action. Recommended doses are 0.1 to 0.2 mL locally spread or injected into the surgical area. It produces 15 to 30 minutes of analgesia in rats.

Iontocaine

In the commercially prepared solution of iontocaine (1% to 2% lidocaine HCl with 1:100,000 or 1:50,000 epinephrine, 50 mL/bottle, Henry Schein), epinephrine was added to lidocaine to increase the depth and duration of anesthesia, presumably because of its vasoconstrictor activity, which decreases the rate of removal of lidocaine from the site of administration. Recommended doses are 0.1 to 0.2 mL applied locally to the surgical area. It produces 30 to 60 minutes of analgesia.

Bupivacaine HCl

Bupivacaine (commercially prepared 0.25% to 0.75%, 30 to 50 mL/bottle, Henry Schein) is an analog of lidocaine with a long-acting duration of local anesthesia, and it very effectively relieves postoperative pain. It can be combined with epinephrine to further increase the anesthesia time (50 mL/bottle, 0.25% bupivacaine and 1:200,000 epinephrine).

Recommended doses are 0.1 to 0.3 mL applied directly to the surgical area, where it produces 4 to 6 hours of analgesia.[37]

TABLE 8.5
Current Recommendations for Analgesia in Rats[31–36]

Drugs	Dosage	Route of Administration	Frequency	Comment
Bupivacaine (0.25% to 0.75%)	0.1 to 0.3 mL	Local	qd to tid	Recommended for stroke study
Aspirin	20 mg/kg	s.c.	qd or bid	For mild pain; neuroprotective
	100 to 400 mg/kg	Oral	qd or bid	
Meloxicam	1.0 mg/kg	Oral or s.c.	qd	Neuroprotective
Buphenorphine	0.01 to 0.02 mg/mL in drinking water	Oral	qd	Controlled substance, need approval; neuroprotective
	0.5 mg/kg in Jello		qd	
	0.05 to 0.1 mg/kg	i.m., s.c., i.p.	qd	
Butorphanol	1 to 2 mg/kg	s.c.	qd	Controlled substance, need approval
Morphine	5 mg/kg	s.c.	Every 2 to 4 hours	Controlled substance, need approval; neuroprotective
Carprofen	5 to 10 mg/kg in water or Jello	Oral or s.c.	Once postoperation	Neuroprotective
Ibuprofen	15 mg/kg or 2.35 mL Children's Motrin in 500 mL water	Oral	qd	Neuroprotective
Ketoprofen	5 mg/kg	Oral or s.c.	qd	Neuroprotective
Flunixin meglumine (Banamine)	2.5 mg/kg	s.c.	qd or bid	No data available
Xylazine	5 to 12 mg/kg	s.c.	q 2 hour	Commonly combined with ketamine; neuroprotective

Notes: bid, two times a day; i.m., intramuscular; i.p., intraperitoneal; s.c., subcutaneous; tid, three times a day; qd, once a day; q 2 hour, once every 2 hours.

ANALGESICS

Because most analgesics have neuroprotective effects, these agents should be avoided when testing new compounds or drugs for neuroprotective efficacy in stroke models. Thus, the choice of analgesic is fully dependent on the purpose of the study. Administration of local anesthetics of long duration is highly recommended; for example, 0.1 to 0.3 mL of bupivacaine can be locally administered to an open wound, and the effect can last 4 to 6 hours and be repeated. A summary of commonly used analgesics is given in Table 8.5. The neurological effects of each analgesic are included in the table.

MANAGEMENT DURING ANESTHESIA

Anesthesia must be induced before surgery and maintained for a certain length of time and of sufficient depth while keeping the physiological state of the animal as near normal as possible. During complicated surgical procedures, metabolic and respiratory acidosis, blood loss, hypothermia, and anesthetic overdose may contribute to cardiorespiratory collapse, which can be fatal unless it is detected and corrected early. Techniques are available for monitoring anesthesia and providing supportive care during surgery.

TABLE 8.6
Anesthetic Reversal Agents (Antagonists)

Antagonists	Anesthetics	Doses and Administration	Comments
Atipamezole (α_2-adrenergic antagonist)	Combinations of xylazine or medetomidine	1 mg/kg i.m., i.p., s.c., or i.v.	Highly specific antagonist
Buprenorphine (partial opioid agonist at μ-receptors)	Opioid (fentanyl) combinations	0.05 mg/kg i.p. or s.c.	Slower onset than naloxone and nalbuphine, longer duration of action
Doxapram (central nervous system stimulant)	All anesthetics	5 to 10 mg/kg i.m., i.p., or i.v.	General respiratory stimulant
Nalbuphine (μ-receptor antagonist)	Opioid combinations	1 to 2 mg/kg i.p. or s.c.	Almost as rapid as naloxone; maintains postoperative analgesia
Naloxone (μ-receptor depression)	Opioid combinations	0.01 to 0.1 mg/kg i.v., i.m., or i.p.	Reverse analgesia and respiratory antagonist
Yohimbine (α_2-adrenoceptor antagonist)	Combination of xylazine or medetomidine	0.5 mg/kg i.p.	Nonspecific antagonist

Notes: i.m., intramuscular; i.p., intraperitoneal; i.v., intravenous; s.c., subcutaneous.

Reversal of Injectable Anesthetic Regimens

There are some specific antagonist drugs available to reverse the problems associated with overdosing and varying responses to the injection of anesthetics. The advantages of these drugs are as follows:

1. Most are available commercially, and their responses are well documented.
2. They can partially reverse some anesthetic complications by reducing sleep time.
3. Many anesthetics depress the cardiorespiratory system, but antagonists can effectively reverse this.
4. Some combinations of agonist/antagonist can reverse the undesirable effects of anesthetics, but will extend by several hours the period of postoperative analgesia.

Table 8.6 lists anesthetic antagonists for use in rats.[9,38,39]

Temperature

Two thermoregulatory aspects need to be considered during surgical and anesthetic procedures in rat stroke models: (1) Due to depression of the central heat-regulating mechanism during anesthesia, small animals such as rats and mice lose a great deal of heat from their body, causing a rapid fall in body temperature to as low as 33°C within 20 minutes and to below 30°C after 1 hour. (2). On the other hand, induction of stroke (especially by means of the filament model) can cause a temporary or persistent temperature increase just after the procedure. A possible cause is ischemic stimulation of the neural centers for temperature regulation (hypothalamus and preoptic areas). In our experience with the filament model of transient cerebral ischemia in the rat, the body temperature under isoflurane anesthesia increased to 38°C to 40°C in 1 to 2 hours after MCAO and recovered to normal in about 12 to 24 hours after reperfusion. Because rats are responsive to temperature changes in the environment, 21°C to 26°C is the optimal temperature for the environment, but if this is over 37°C in the environment, most rats subjected to filament-induced stroke will die. Note that for a general surgery procedure, people will place the operated-animal in a warm intensive care unit box for postoperative care. However, if the environmental temperature is lower than 20°C, the body temperature in stroked rats will decrease very quickly. To avoid this consequence, the body

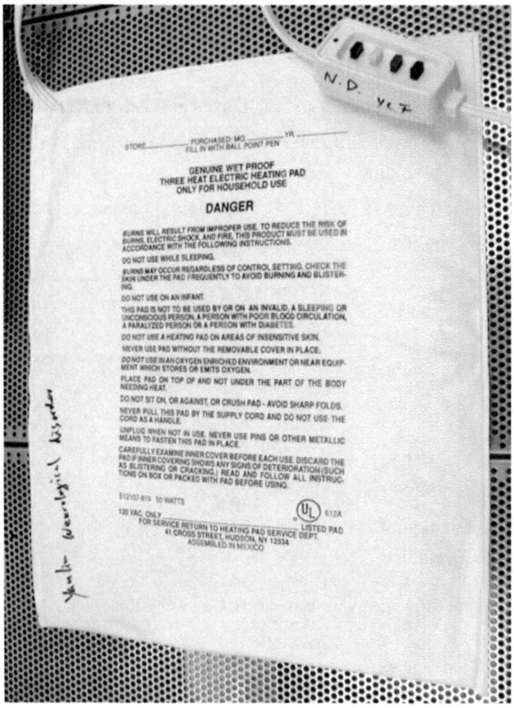

FIGURE 8.13 Heating pad purchased from CVS with three settings to control the temperature. The low setting is recommended for stroke surgery. Careful attention should be paid to the animal's body temperature. It is necessary to turn the heating pad off if the animal's body temperature increases above 37°C.

temperature can be maintained at 36°C to 37°C by placing the rat on a thermostatically regulated heating pad (available in CVS stores; Figure 8.13) during the surgical procedure. The rectal temperature can be monitored with a rectal thermistor probe inserted 2 cm into the rectum.

Eye Protection

Rats under anesthesia lose their eye blink function and the resultant inability to moisten their corneas can lead to permanent damage, even blindness, especially after long anesthetic and surgical procedures. A suitable eye cream (such as Triple Antibiotic Ophthalmic Ointment, Cat. No. 900862, J.A. Webster) should be used to protect against dry eye.

ACKNOWLEDGMENTS

We thank James Hastings and Dr. Sally Wixson from Johnson & Johnson, PRD, for their technical assistance and advice.

REFERENCES

1. Salo, M., The relevance of metabolic and endocrine responses to anesthesia and surgery, *Acta Anesthesiol Belg*, 30(3), 133, 1988.
2. Strommer, L. et al., Early impairment of insulin secretion in rats after surgical trauma, *Eur J Endocrinol*, 147(6), 825, 2002.
3. Grundy, P.L. et al., The hypothalamo-pituitary-adrenal axis response to experimental traumatic brain injury, *J Neurotrauma*, 18(12), 1373, 2001.

4. DeKeyser, F.G., Leker, R.R., and Weidenfeld, J., Activation of the adrenocortical axis by surgical stress: involvement of central norepinephrine and interleukin-1, *Neuroimmunomodulation*, 7(4), 182, 2000.
5. Reis, F.M. et al., Alterations in plasma prolactin and glucose levels induced by surgical stress in hyperprolactinemic female rats, *Braz J Med Biol Res*, 29(6), 811, 1996.
6. Johnson, T.J. et al., Glucagon, stress, and portal hypertension. Plasma glucagon levels and portal hypertension in relation to anesthesia and surgical stress, *Dig Dis Sci,* 40(8), 1816, 1995.
7. Hoover-Plaw, J.L. and Clifford, A.J., The effect of surgical trauma on muscle protein turnover in rats, *Biochem J,* 176, 137, 1978.
8. Fagin, K.D., Shinsako, J., and Dallman, M.F., Effect of housing and chronic cannulation on plasma ACTH and corticosterone in the rat, *Am J Physiol*, 245, E515, 1983.
9. Waynforth, H.B. and Flecknell, P.A., *Experimental and Surgical Technique in the Rat*, 2nd ed., Academic Press, San Diego, CA, 1992, chap. 3, p. 100.
10. Lovell, D.P., Variation in pentobarbitone sleeping time in mice. 1. Strain and sex differences, *Lab Anim*, 20, 85, 1986.
11. Zhang, P.B. et al., Effects of ketamine-midazolam anesthesia on focal cerebral ischemic injury in rats, *Di Yi Jun Yi Da Xue Xue Bao,* 24(12), 1337, 2004.
12. Brunson, K.L. et al., Effect of the noncompetitive NMDA antagonists MK-801 and ketamine on the spastic Han-Wistar mutant: A rat model of excitotoxicity, *Dev Neurosci*, 23(1), 31, 2001.
13. Spandou, E. et al., Effect of ketamine on hypoxic-ischemic brain damage in newborn rats, *Brain Res*, 20, 819(1–2), 1, 1999.
14. Nagels, W. et al., Evaluation of the neuroprotective effects of S(+)-ketamine during open-heart surgery, *Anesth Analg*, 98(6), 1595, 2004.
15. Kohn, D.F., Wixson, S.K., White, W.J., and Benson, G.J., *Anesthesia and Analgesia in Laboratory Animals*, Academic Press, New York, 1997, chap. 9, p. 165.
16. National Institute for Occupational Safety and Health, Criteria for a Recommended Standard Occupational Exposure to Waste Anesthetic Gases and Vapors (77–140), U.S. Government Printing Office, Washington, DC, 1977.
17. National Institute for Occupational Safety and Health, NIOSH Alert: Request for Assistance in Controlling Exposures to Nitrous Oxide during Anesthetic Administration (94–100), U.S. Government Printing Office, Washington, DC, 1994.
18. Boswell, M.V. and Collins, V.J., Pharmacology of inorganic gas anesthetics. In: Colins, V.J., *Physiologic and Pharmacologic Bases of Anesthesia*, Williams & Wilkins, Baltimore, MD, 1996, pp. 712–725.
19. Moyle, J.T.B., Darey, A., and Ward, C., *Ward's Anesthetic Equipment,* 4th ed., Saunders, London, 1998.
20. Fenwick, D.C. and Blackshaw, J.K., Carbon dioxide as a short-term restraint anesthetic in rats with subclinical respiratory disease, *Lab Anim*, 23, 220, 1989.
21. Svendsen, P., Experimental design and statistical analysis. In: Svendsen, P. and Hau, J., eds., *Handbook of Laboratory Animal Science, Vol. 1, Selection and Handling of Animals in Biomedical Research*, CRC Press, Boca Raton, FL, 1994, pp. 311–337.
22. Smiler, K.L. et al., Tissue response to intramuscular and intraperitoneal injections of ketamine and xylazine in rats, *Lab Anim Sci*, 40, 60, 1990.
23. Engelhard, K. et al., The effect of sevoflurane and propofol on cerebral neurotransmitter concentrations during cerebral ischemia in rats, *Anesth Analg,* 97(4), 1155, 2003.
24. Cross, A.J. et al., Neuroprotective activity of chlormethiazole following transient forebrain ischemia in the gerbil, *Br J Pharmacol*, 104(2), 406, 1991.
25. Chen, L., Gong, Q., and Xiao, C., Effects of propofol, midazolam and thiopental sodium on outcome and amino acids accumulation in focal cerebral ischemia-reperfusion in rats, *Chin Med J (Engl)*, 116(2), 292, 2003.
26. Fischer, S. et al., Barbiturates decrease the expression of vascular endothelial growth factor in hypoxic cultures of porcine brain derived microvascular endothelial cells, *Brain Res Mol Brain Res*, 18, 60(1), 89, 1998.
27. Hsu, W.H., Bellin, S.I., Dellmann, H.D., and Hanson, C.C., Xylazine-ketamine induced anesthesia in rats and its antagonism by yohimbine, *J Am Vet Med Assoc,* 189, 1040, 1986.
28. Maier, C., Steinberg, G.K., Sun, G.H., Zhi, G.T., and Maze, M., Neuroprotection by the alpha 2-adrenoceptor agonist dexmedetomidine in a focal model of cerebral ischemia, *Anesthesiology*, 79, 306, 1993.
29. Tolkkonen, J., Pnurunen, K., Koistinaho, J., et al., Neuroprotection by the alpha 2 adrenoreceptor agonist, dexedemidine, in rat focal cerebral ischemia, *Eur J Phamacol,* 372, 31, 1999.

30. Johansen, O. et al., Increased plasma glucose levels after Hypnorm anesthesia, but not after pentobarbital anesthesia in rats, *Lab Anim*, 28(3), 244, 1994.
31. Tsai, S.K. et al., The effect of desflurane on ameliorating cerebral infarction in rats subjected to focal cerebral ischemia-reperfusion injury, *Life Sci*, 2, 74(20), 2541, 2004.
32. Couturier, J.Y. et al., 3-Aminobenzamide reduces brain infarction and neutrophil infiltration after transient focal cerebral ischemia in mice, *Exp Neurol*, 184(2), 973, 2003.
33. Siverman, J. and Muir, W.W., A review of laboratory animal anesthesia with chloral hydrate and chloralose, *Lab Anim Sci*, 43(3), 210, 1993.
34. Beland, F.A. et al., Effect of ethanol on the tumorigenicity of urethane (ethyl carbamate) in B6C3F1 mice, *Food Chem Toxicol*, 43(1), 1, 2005.
35. Yokoyama, O. et al., Influence of anesthesia on bladder hyperactivity induced by middle cerebral artery occlusion in the rat, *Am J Physiol*, 273(6 Pt 2), R1900, 1997.
36. Wang-Fischer, Y.L. et al., Refined technique for inducing and grading middle cerebral artery occlusion in rat stroke model, *American Association of Laboratory Animal Science 54th National Meeting*, Seattle, platform sessions speaker PS 43, 2003, October 11–16, and *AALAS Tri-Branch Symposium, Biotechnology in the 21st Century and Beyond*, presenting poster 10, Philadelphia, June 8–10, 2003.
37. Porter, C.J. and Frizelle, F.A., Use of local anesthetic agents among New Zealand plastic surgeons—Their practices and philosophies [review], *Med Sci Monit*, 6(1), 194, 2000.
38. Cruz, J.I., Loste, J.M., and Burzaco, O.H., Observations on the use of medetomidine/ketamine and its reversal with atipamezole for chemical restraint in the mouse, *Lab Anim*, 32, 18, 1998.
39. Pao, L.H. et al., *In vitro* and *in vivo* evaluation of the metabolism and pharmacokinetics of sebacoyl dinalbuphine, *Drug Metab Dispos (DMD)*, 33, 395, 2005.

9 General Principles of Microsurgery on Animals

Yanlin Wang-Fischer, Brant D. Watson, and Lee Koetzner

CONTENTS

THE MICROSURGICAL SURGEON

The practice of microsurgery demands extensive training along with an abundance of mental and physical energy. A surgeon must develop the ability to concentrate on the visual field, as seen through the microscope, and become familiar with how objects change in their appearance as they are magnified. Good knowledge of anatomy and firsthand experience are also essential.

Some suggestions to improve performance of microsurgical procedures are as follows:

1. Avoid mental stress; one needs to be able to devote all necessary time and attention to the exercise.

2. Sleep well and avoid heavy physical exertion during the 24 hours preceding the surgical exercise as the latter will interfere with fine muscular control and increase hand tremor.
3. Do not change any habits relating to coffee intake as either a radical increase or decrease will increase hand tremor.
4. Do not work for too long at a stretch. If possible, take a 10-minute break every hour; otherwise, one will lose concentration, followed by loss of dexterity. Sometimes, muscle or joint pain may result from an overlong surgical session.
5. A comfortable chair and an adjustable surgical table can also help surgical performance. Good posture and hand support facilitate dexterity.

ASEPSIS

Asepsis refers to achievement of a germ-free condition. The misconception that rodents have an innate resistance to bacterial infection has not been scientifically substantiated. Infections, which may not be apparent on casual observation, may cause loss of vessel cannulations[1] and numerous changes in physiological parameters.[2] Remie et al.[3] in 1990 reported use of a sterilized catheter and surgical instruments to cannulate a jugular vein and found that the catheter could be kept patent for up to 6 months, but with use of an unsterile catheter and instruments, the period of patency was drastically reduced to 1 to 2 weeks. Further, the time needed to recover to the preoperative weight was extended. In accordance with good scientific practice and standards set forth in the *Public Health Service* (PHS; now it is the NIH) *Guide for the Care and Use of Laboratory Animals* and the federal Animal Welfare Act,[4] aseptic surgical procedures must be used, especially when dealing with survival surgical procedures on all mammalian species. Many supplies such as gloves, surgical blades, and suture materials are commercially available as sterile packs. However, it is frequently necessary to sterilize in house such items as surgical instruments, drapes, gowns, and so on.

STERILIZATION VERSUS DISINFECTION

In considering methods for sterilization procedures, it is important to differentiate between *sterilization* and *disinfection*. Sterilization results in the destruction of all forms of microbial life, while disinfection specifically results in the destruction of pathogenic microorganisms and thus only reduces the number of viable microorganisms. Accordingly, disinfection is faster and less expensive. High-level disinfection will not kill the more resistant bacterial spores. Some investigators substitute high-level disinfection for sterilization of medical instruments. Commonly used disinfectants such as alcohol, iodophors, quaternary ammonium, and phenolic compounds are not effective sterilants and therefore are not acceptable for sterilization of surgical tools in survival surgical procedures.

An object should be disinfected or sterilized depending on its intended use. Critical objects (those that enter tissues or the vascular system or permit conduction of blood, such as implanted medical devices or catheters) require sterilization before use.

Items to be in contact with mucous membranes or non-intact skin, such as endoscopes, respiratory therapy equipment, and diaphragms, require high-level disinfection.

METHODS OF STERILIZATION/DISINFECTION

Methods of sterilization fall into the following categories:

1. Chemical sterilization
2. High-temperature/high-pressure sterilization (by autoclave), which is very commonly used in conjunction with laboratory animal studies
3. Radiation sterilization, which is not often associated with laboratory animal surgeries

TABLE 9.1
Sterilization Methods

Methods	Common Used	Others
Chemical	Ethylene oxide (EtO)	Solution glutaraldehyde
	Sterrad	(Cidex and Cidex Plus)
	Steris System 1	VHP MD series
	Cidex OPA	Chlorine dioxide
	Chlorhexidine	Ozone
High temperature/pressure	Steam autoclave	Dry autoclave
Radiation	Gamma	Electron beam (E-beam) x-ray

Table 9.1 summarizes the various sterilization methods available within each category.[5]

Some commonly used sterilization methods are chemical sterilization, high-temperature/high-pressure sterilization, and radiation.

Chemical Sterilization

Ethylene Oxide Gas

Ethylene oxide (EtO) gas was introduced in the 1950s and is an effective, low-temperature chemical sterilization method. It also takes *longer* than steam sterilization, typically 16 to 18 hours for a complete cycle. Temperatures reached during sterilization are usually in the 50°C to 60°C range.

EtO is a chemical agent that kills microorganisms, including spores. EtO gas must have direct contact with microorganisms on the items to be sterilized. Because EtO is highly flammable and explosive in air, it must be used in an explosion-proof sterilization chamber in a controlled environment. Items sterilized by this process must be packaged with wraps and be aerated.[6] EtO-permeable sterilization envelopes are commercially available. Figure 9.1 shows a popular low-temperature sterilization system based on EtO use.

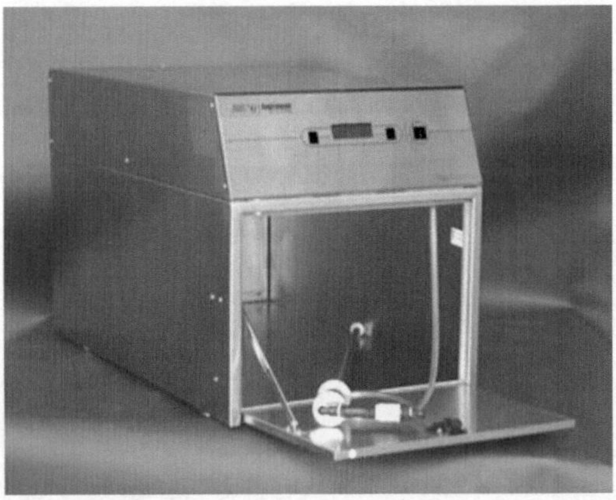

FIGURE 9.1 A sterilizer based on ethylene oxide (EtO), AN74j/Anprolene Sterilizer (Andersen Sterilizers). This system gently sterilizes a wide variety of delicate items that are sensitive to heat and moisture. A combination of ventilation and purge systems ensures that EtO exposure levels are below the standard limits for safety. It is easy to install and requires only a 110 V alternating current outlet and an exhaust line to the outside. (Courtesy of Andersen Sterilizers, Haw River, North Carolina.)

In general, EtO gas is a reliable and safe agent for sterilization when handled properly. EtO is used to sterilize items that are heat or moisture sensitive, for example, polyethylene tubing for implantation into blood vessels or delicate cellulose fiber implants.

The disadvantages of EtO gas are several. It can leave toxic residues on sterilized items, and it can present several health hazards to personnel that merit special attention.

Sterrad: Low-Temperature Hydrogen Peroxide Plasma

The dry, low-temperature sterilization technique based on Sterrad (a Johnson & Johnson product) is easy to install and use (Figure 9.2). This low-temperature plasma sterilization was introduced to fill the gap between the autoclave method of high-temperature steam sterilization (safest, fastest, and least expensive) and EtO gas sterilization, which leaves toxic residuals. Sterrad is a low-temperature, nontoxic, but fairly expensive sterilization method. In this process, hydrogen peroxide is activated to create a reactive plasma or vapor.

Note that plasma is an ionized gas made up of ions and electrons and is distinguishable from solids, liquids, and gases. Plasma is often referred to as the fourth state of matter.

The Sterrad system is a hydrogen peroxide gas plasma sterilization system with an operating temperature range of 45°C to 50°C. Operating cycle times range from 45 to 70 minutes, depending on system size.

This sterilization system uses a combination of hydrogen peroxide and low-temperature gas plasma to quickly sterilize most surgical instruments and materials without leaving any toxic residues. Hydrogen peroxide is a known antimicrobial agent capable of inactivating resistant bacterial spores. Sterilization by this method occurs in a low-moisture environment.

This system is best suited for sterilizing heat-sensitive equipment such as nylon monofilament sutures and polyethylene tubing for implantation into blood vessels.

FIGURE 9.2 The Sterrad family of dry and low-temperature sterilizers is easy to install and simple to use. No plumbing, water source, heat booster, or other provisions are required. The absence of hookups allows for superior portability of the Sterrad System as it can be moved to any room with an approved electrical outlet.

Steris System 1 Sterile Processing System

The System 1 (Steris Corporation) Sterile Processing System is a proven low-temperature sterile processing system for immersible surgical instruments or devices. It uses the Steris 20 Sterilant Concentrate, which combines peracetic acid, a chemical biocidal agent, with a proprietary anticorrosion formulation to kill microorganisms at low temperatures (50°C to 56°C).

The Steris 20 Sterilant Concentrate is mixed with sterile water to create the solution that flows into the sterilization chamber. The sterilization time is 12 minutes. This is followed by repetitive sterile water rinses to complete the process. The entire process is completed in less than 30 minutes for a standard cycle.

Peracetic acid by itself is an oxidant and disinfecting agent for liquid immersion. It maintains its effectiveness when high levels of organic debris are present. It is also known as acetyl hydroperoxide and reacts with most cellular components to destroy cells.

Cidex OPA Solution: Alternative to Glutaraldehyde

Cidex® OPA Solution (a Johnson & Johnson product) is a high-level disinfectant (HLD) for use in reprocessing heat-sensitive medical devices. Cidex OPA Solution provides high-level disinfection in 12 minutes at room temperature (20°C) and is particularly active against mycobacteria, including glutaraldehyde-resistant strains of *Mycobacterium chelonae*. Cidex OPA Solution has the broad materials compatibility of glutaraldehyde (but does not contain glutaraldehyde), requires no activation, and has minimal odor (Figure 9.3). Cidex OPA Solution is replacing Cidex and Cidex Plus owing to their toxicity concerns.

Chlorhexidine

Chlorhexidine [biguanide, 1,1′-hexamethylenebis 5-(*p*-chlorophenyl)-, diacetate] is a liquid disinfectant and is effective against Gram-negative and some Gram-positive bacteria, viruses, and yeasts. It adheres to the epithelial surfaces of the skin, thus creating a persistent and residual effect not seen

FIGURE 9.3 Cidex OPA solution (1 gallon, Johnson & Johnson). This is a high-level disinfectant (HLD) for use in reprocessing heat-sensitive medical devices. Cidex OPA solution is the first new HLD available in the past 35 years with the broad materials compatibility of glutaraldehyde. Cidex OPA solution will provide rapid HLD in 12 minutes at room temperature (20°C) and is particularly active against mycobacteria, including glutaraldehyde-resistant strains of *M. chelonae*. Cidex OPA requires no activation and has minimal odor.

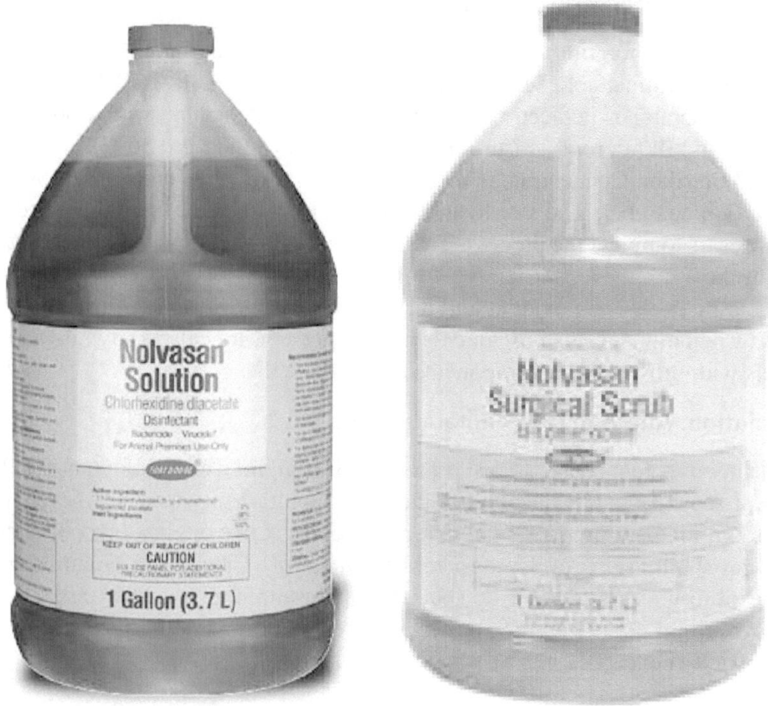

(a) (b)

FIGURE 9.4 Nolvasan Surgical Solution (a) and Scrub (b) are 2% chlorhexidine acetate in a stable detergent base. They possess a wide range of antimicrobial activity and a rapid and residual antimicrobial effect for the cleaning and nonirritating treatment of wounds and preparation for surgical procedures.

with iodine or other antiseptics. Chlorhexidine acetate is commercially available as both a 2% surgical scrub (Nolvasan Surgical Scrub) and 2% solution (Nolvasan Solution, Aveco/Fort Dodge) (Figure 9.4). The scrub is used undiluted as an antiseptic. Chlorhexidine scrub is excellent for use as a surgical hand scrub, preoperative skin preparation, skin wound cleanser, and hand wash. The chlorhexidine solution is used as a disinfectant by diluting 3 ounces in a gallon of water. Nolvasan solution is not effective against Gram-positive cocci or *Pseudomonas aeruginosa* on inanimate objects. This solution can be used to sterilize instruments between surgical procedures on multiple animals. It should not be used as the initial sterilant. When performing multiple consecutive surgeries, placement of the instruments in these solutions for at least 20 minutes between animals is permissible to eliminate potential cross contamination. Instruments should be free of all organic debris prior to placement in solution. This product is fragrance free to reduce possible irritation. Once removed from the chemical solution, it is extremely important that instruments be thoroughly rinsed with sterile water prior to use as the chemical solutions can be very irritating to healthy tissues. Chlorhexidine is neutralized by alcohols; therefore, the two agents should not be used in combination. It can be purchased from J.A. Webster (Cat. No. 255914) or Henry Schein (2% chlorhexidine acetate, Cat. No. 9953226).

High-Temperature/High-Pressure Sterilization

Steam Autoclave

Steam autoclaving is the oldest, safest, and most cost-effective method of sterilization in laboratory animal surgery. The steam reaches 121°C to 148°C (250°F to 300°F) in the pressure chamber at 15 pounds per square inch. The sterilization period is dependent on the temperature, pressure, wrapping, and size of load and can range from 10 to 60 minutes; for example, a small surgical instrument package (8 × 10 square inches) can be autoclaved in 15 minutes.

Dry Heating Bead Sterilization

A hot, dry glass bead sterilizer (Figure 9.5) is a simple, inexpensive way to quickly decontaminate microdissection instruments between procedures and animals. It kills all viruses, aerobic and anaerobic bacteria, yeasts, and spores. The unique stainless steel glass bead bath (1.5-mm lead-free glass beads) remains at a constant 233°C to 260°C (over 500°F), allowing insertion of the tips of the instruments whenever decontamination is needed. Most microdissecting instruments are decontaminated within 15 seconds (see Figure 9.5; to decontaminate surgical scissors, just place the scissors into the glass bead bath or well for 15 seconds). The instrument chassis stays cool, even at the end of a full day of operation. The air above the unit is cool enough to insert an instrument into the well without discomfort. When inserting the surgical instrument into the well or bath, sterilization of the inserted parts takes only 2 to 5 seconds. The handles of the instruments do not get hot. Within 30 seconds after removal, the surgical instruments are cool enough to use. The beads will, after extensive use, eventually be worn or broken to powder, which will settle to the bottom of the well. When it becomes difficult to push the instruments in deep enough, the beads must be replaced.

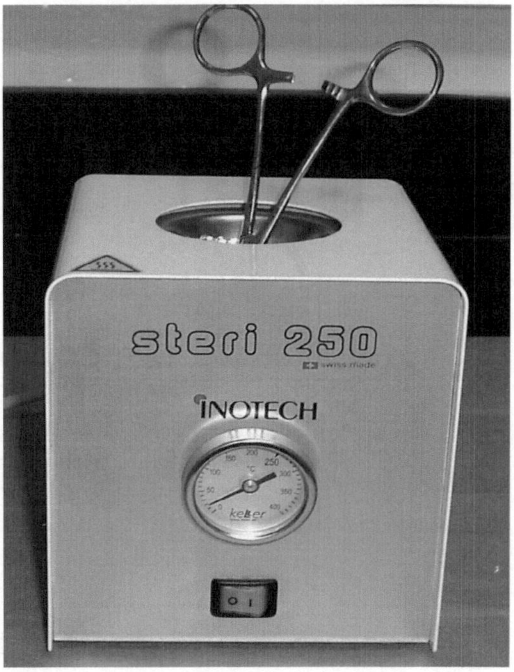

FIGURE 9.5 A dry sterilizer is a convenient, effective way to quickly decontaminate microdissection instruments between animals.

Note that this dry hot bead sterilizer cannot substitute for a steam autoclave for surgical instrument sterilization because it sterilizes only the tips of surgical instruments. It is used to decontaminate surgical tools between animals and procedures. For example, if one is performing surgery on a group of six rats, surgical tools initially were sterilized by a steam autoclave; after surgery on the first one or two rats, the tips of surgical tools can be decontaminated with the dry hot bead sterilizer.

Common Radiation Sterilization

Gamma Sterilization

Irradiation is an effective sterilization method. The product to be sterilized is exposed to radiation for 10 to 20 hours, depending on the strength of the source. The highest temperatures reached in gamma sterilization are usually 30°C to 40°C. Gamma radiation is popular for sterilizing before shipment and can be done through the packing material. A dose of 2.5 megarad is generally selected for many items.

Other Chemical Sterilization and Disinfectant Methods

Chlorine Dioxide

The use of chlorine dioxide (ClO_2) is a chemical liquid sterilization process. The best operating temperature range for this process is 25°C to 30°C while using low concentrations of ClO_2. The

TABLE 9.2
Sterilization Comparison

	T (C)	Cycle Time (minutes)	Type	Sterilize/ Disinfect	Advantage	Disadvantage
			High Temperature (C)/Pressure			
Autoclave	121 to 148	10 to 60	Steam	Sterilize	Safe, effective; economical; reliable; fast	Wet and hot, not for heat-sensitive materials
			Chemical			
Sterrad	45 to 50	55 to 70	H_2O_2 vapor Gas plasma	Sterilize	Safe, effective; no toxicity; no heat; no moisture	Damage to nylon material; expensive
Steris System 1	50 to 56	30	Peracetic acid Proprietary Anticorrosion	Sterilize	Safe; nontoxic; short time	For small items, soak in solution
Ethylene oxide (EtO) gas	50 to 60	16 to 18 hours	EtO gas	Sterilize	Effective; no heat; no moisture; reliable	Long cycles; toxic; carcinogenic; long aeration
Cidex OPA	20	12	Orthophthal-aldehyde	Disinfect	Quick; less odor; no heat	Costly; disinfectant; nonsterilant
Chlorhexidine	15 to 30	20	Unknown	Disinfect	Quick, odorless	Disinfectant; nonsterilant
			Radiation			
Gamma	30 to 40	10 to 20 hours	Radiation	Sterilize	Simple, reliable, no heat	Radioactivity

process requires 6 hours of contact time to achieve sterilization. The presence of organic matter reduces its activity. A processor converts a compound of dilute chlorine gas with sodium chlorite to form ClO_2 gas, and this gas is then exposed to the equipment in a sterilizing chamber.

Betadine Scrub: Povidone-Iodine Scrub

Povidone-iodine surgical scrub (Betadine®) is a highly effective germicidal cleanser for pre- and post-operative skin washing and is also effective as a cleanser for bacterial and fungal skin infections.

A comparison of different sterilization methods is summarized in Table 9.2.

STANDARDS FOR ASEPTIC PROCEDURES

The National Institutes of Health (NIH) has established recommendations for conducting major rodent survival surgical procedures,[7] and, along with the *Public Health Service (PHS) Policy on Humane Care and Use of Laboratory Animals*[8] and the *Guide for the Care and Use of Laboratory Animals* (the guide),[8] specified that procedures conducted on rodents are to be performed using an aseptic technique. The following definitions should be considered when determining whether the procedures employed must meet these requirements. Note that the institutional animal care and use committees (IACUCs) require that *all* investigators utilizing rodents for survival surgical procedures attend a mandatory hands-on or didactic orientation session prior to commencement of their studies. The following definitions are related to aseptic procedure standards:

Survival procedure: A procedure in which the animal recovers from the anesthetic, even if for a short time.

Major surgical procedure: A survival surgical procedure in which the surgical intervention penetrates a body cavity or has the potential for producing a permanent handicap in an animal that is expected to recover. Procedures including, but not limited to, *laparotomy, thoracotomy, craniotomy,* and *orthopedic manipulations* are all considered major surgical procedures.

To achieve a high level of aseptic technique in stroke studies *in vivo,* the investigator should adopt the following standards for aseptic procedures:

1. Sterilization of surgical instruments and all materials that are permanently implanted in the animal. This can be achieved by the above methods based on the type of material and laboratory equipment availability.
2. Use of sterile sutures, needles, and syringes: Selection and use of appropriate suture materials is imperative for successful wound closure and healing. Sutures are either absorbable or nonabsorbable depending on the materials from which they are manufactured. Sutures are made from natural materials such as silk, cotton, or catgut or synthetics such as nylon, polygalactin, or stainless steel. For survival surgical manipulations, it is imperative that suture materials are sterile at the time of use since they are foreign materials and that they provide a substrate where bacteria may proliferate. Most rodents will continue to exhibit growth at the locations of any externalized sutures, so a buried suture line (subcuticular sutures) or stainless steel wound clips/staples are recommended for skin closure. Wound clips, as with sutures, should be removed between 10 and 14 days after placement. Cyanoacrylate surgical adhesives, such as Vetbond® or Nexaband®, may be used to close incisions.
3. Use of sterile solutions can be achieved through germ filters (0.22-μm cellulose acetate 50-mL tube-top filter from VWR, Cat. No. 430320, or small filters for syringes).
4. A separate surgical room used primarily for aseptic procedures is desirable; however, it is appropriate to perform survival rodent surgical procedures in a conventional laboratory setting using aseptic technique. A clean, uncluttered work area and a sanitized work surface should be utilized. The work area should be located to minimize laboratory traffic not related to the surgical procedure and dedicated exclusively for surgery when in use. Consideration should also be given to locate the surgery away from potential sources of contamination, such as open windows, fans, or fume hoods, which can blow dust into the area and increase desiccation of exposed tissues.
5. The surface on which the procedure will be performed should be disinfected (2% chlorhexidine, 10% povidone-iodine, or a quaternary ammonium compound) before use. The maintenance of asepsis is enhanced if a sterile drape is then applied (this can be obtained from J.A. Webster, Cat. No. 904409, sterile poly-lined disposable drapes, 18 × 26 inches).
6. The site of surgical incision is prepared by first removing hair from it. Skin asepsis begins with shaving the hair using a #40 clipper blade. The size of the shaved area should be three to four times larger than the incision area. All loose hair should be vacuumed or carefully dusted away to prevent translocation into the incision. Once the site is free of all hair, surgical preparation of the skin may commence. The use of either a povidone-iodine scrub (Betadine Scrub) or a chlorhexidine scrub (Nolvasan) is recommended. Both of these agents have good bactericidal activity and contain a detergent. By means of 3 × 3 gauze squares, cotton-tipped applicators, or the equivalent, the area should be scrubbed beginning at the center of the incision site and working out toward the perimeter. After reaching the perimeter, a new gauze square or applicator should be selected and the process repeated. After completing the above preparation, the area should be wiped with a 3 × 3 gauze soaked in 70% isopropyl alcohol or 70% ethanol. The final step prior to making the incision is to paint or spray the surgical site with a 10% povidone-iodine (Betadine solution) or chlorhexidine solution (Nolvasan). Do not saturate other areas of the

body with disinfectant since this enhances hypothermia, which is a common postoperative complication in rodents. Covering the animal with sterile drapes to avoid contamination of the incision, instruments, and supplies is strongly recommended. Opaque and nonopaque materials have been utilized. Clear materials have the advantage of allowing the investigator to monitor respirations and perfusion through the drape. Autoclavable plastic and sterile adhesive dressings are available for use.

7. Surgeon preparation: The surgeon and any assistant must wear a clean three-quarter-length laboratory coat or a surgical scrub shirt. Surgical scrubbing of the surgeon's hands by the appropriate disinfectant, use of sterile gloves, and wearing a surgical mask by the surgeon and any assistants working in the immediate area must be performed.

CONSIDERATIONS WHEN PERFORMING SERIAL SURGERIES: MULTIPLE SURGERIES IN A SINGLE SESSION

It is often necessary to surgically prepare several different animals during one session using one sterile pack. This is appropriate, providing care is taken to maintain sterility of the instruments. The following considerations should be made:

1. Use two sets of instruments. One set is used for incising and manipulating the skin, which is considered a potentially contaminated site because of resident microbial flora. A second set of instruments, which is sterile, is used to manipulate the deeper tissues.
2. To maintain sterility between operations on different animals, the instruments can be soaked in cold sterilant solution for at least 10 minutes to kill vegetative bacteria prior to reuse. *Instruments must be rinsed thoroughly with sterile saline or sterile water prior to use with the next animal.* Alternatively, a glass hot bead sterilizer may be used to sterilize the tips of surgical instruments between animals. Recognize that only the instrument tips are sterilized. Care must be taken to avoid touching tissues with hot instruments, instrument handles, or other nonsterile instrument parts. Because nonsterile instrument handles are held with gloved fingers, contact of tissues with the fingers (which tend to be more easily contaminated) must also be avoided.
3. Manipulate the tissues with only the tips of the instruments.
4. Consider using a separate sterile pack if more than six animals must be processed.

PREOPERATIVE CONSIDERATIONS OR CARE

Animal Health/Selection

A healthy animal is important to a successful surgery. Animals undergoing clinical or subclinical disease often experience anesthetic complications and are not appropriate for surgery. It is recommended that animals purchased for surgery be barrier housed prior to surgery to ensure the absence of rodent diseases. Also, in general, a minimum of *48 hours* is required for an animal to recover from the stress of shipping; therefore, surgery should not be performed immediately on arrival.

Preoperative Withholding of Food

While it is common in larger species to withhold food prior to surgery to prevent the possibility of aspiration pneumonia after regurgitation, this practice is not necessary in rats. Fasting for 4 hours before surgery, however, may promote the absorption of intraperitoneally administered anesthetics.[9] Water should never be withheld.

Pre-/Postoperative Antibiotics

Administration of prophylactic antibiotics is not a substitute for the practice of aseptic surgery. If surgical induction of experimental stroke does not involve intestinal tract exposure and proper aseptic technique is utilized, antibiotics should not be necessary. In fact, antibiotics are contraindicated in hamsters and guinea pigs owing to the frequent development of fatal clostridial enteritis (because the antibiotics can kill the normal intestinal bacteria). If the interior of the intestinal tract is exposed, however, antibiotics are commonly administered. To have the desired effect, antibiotics should be administered prior to surgery to provide adequate blood/tissue levels at the time of surgery.

REFERENCES

1. Popp, M.B. and Brennan, M.F., Long-term vascular access in the rat: Importance of asepsis, *Am J Physiol*, 241, H606, 1981.
2. Bradfield, J.F. et al., Behavioral and physiologic effects of unapparent wound infection in rats, *Lab Anim Sci*, 42(6), 572, 1992.
3. Remie, R. et al., General principles of microsurgery. In: *Manual of Microsurgery on the Laboratory Rat*, Elsevier Science, Amsterdam, the Netherlands, 1990, part 1, p. 13.
4. Institute for Laboratory Animal Research (ILAR), *Public Health Service (PHS) Policy on Humane Care and Use of Laboratory Animals*, National Academy Press, Washington, DC, 1996.
5. Patel, M., Medical Sterilization Methods, white paper, Design Engineer Lemo USA, (www.lemousa.com/pdfs/whitepaper/P_series_whitepaper.pdf), 2003.
6. Strain, P. and Young, B., Emphasis on Ethylene, Methods to Reduce Sterilization Process Time (www.sterigenics.com), 2004.
7. NIH Animal Research Advisory Committee, *NIH Intramural Research Program Guidelines for Survival Rodent Surgery*, 1999. Available at: http://oacu.od.nih.gov/ARAC/surguide.htm.
8. ILAR, *Guide for the Care and Use of Laboratory Animals*, National Academy Press, Washington, DC, 1996.
9. Flecknell, P.A., *Laboratory Animal Anesthesia: An Introduction for Research Workers and Technicians*, 2nd ed., Academic Press, New York, 1996.

10 Microsurgical Instruments for Stroke Studies

Yanlin Wang-Fischer and Lee Koetzner

CONTENTS

GENERAL SURGICAL INSTRUMENTS

Stroke surgery on small rodents requires dedication. It is essential to have one's own set of surgical tools. These should be of good quality and be well maintained. Old, worn out, or obsolete tools must not be used. Surgical instruments used in pediatric or ophthalmic surgery are sized appropriately for rodent surgery. An appropriate set for stroke surgery should consist of common surgical instruments (Figure 10.1) and microsurgical instruments (Figure 10.2).

COMMON SURGICAL INSTRUMENTS

Skin Scissors

There are many different commercially available scissors. A good pair of scissors for cutting skin is important (Figure 10.1A). The varieties known as Super Cut Scissors, Operating Scissors, Veterinary Heavy pattern, or Tough Cut Scissors have in common one blade with serrations (to avoid tissue slippage during cutting), and the other is honed to razor sharpness. These can cut through the

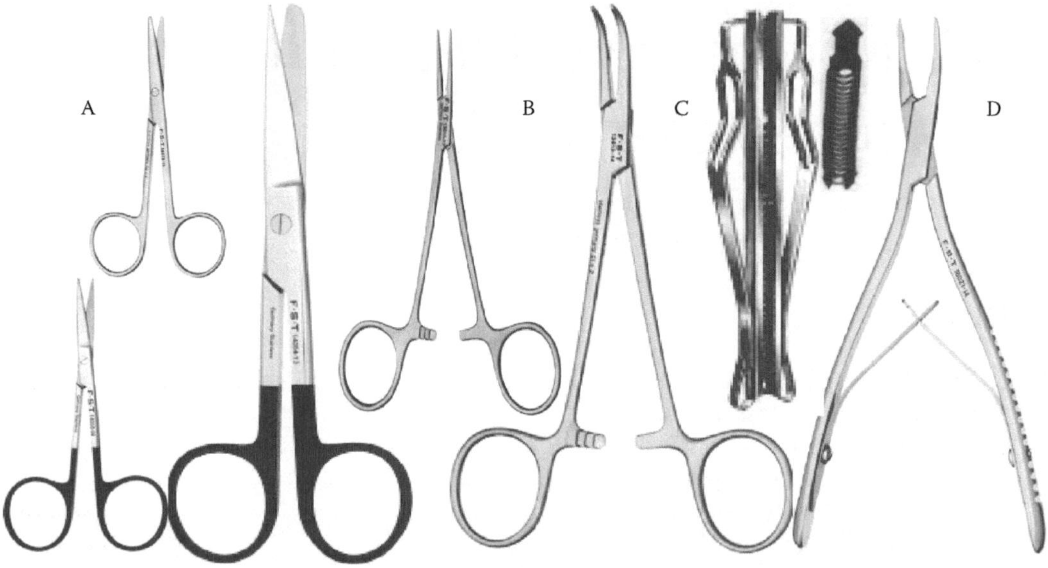

FIGURE 10.1 General surgery tools. (A) Different sizes of scissors. (B) Different sizes of delicate and nondelicate hemostatic forceps. (C) Reflex wound clip applier and 9-mm clips. (D) Microbone rongeur.

toughest tissue with exceptional ease and accuracy and are more durable than general scissors, lasting up to five times longer. They can be obtained from Roboz, Biomedical Research Instruments, Fine Surgery Tools, or other surgical instrument supply companies.

Regarding scissors for other tissues (Figure 10.1A), also needed are one pair of microdissecting scissors (about 9 to 11 cm or 3.5 to 4.5 inches long) with sharp tips, straight or curved, and another pair of microdissecting scissors with blunt tips. These can obtained from Roboz or Biomedical Research Instruments. Tough Cut fine iris scissors from Fine Surgery Tools are also good instruments for stroke surgery.

Hemostatic Forceps

There are different kinds and sizes of forceps on the market: delicate, nondelicate, straight or curved, serrated or smooth, hemostatic, and mosquito. One pair of hemostatic forceps and one pair of mosquito forceps should be included in a surgery package. Figure 10.1B shows a delicate hemostatic forceps; a straight and a curved mosquito forceps 12.5 cm (5 inches) long.

Clamp-applying forceps are designed to hold and apply small or microvessel clamps.

Sponge-holding forceps with curved smooth jaws help to apply disinfection gauzes during preoperative preparation or to mop up blood and other fluids during surgery.

Suture Instruments

The Autoclip wound clip applier and Reflex 9-clip applier with 9-mm clips (Figure 10.1C) are convenient to use. They are autoclavable and durable. A staple remover helps to remove the clips after 10 to 14 days.

Bone Rongeur or Microrongeur

Figure 10.1D shows a fine-tipped rongeur, which is especially suited for delicate stroke surgery on rats.

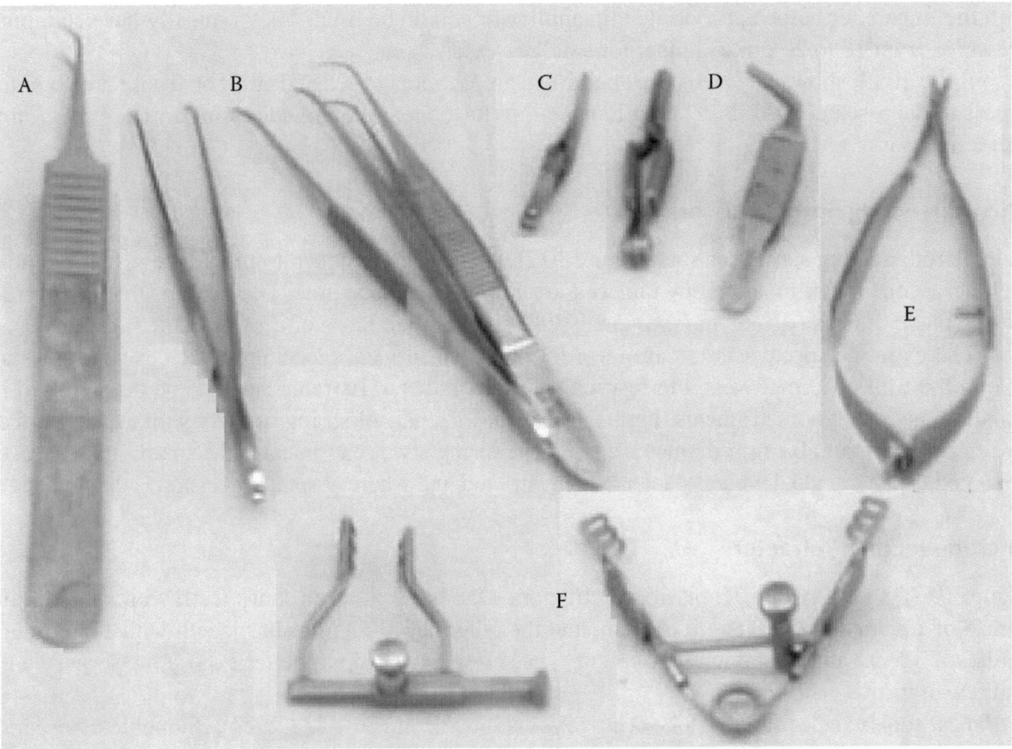

FIGURE 10.2 Some tools for microsurgery. (A) Microdissecting tweezers. (B) Different microdissecting forceps. (C) Vascular clamps. (D) Microdissecting spring scissors. (E) Dietrich microbulldog clamps, both curved and straight. (F) Microdissection retractors.

Autoclavable Trays or Instrument Cases

Autoclavable trays or instrument cases can be found at Fine Surgery Tools or other surgical supply companies (www.finescience.com).

Microsurgical Instruments

Microdissecting Tweezers

Figure 10.2A shows a pair of microdissecting tweezers with 0.05 × 0.01 mm tips curved at 45° for improved visibility. It is an ideal tool for precise procedures on small vessels.

Microdissecting Forceps

Figure 10.2B shows microdissecting forceps that have microserrated surfaces on the inside of the tips. They are 10 cm (4 inches) long, half-curved or fully curved, standard pattern, with 0.5- to 0.8-mm-wide tips. A large number of small teeth (for example, DeBakey pattern) reduce the clamping force required to manipulate a vessel; for this reason, they are called nontraumatic forceps.

Vascular Clips

Appropriate clip size and clip pressure are based on the size of the artery. Detailed information can be found at www.finescience.com. The clips shown in Figure 10.2C are suitable for common carotid artery and internal carotid artery application in small rodents. These clips should never be applied with the fingers; only the appropriate clip applicator should be used. They typically have clamping pressures from 10 to 20 g/mm^2, ideal for smaller vessels.

Figure 10.2E shows Dietrich microbulldog clamps, both curved and straight. These are good for access to deep-seated vessels. The bulldog clamps have an approximate closing force of 50 g/mm^2 and weigh 3 g.

Microdissecting Spring Scissors

The microdissecting spring scissors (Figure 10.2D) have been designed to maximize visibility under the microscope by eliminating any unnecessary bulk in the blade area. The result is that the overall cutting blade dimensions are approximately 30% smaller than those of the "industry standard" scissors. The extrafine spring scissors are ideal for precision surgical procedures, such as cutting a tiny hole in the wall of a small vessel to insert a suture or catheter. Extreme care should be exercised in handling any of these instruments. Figure 10.2D shows a pair of spring scissors with a 2-mm effective cutting edge and 0.1-mm-diameter tips. These spring-style scissors have the smallest blades ever produced. They are ideal where space is very restricted and where absolute precision is demanded.

Microdissecting Retractors

Figure 10.2F shows two different microretractors. One has 3 × 3 mm sharp teeth with a maximum spread of 1.8 cm and total length of 3 cm, and the other has 3 × 3 mm sharp teeth with a maximum spread of 3.5 cm and a total length of 4 cm. These retractors help to better expose the surgery area and thus improve visibility.

These tools can be found at Roboz (http://www.roboz.com, 800-424-2984), Biomedical Research Instruments (http://www.biomedinstr.com, 800-327–9498), Fine Science Tools (http://www.finesciencetools.com, 800-521-2109), or other surgical instrument companies.

STANDBY EQUIPMENT

Electric Hair Clipper

Several different hair clippers are commercially available (Figure 10.3). Figure 10.3a shows a clipper with a vacuum system. All units and attachments are ultraquiet and produce a constant airflow to cool the blades, eliminating the use of unhealthy solvent sprays and thus extending the blade life by 80%. This system helps protect the surgeon from respiratory problems and skin diseases. A

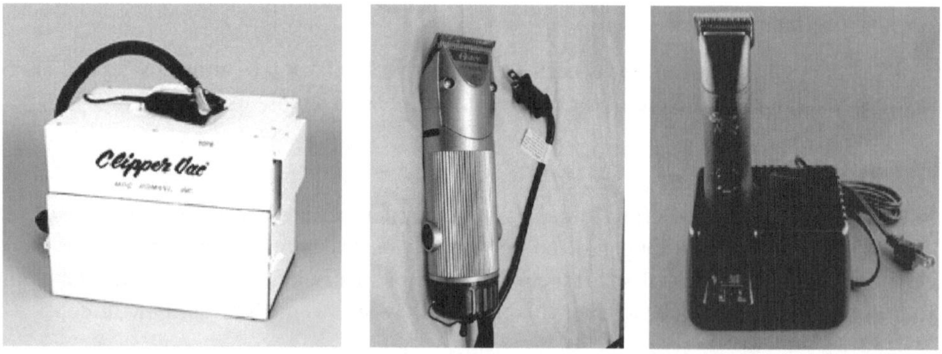

(a) (b) (c)

FIGURE 10.3 Different kind of hair clippers. (a) A portable, one-operator vacuum unit with 6 gallons of hair storage. A top-mounted carrying handle renders this 27-pound unit easy to move. The clipper attachment is included. (b) Simple electromagnetic hair clippers for heavy-duty use with rats. (c) A cordless clipper.

patented regulator on the clipper attachment controls the vacuum flow for multiple blade lengths. The clipper attachments are extremely lightweight and are economically designed for each style of clipper to produce an even cut and can be properly fitted (blade size #40 is good for rats). This system can be found at M.D.C. Romani (www.clippervac.com, 2860 West Pike Road, Indiana, Pennsylvania 15701, 724-463-6101, e-mail: info@clippervac.com).

Other hair clippers (Figure 10.3b) can be found at VWR (Cat. No. 100231-062), Rattus, Webster Veterinary Supply (800-225-7911), Stoelting, or other animal supply companies. Some clippers are cordless, rechargeable, and convenient to use. Figure 10.3c shows one of these; it is as powerful and uses the same blade set as the clippers powered by wall current. It can shave any size animal for surgery and work all day on a single charge. A fast recharger stand can recharge clippers in 1 hour or less but will not damage its batteries when left on charge for longer periods. The battery pack can be removed from the clipper for recharging, and a spare battery pack can be used for high-volume applications.

Light Source

Proper lighting is of enormous help for surgery. A cold-illumination lighting system with adjustable intensity is effective. Moreover, the flexible fiber-optic arm can be equipped with a small focusing lens to obtain diffuse or pinpoint lighting. Halogen lamps can also be used, although these commonly do not have heat filters, which necessitates more frequent moistening of the operating area. Figure 10.4 shows some surgical lights used in surgery laboratories. The ST9W series from

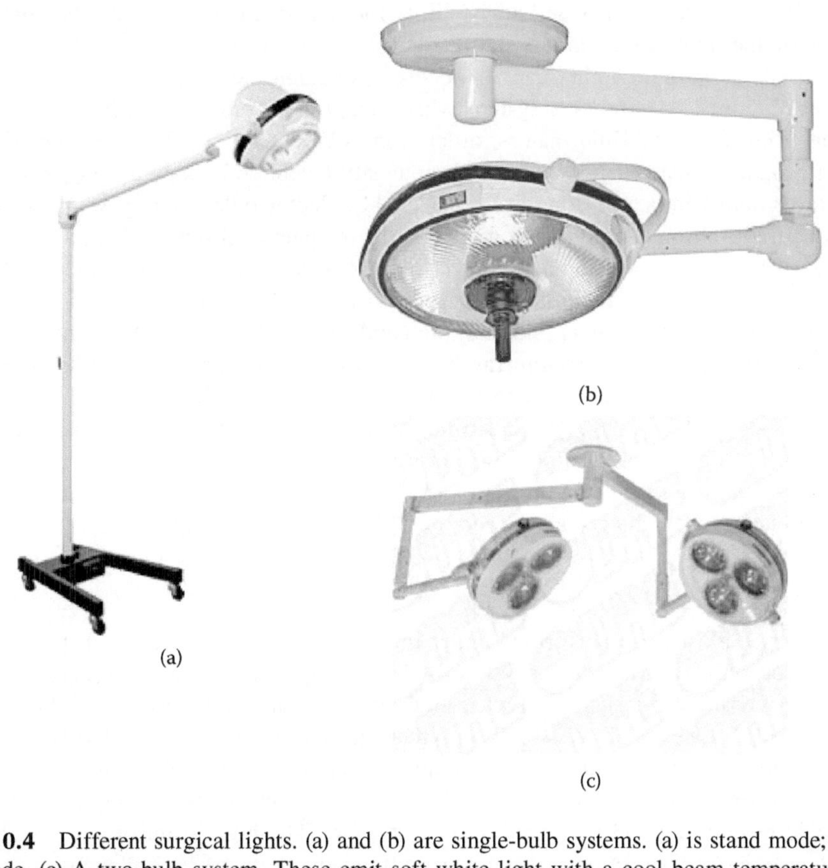

FIGURE 10.4 Different surgical lights. (a) and (b) are single-bulb systems. (a) is stand mode; (b) is wall-mount mode. (c) A two-bulb system. These emit soft white light with a cool beam temperature, superior shadow control, excellent maneuverability, and adjustable friction points for drift-free performance. The light intensity is adjustable.

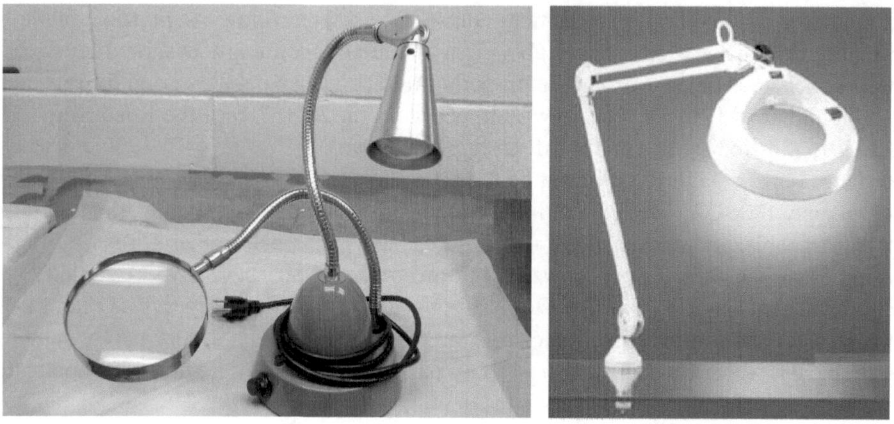

(a) (b)

FIGURE 10.5 Different lights with magnifiers. (a) A table-stand-based light magnifier system. (b) An illuminator magnifier system from Luxo. These can be ordered from VWR in the United State (800-932-5000).

Skytron has stand, ceiling, or wall-mounted modes with a 9-inch-diameter light head. These are high-intensity single- or two-bulb systems with cool beam temperature and glare reduction. This light has excellent maneuverability and adjustability for drift-free performance. The intensity control is located on the light head, and it is simple to use.

If these kinds of surgery lights cannot be made available, alternative light sources can be recommended. Figure 10.5a shows a system with a magnifier combined with one or two lights (Dyna-Lume, High Intensity Illuminators, order from VWR, Cat. No. 36518-971 or 36518-973). This system produces electronically adjustable intensity levels with exacting precision. Features include a power on/off indicator, long-life replaceable reflector bulbs, and a weighted base. It is finished with durable chrome, with a high-gloss coating of enamel on steel and anodized aluminum bulb guards. An all-angle swivel and a 28-cm (11-inch) long chrome flex arm create adjustable light heads. A built-in socket for the magnifier accessory kit allows the lamps to be used to view "mini" to "macro" sizes. The base is 4 cm (1.5 inches) high and 14.5 cm (5.75 inches) in diameter. The heat shield accessory protects objects from infrared energy (heat) emitted from the bulbs. Figure 10.5b shows a system with an illuminated magnifier and base (Luxo, VWR, Cat. No. 53514-000). It is designed to provide brilliant light. The magnifier provides shadow-free inspection. The system is constructed with a weighted base support that permits freestanding movement. The adjustable 12.7-cm (5-inch)-diameter arm moves the three-diopter-lens ring fixture over a wide area. The lamp has a 76.2-cm (30-inch) range. Objects are viewed through the bulb assembly ring.

The Operating Board

As stroke surgeries mostly are performed under inhalation anesthesia, a downdraft table is often used for this procedure. To avoid unwanted movements, proper and stable placement of the rodent during surgery is critical for success. Figure 10.6 shows a surgery board from VWR (Cat. No. 10718-012, in sizes from 28 × 11.4 to 41 × 21 cm) that can be used for animals 23 to 31 cm long. The board features quick setup and release with flexible nylon tie-downs hooked into locking grooves along the side of the board to hold the animal securely without injury. An angled body channel provides stable positioning and increased animal comfort. The boards are of autoclavable Noryl (polyphenylene oxide) construction and stand up well to detergents, bases, and acids.

A soft pad about 40 × 45 cm and covered with a sterile drape is an alternative to the board (see Figure 8.11 in Chapter 8).

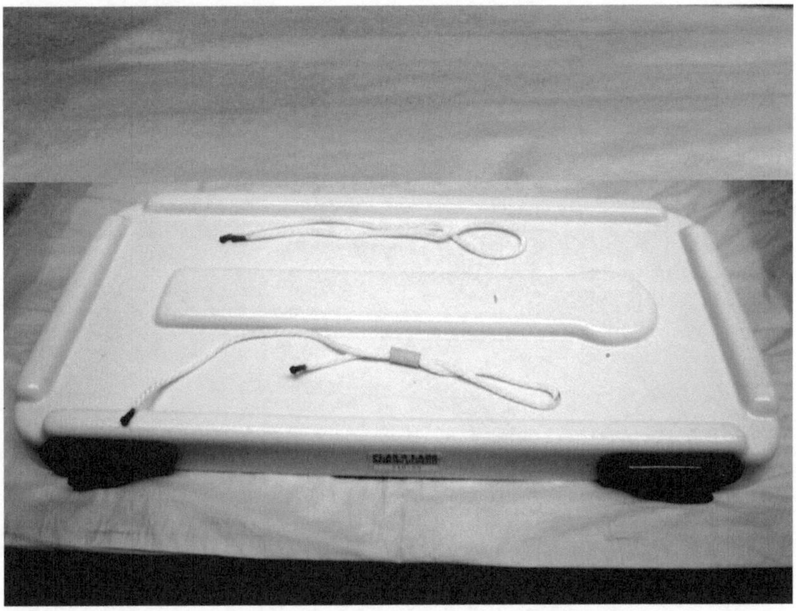

FIGURE 10.6 Rodent surgery boards. These boards feature quick setup and release and flexible nylon tie-downs that hook into locking grooves along the side of the board to hold the animal securely without injury. An angled body channel provides stable positioning and increased animal comfort. It is made of autoclavable Noryl (polyphenylene oxide) construction and can stand up well to detergents, bases, and acids.

Rectal Thermometer

Maintaining the animal's body temperature in the normal range is very important in stroke studies. Body temperature can be influenced by the type of anesthesia as well as by surgical procedures. A rectal thermometer can check an animal's body temperature before, during, or after surgery. Figure 10.7 shows a thermometer from Welch Allyn (SureTemp Plus 690 Electronic Thermometer). This single thermometer features speed, accuracy, security, and convenience. The Welch Allyn SureTemp is the fastest oral/axillary/rectal thermometer on the market. It provides a 4-second oral, 10-second rectal, and 15-second axillary response to temperature changes. Our laboratory has adopted this system for measurement of a rodent's rectal temperature. This system has several helpful features:

1. Used probe covers may be ejected automatically.
2. Large LCD (liquid crystal digital) displays temperatures in Fahrenheit or Celsius.
3. It can recall previous temperature measurement.
4. It has a waterproof, stainless steel probe shaft.
5. It has convenient storage housing for 25 probe covers.
6. There is a color-coded, interchangeable, and removable probe well to minimize the risk of cross-contamination.
7. A user-selectable icon or words identify thermometer modes.
8. There is a battery life indicator.

Note that the probe cover must be lubricated (lubricant supplied) before insertion into the rat's rectum.

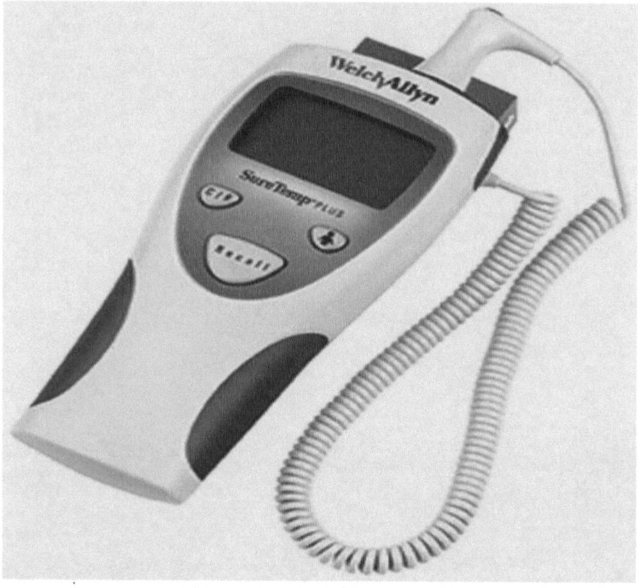

FIGURE 10.7 A thermometer from Welch Allyn (SureTemp Plus 690 Electronic Thermometer). It provides 10-second response time for rectal temperature determination in rats.

Operation Microscopes

There are many types of surgical microscopes or magnifiers (Figure 10.5) available on the market. For simplicity, the whole group falls under the term *operation microscopes.*

In 1920, Carl Zeiss created his well-known Zeiss surgical loupes, which consist of a pair of ×2 magnifiers mounted in a spectacle frame and are still used today in surgical settings (Figure 10.8). But, this device is not sufficiently powerful to be of help in very delicate microsurgeries, such as operations on eyes and on brain vessels in small rodents. In 1952, Zeiss developed a more powerful microscope called the OPMI 1, which increased the magnification power and was more efficient to use. Figure 10.9 shows an OPMI 1FC S21 floor-stand microscope with mobility and stability in one device. This floor-stand device has been specially designed for use in delicate surgery and

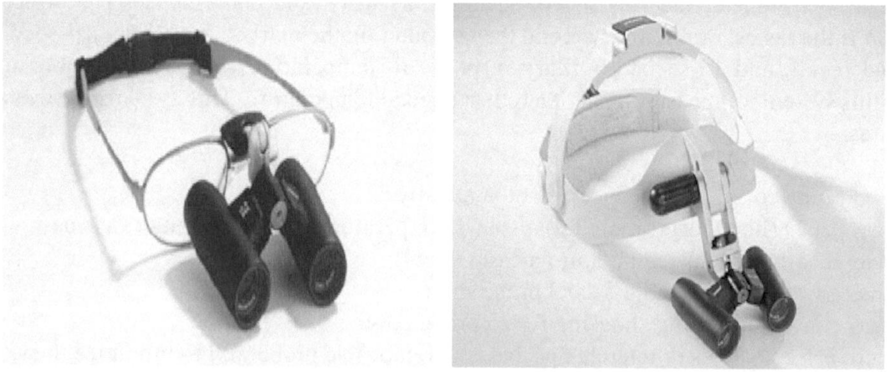

(a) (b)

FIGURE 10.8 Two different loupes from Zeiss. (a) A Prism Loupe KF; (b) a Prism Loupe KS. Both are easy to use, facilitating ergonomically correct body posture and optimum working distance. These feature (1) a lightweight titanium frame and soft bridge for maximum comfort; (2) simple adjustments, with height and tilt easy to set with a Torx key; and (3) ease of setting for each eye using the folding bridge.

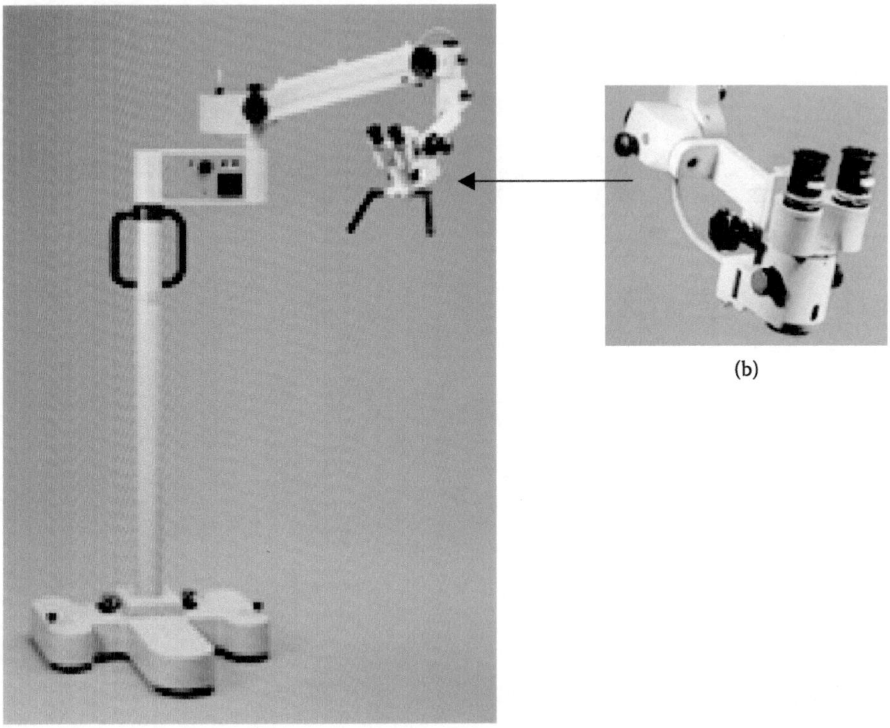

(b)

(a)

FIGURE 10.9 (a) An OPMI 1FC S21 floor-stand microscope. It is easy to position and offers five magnifications from ×3 to ×21. (b) An enlarged picture of the optical system. This microscope can also be integrated with a video camera for documentation.

permits easy, exact positioning of the OPMI 1FC. The friction of all joints and the upward/downward motion of the microscope can be set as required by the surgeon. A lamp module with a 12 V/100 W halogen reflector lamp is integrated. A backup lamp module allows the lamp to be changed in a matter of seconds, also during surgery. This microscope system can also be interfaced with the latest developments in computer-digital-video technology. Figure 10.10 shows an OPMI® pico floor stand. Except for the magnification function, it offers many of the new integrated features, such as documentation accessories and video and digital camera connection to a computer or a monitor. With these functions, still images can be created at the push of a button. Video recording of procedures can be easily and clearly displayed on the monitor or stored in a computer for teaching and data collection purposes. All OPMI microscopes offer five magnifications (×3, ×5, ×8, ×13, and ×20) at a 250-mm working distance. These microscopes can be purchased from Zeiss or from Seiler Instruments. Detailed information can be found in their Web sites at www.zeiss.de or www.seilerinst.com, respectively.

Stereotaxic Apparatus (Stereotaxic Head Holder)

A stereotaxic apparatus is ideal for researchers in need of a versatile, reliable instrument for sensitive, highly skilled procedures with small animals. Precision alignment of the stereotaxic instrument ensures accurate placement of electrodes, micropipettes, and other devices such as brain ventricular or local brain injection. There are many kinds of stereotaxic instruments available. Figure 10.11 shows a Lab Standard Stereotaxic Instrument (Single and Dual Manipulator Models)

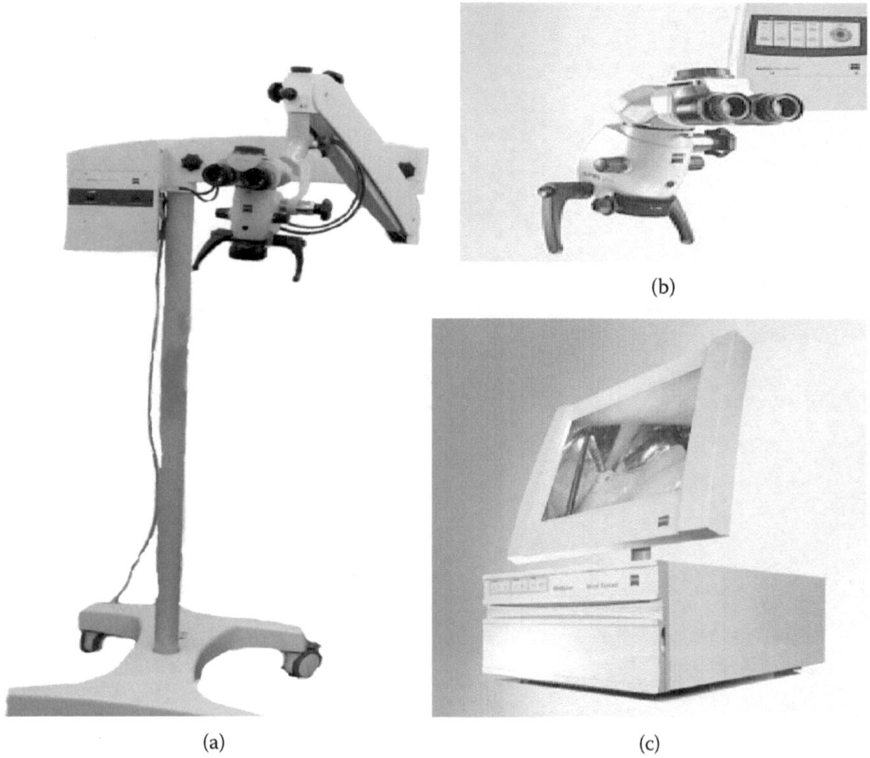

FIGURE 10.10 An OPMI pico microscope floor stand (a) with a new integrated video camera connecting to a computer or a monitor (c). (b) An enlarged view of the optical system with its control board.

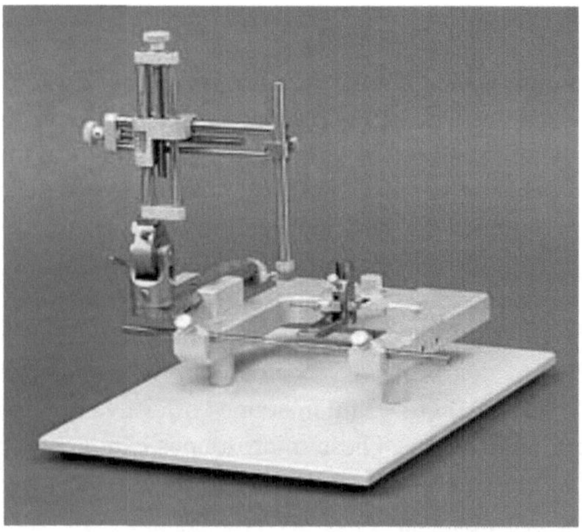

FIGURE 10.11 A stereotaxic instrument, sold by Stoelting, with a single manipulator.

sold by Stoelting. You can find detail information at www.stoeltingco.com. The Lab Standard Stereotaxic offers several advantages:

1. Easy-to-read scales: All scales are oriented to be read easily from the open end of the U. This is the position from which most scientists prefer to work. The numerals on the scales are larger and therefore easier to read. The scale lines are laser engraved to ensure the finest possible permanent marking of scales on all three axes. Precise alignment with facing vernier scales gives resolutions accurate to 0.1 mm.
2. Smooth movements: The Lab Standard's™ exclusive triple lead screws allow the fastest positioning possible, consistent with aligning the scales easily with a given coordinate.
3. Versatile positioning: The manipulator arm controls mediolateral and vertical positioning via lead screws and anteroposterior movement via a dovetail slide, with 80 mm of travel possible in each direction. A universal joint allows the investigator to change the angle of the probe up to 90° in either the anteroposterior or mediolateral planes. The improved locking mechanism on the Lab Standard will hold any angle position without slippage. And, of course, it also provides an absolute lock at 90° vertical. In addition, a swing joint allows the investigator to conveniently swing the manipulator arm and probe out of the way for performing a procedure, then reliably return the probe to the same point.
4. Convenient for electrophysiology: Integral brass bushings in the manipulator arm allow grounding directly to the closest metal on the manipulator arm—even the probe holder.
5. Dual Manipulator Model: The Dual Manipulator Lab Standard has three-dimensional manipulator arms on both sides, with the second manipulator arm properly reversed regarding handedness. Dual manipulators are useful in studies in which double injections or simultaneous stimulation and recording are necessary.

ACKNOWLEDGMENT

We thank James Hastings from Johnson & Johnson, PRD, for his technical assistance and advice.

11 Postoperative Care

Yanlin Wang-Fischer and Lee Koetzner

CONTENTS

NATIONAL INSTITUTES OF HEALTH (NIH) RECOMMENDATIONS FOR POSTOPERATIVE CARE FOLLOWING RODENT SURVIVAL SURGERY

Postoperative care guidelines have been recommended for application to major survival surgical procedures in rodents in the *Guide for the Care and Use of Laboratory Animals* established by NIH,[1] concomitant with the *Public Health Service (PHS) Policy on Humane Care and Use of Laboratory Animals.*[2] The postoperative recovery period is a *critical* time for rodents undergoing surgery. Providing appropriate postoperative care can promote the optimal recovery of an animal from surgery and anesthesia. Owing to their higher metabolic rates and increased surface area-to-volume ratios, rats and mice lose a considerable amount of body heat during and after surgical procedures.

Animals may exhibit systemic hyperthermia or hypothermia after stroke surgery, depending on the type of model. The classic filament suture model and embolic stroke model cause hyperthermia in rats, whereas models of stroke induced by distal middle cerebral artery occlusion may cause hypothermia because distal middle cerebral artery occlusion does not involve the hypothalamic artery (refer to Chapters 4, 12, and 13). Hyperthermic animals may become dehydrated owing to water loss, while hypothermic animals may become hypotensive due to shunting of blood toward vital organs. Ultimately, both hyperthermia and hypothermia may result in decreased drug clearance, prolonged anesthesia, and increased mortality.[3,4] Postoperative care programs should be considered and designed before commencing any experimental procedures. The following essential guidelines should be routinely incorporated into the postsurgical care protocol.

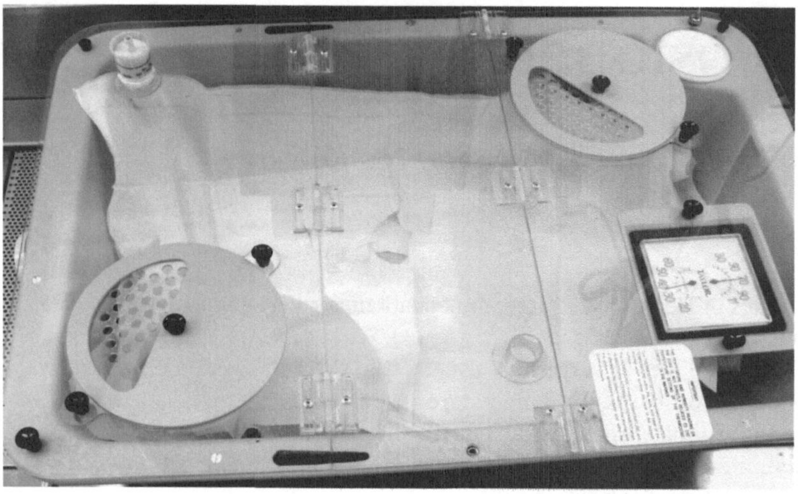

FIGURE 11.1 A commercially available intensive care unit (ICU). This Thermocare Intensive Care System Unit supports (1) heat safety: Animals are in a "safe heat" environment free of "hot spots"; (2) humidity requirements, which can be preset as desired.

GUIDELINES FOR SPECIAL STROKE SURGERIES

MAINTAINING ANIMAL BODY TEMPERATURE

In focal models of distal middle cerebral artery territory stroke (based on filament occlusion), postoperative animals should be kept warm in an intensive care unit (see Figure 11.1), and a layer of tissue paper should be placed in the box to absorb urine. Temperature in this unit is controlled between 27°C and 32°C with humidity about 50% to 60%. This system can be ordered from Thermocare (www.thermocare.com, 800-262-4020). Animals should be moved into their own cages after recovery from anesthesia.

In the absence of a dedicated intensive care unit, the room in which the animals are to recover must be warm (22°C to 28°C) and quiet. An incandescent lamp (50 to 75 W) can be placed 12 to 14 inches away from the animal to provide supplemental heat. Cold or noisy conditions will stress an animal, especially those subjected to stroke. The resultant increases in catecholamine and corticosteroid levels may confound data interpretation and analysis.

In the filament suture stroke model or classic embolic stroke model, animals should be kept on a soft pad or a surgical board without heating because these models cause hyperthermia (refer to Chapters 13 and 14 on surgical procedures).

PROTECTING ANIMALS FROM AIRWAY OBSTRUCTION

Animals recovering from anesthesia should be rotated from side to side until they are able to maintain sternal recumbency. They should not be left unattended until they have fully regained consciousness. If an animal is allowed to recover in its own cage with bedding materials, it should not be placed directly onto contact bedding as the animal may inhale or ingest bedding particles. Placing the animal onto a paper towel or equivalent will prevent airway obstruction.

MAINTAINING FLUID BALANCE

Hypovolemia causes cardiovascular failure, eventually leading to animal death because blood and fluid loss can occur, especially during a prolonged (more than 30 minutes) surgical procedure, possibly due to operator inexperience. The total amount of circulating blood in rat is about 6 mL/100

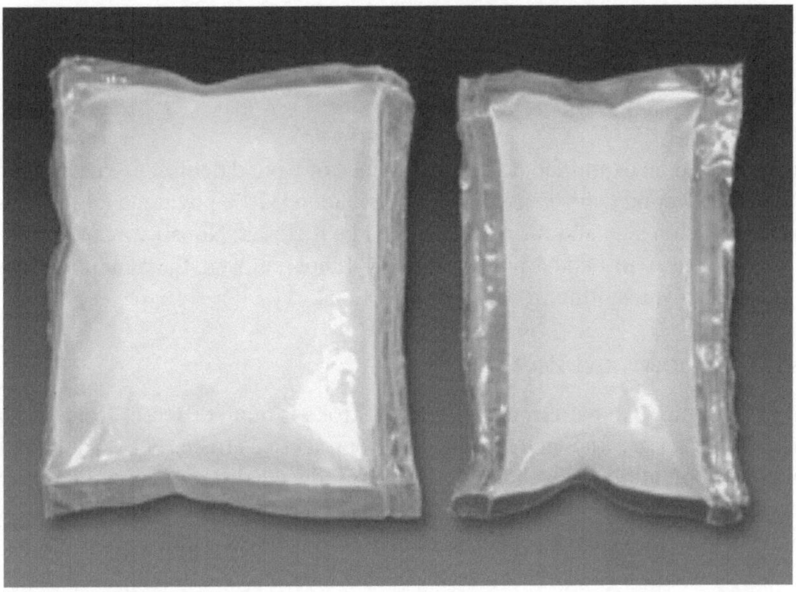

FIGURE 11.2 Napa Nectar is a "water gel pack" available in two convenient sizes, 4 ounces (113 g) and 8 ounces (226 g), for daily water and nutrition needs. This product is beneficial for stroked animals, especially in the first 1 to 2 days after surgery.

g body weight. Administration of warmed lactated Ringer's solution or normal saline (0.9%) by intraperitoneal or subcutaneous injection (with not more than 1.5 mL at one place) also can assist a rodent's recovery from anesthesia. Usually, fluid support of 1 to 2 mL/mouse or 3 to 4 mL/rat immediately postsurgery will help to prevent the inevitable dehydration that occurs during the first 24 to 48 hours as animals are most likely to be anorexic during this postoperative time period.

Although the chance of fluid loss from a small craniectomy or ventral neck area is lower than that during surgery of the abdominal cavity, stroked animals are often hemiparalyzed, with difficulty in eating and drinking, especially during the first 1 to 2 days. Administration of fluid is recommended, especially if multiple surgical procedures were performed on a rat and the procedures lasted more than 20 minutes with unexpected bleeding.

Supplying Adequate Nutrition

Adequate nutrition is necessary in postoperative animals. It is recommended to give soft food or gel food to the animal on the first day after surgery. Such food has the same nutritional components as rodent chow but is soft and rich in fluid content. Figure 11.2 shows a kind of gel food called Napa Nectar™, packaged in an inert polypouch. It is a sterile product containing ingredients approved by the U.S. Food and Drug Administration (FDA). The nonrefrigerated shelf life is 18 months, and moisture content is 99.3% by volume. It can be directly ordered from an animal supply company, SE Lab Group (www.selabgroup.com).

Antibiotics

It is advised that rats should be administrated an antibiotic to protect against postoperative infection, despite the fact that rats are remarkably resistant to the development of wound infection. Furazolidone aerosol powder (4%) from Henry Schein, spread locally around the surgical area, has been found quite effective in our stroke surgeries. Alternatively, a single subcutaneous dose of 150 mg/kg amoxicillin or 150 mg/kg ampicillin given just after the surgery will also give adequate protection against infection. Using antibiotics for long-term postoperative care is not recommended in rodents.

Care of Incision Sites

The incision site should be examined daily for evidence of wound dehiscence or infection until it is completely healed. If wound dehiscence or infection is found, 10% povidone iodine solution should be applied on the incision site, and the wound should be reclosed. Nonabsorbable sutures or wound clips should be removed 10 to 14 days postoperatively. If not removed, the remaining sutures or clips can act as a foreign body, resulting in bacterial infections.

Management of Postoperative Pain

Management of postoperative pain in the context of stroke is quite different from that in other surgeries because analgesics are known to produce neuroprotective effects. This would make it impossible to evaluate the potential of novel compounds to produce neuroprotection in animal models of stroke. Bupivacaine (0.25%, 0.1 to 0.2 mL), however, is accepted for local use to decrease postoperative pain. Bupivacaine (from Henry Schein, United States) is an extended activity form of lidocaine, lasting 4 to 6 hours; it exerts its analgesic effect by blocking peripheral nerve activity.

The evaluation of postoperative pain in stroked rodents follows the procedures used in general surgery. Clinically, the extent and course of postoperative pain varies with the type of procedure performed, the amount of tissue handling, the level of surgical skill, and the presence of inflammation or wound infection. It is also important that the investigator be cognizant of the "normal" amount and duration of pain associated with a given procedure. Severe, unexpected pain behaviors may indicate serious complications, such as brain infection, sepsis, and the like, that require rapid intervention with appropriate treatment or euthanasia. Useful parameters for assessing postoperative pain in rats are activity, appearance, vocalization, and body weight and feeding behavior changes.

Activity

The activity level of a stroked animal experiencing pain is usually reduced, but the animal sometimes may show unusual restlessness or limb spasm.

Appearance

- Animals may have a hunchback and huddle in the corner of the cage.
- Pain may result in cessation of grooming, which may lead to an unkempt or ruffled hair coat and soiling of the perineal region.
- An encrusted pigmented discharge (porphyrin pigment) may also accumulate around the eyes, nose, and mouth of rats (occasionally mice). Although this porphyrin staining is a nonspecific stress response, its presence should alert the observer to the possibility that the stress involved may be pain related.

Vocalization

- Acute pain may cause an animal to vocalize (make noise).
- When assessing rodents, however, it is important to appreciate that their vocalizations are often at higher frequencies and thus may be inaudible to humans, so careful observation of other signs is important for assessing postoperative pain.

Body Weight and Feeding Behavior Changes

- While subtle changes in activity or appetite may not be observed at the cage side, changes in weight can be quickly detected, allowing appropriate intervention. During the early postoperative period (days 1 to 3), it is highly recommended that animals be weighed on a daily basis to assess changes in body weight. Our unpublished data showed that stroked rats lose much more body weight than sham surgery rats (17% to 20% versus 5% to 10%) compared to presurgical baseline values.
- Decreased food and water consumption are useful indicators of postoperative pain in rats. Food and water intake is often markedly reduced when an animal is in pain. Severe pain may be associated with complete cessation of eating and drinking.
- Supplying a softer, more palatable, easily accessible diet (e.g., a dough diet) may encourage the animal to resume eating after surgery. If the problem cannot be resolved within 48 hours, the sick animal should be euthanized.

Record Keeping

Appropriate records must be kept of postoperative care, evaluation, and treatment. For rodents, these can be kept on the surgery card attached to the animal's cage or in the laboratory notebook, which must be available for inspection on request. It is recommended to place rats in separate, individual cages after stroke surgery.

ACKNOWLEDGMENTS

We thank James Hastings and Dr. Sally Wixson from Johnson & Johnson, PRD, for their technical assistance and advice.

REFERENCES

1. NIH Animal Research Advisory Committee, NIH Animal Research Advisory Committee, *NIH Intramural Research Program Guidelines for Survival Rodent Surgery*, 1999. Available at: http://oacu.od.nih.gov/ARAC/surguide.htm.
2. The Office of Laboratory Animal Welfare at NIH. *Public Health Service (PHS) Policy on Humane Care and Use of Laboratory Animals*, Bethesda, MD, 1996.
3. Animal Welfare Act (P.L. 94-279), Regulatory Enforcement and Animal Care, APHIS, USDA, Riverdale, MD.
4. Flecknell, P.A., Postoperative care. In: *Laboratory Animal Anesthesia*, Academic Press, London, 1996, p. 131.

12 Surgery Procedure for Distal Middle Cerebral Artery Occlusion Model

Yanlin Wang-Fischer and Lee Koetzner

CONTENTS

SPECIAL INSTRUMENTS

Electric Dental Drill, Cordless Microdrill, and Rechargeable Microdrill

Removing bone from the skull is often a necessary step in preparing an animal for middle cerebral artery occlusion (MCAO). For this purpose, a portable dental drill has been used in many laboratories (Figure 12.1a). In recent years, cordless and rechargeable microdrills have become commercially available. The new cordless microdrill from Stoelting (Figure 12.1b) is designed for research applications that require surgical burrs and trephines. It is made from a lightweight aluminum alloy for balance and control. It is powered by a rechargeable 6-V nickel cadmium battery and can function for up to 8 hours under normal operating conditions. Its average speed is 14,400 rpm (VWR, Cat. No. 100230-730). Its carbide burr drill bits have a shaft diameter of 2.3 mm and are 4.4 cm long. A microdrill from Fine Science Tools (Figure 12.1c) is portable, rechargeable, and high speed, powered by a 7-V nickel cadmium rechargeable battery. This drill operates at approximately 11,000 rpm, runs quietly, and is vibration free. It is provided with its own recharging unit for overnight charging. The drill cannot be overcharged. A sensitive switch allows precise fingertip control of drilling. This microdrill will hold all standard bits and reamers (Cat. No. 18000–17, www.finescience.com). All bits or burrs can be autoclaved, but the drills cannot.

Electrosurgical Generator with Bipolar Forceps Coagulation, Electrocauterization

An electrosurgical generator is needed in order to use a bipolar coagulator. Figure 12.2a shows a generator from Medical Resource USA (Cat. No. Bovie 1250). It can operate in cut, blend, coagulation, fulguration, and bipolar modes for use in stroke surgery in laboratories or hospitals. It features both monopolar and bipolar functions to satisfy surgical demands with safety, flexibility, reliability, and convenience. The Bovie 1250 incorporates 120 W of cutting power and 90 W for blended cutting. It also has two levels of coagulation: pinpoint (80 W maximum power) and fulguration (40 W maximum power) plus bipolar (30 W maximum power). The bipolar coagulator is an indispensable

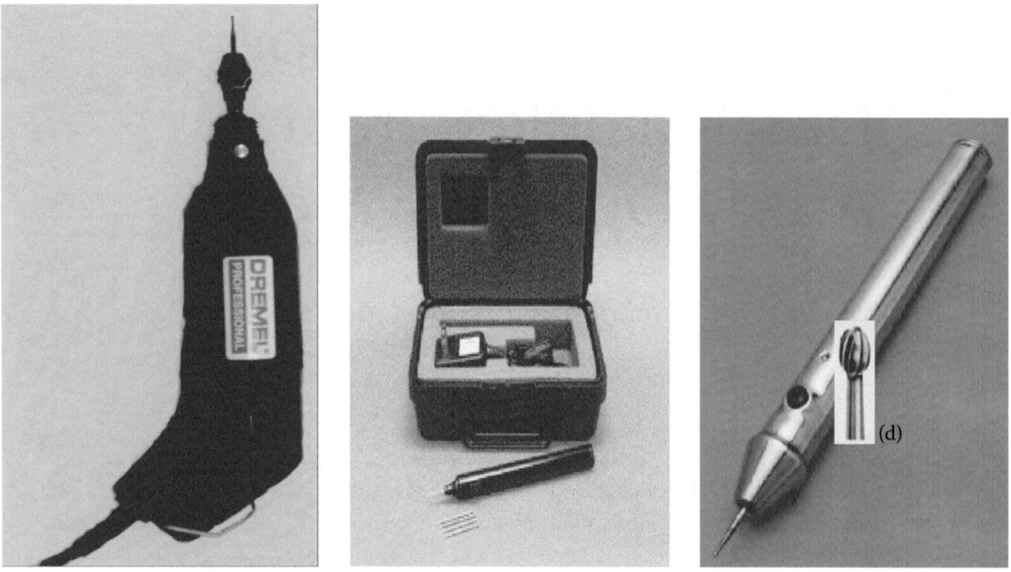

FIGURE 12.1 Different drills used in experimental stroke studies. (a) A portable dental drill from Dremel Professional®. (b) A rechargeable microdrill from Fine Science Tools. (c) A cordless microdrill from Stoelting. These microdrills are powered by rechargeable 6- or 7-V nickel cadmium batteries and can function for up to 8 hours under normal operating conditions at high speed (11,000 to 14,400 rpm) while running quietly, free from vibration. The steel burr has a shaft diameter of 2.3 mm and is 4.4 cm long (D).

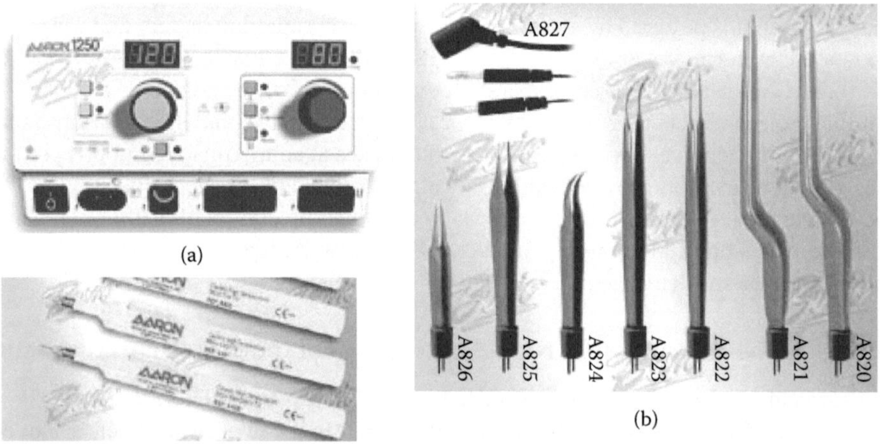

FIGURE 12.2 Different cauterization systems. (a) A multipurpose electrosurgical generator (Bovie 1250). It features both monopolar and bipolar functions to satisfy all surgical demands. (b) A reusable bipolar forceps that is connected to the generator by a cable for cerebral artery occlusion; available in different sizes. (c) High-temperature electrocauterizers, which are popular for many types of procedures, including pinpoint hemostasis, vasectomies, and middle cerebral artery occlusion.

surgical instrument for middle cerebral artery (MCA) coagulation in stroke surgery. It facilitates high-grade hemostasis and increased speed of closure, reflecting a higher efficiency of operation.

Reusable bipolar forceps are available in micro-, 1.0-mm, and 2.0-mm tip sizes and in lengths ranging from 4.5 to 8 inches (Figure 12.2b) (Medical Resource USA, Cat. No. A824). These forceps can be autoclaved for 20 cycles.

If an electrosurgical generator with bipolar forceps is not available, a simple electrocauterizer can be used for stroke surgery (Figure 12.2c). It is inexpensive and easy to use. Within a range of relatively low temperatures (regulated by one or two batteries) and fine tips, the surgeon can be assured of pinpoint hemostasis. All disposable cauterizers are individually packaged and sterile. High-temperature cauterization is popular for many types of procedures, including pinpoint procedures such as hemostasis, vasectomies, and MCAO but may require peripheral cooling (for example, by cauterizing under saline) when necessary to localize the effects of heating. These cauterizers may be obtained from Henry Schein, Medical Resource USA, or other companies.

STEP-BY-STEP SURGICAL PROCEDURE

The step-by-step surgical procedure[1,2] is as follows:

- The rat is weighed and placed in a chamber to induce anesthesia with 3% to 4% isoflurane in medical-grade oxygen or in a gas mixture of 30% oxygen (O_2) and 70% nitrous oxide (N_2O) (see Figure 8.7 in Chapter 8).
- The right side of the head and ventral area of neck are shaved.
- Skin is cleaned locally by 0.5% Betadine and 75% alcohol or 2% chlorhexidine (blue color solution).
- Eye cream should be placed on both eyes to prevent dryness if the procedure is to last longer than 20 minute.
- The rat is then transferred to the surgical board or surgical pad and is anesthetized via a facemask with 2.5% isoflurane in medical-grade oxygen or in a gas mixture of 30% O_2 and 70% N_2O.
- The rat is placed in the supine position; a 10-cc syringe-sized tube can be placed under the neck to accentuate access to the common carotid arteries (CCAs).
- A ventral midline incision is made at the neck area, and the sternohyoid and sternomastoid muscles are blunt dissected to expose the CCAs (Figure 12.3).

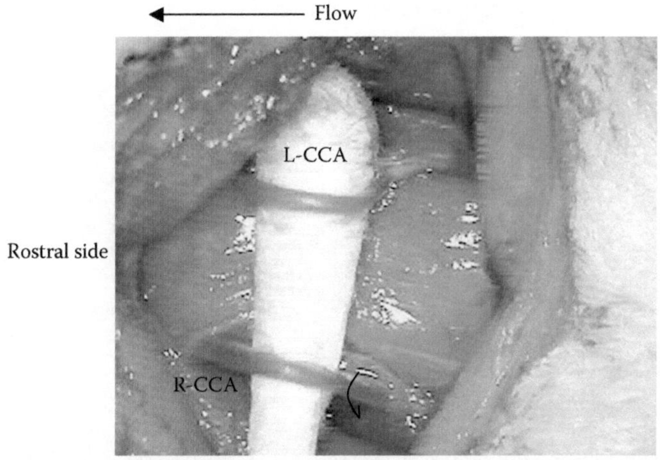

FIGURE 12.3 Both common carotid arteries (CCAs) are exposed. The right CCA is permanently ligated with a 4-0 suture; the left CCA is temporarily blocked for 1 hour.

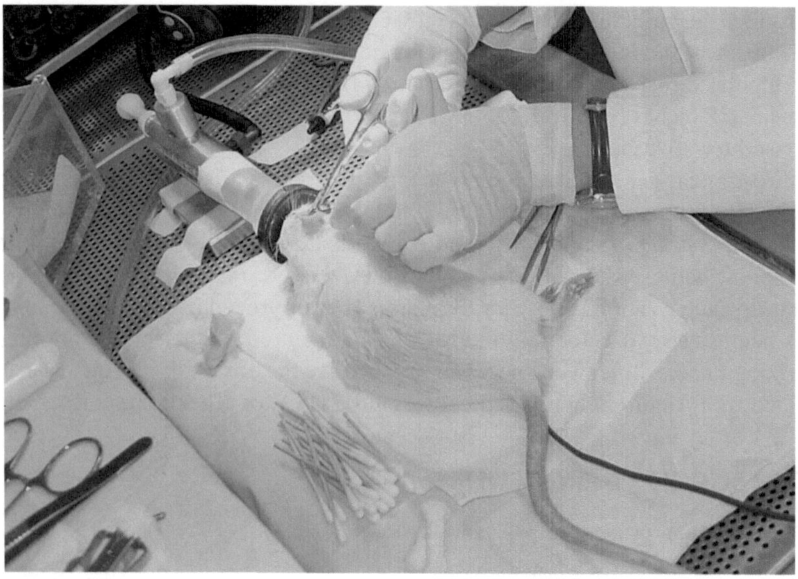

FIGURE 12.4 Rat is placed in a lateral position with right side up.

- Bupivacaine (0.25%, 0.1 to 0.2 mL) is administrated locally around the vessels and nerves to block the carotid sinus reflex.
- The right CCA is permanently ligated with a 4-0 suture, and both CCAs are isolated from their surrounding tissues with a cotton-tipped applicator (Figure 12.3).
- The rat is then turned laterally onto its left side (Figure 12.4). Another incision about 2 cm long is made between the lateral canthus of the right eye and the external auditory canal (Figure 12.5a).
- The underlying temporal muscle is exposed; the temporal nerve, artery, and vein can be seen overlying the muscle (Figure 12.5a). The temporal muscle is blunt dissected and retracted without damaging the nerves and vessels to expose the underlying skull (Figure 12.5b).
- The zygomatic arch is exposed (Figure 12.6a), and most of it is removed by a fine bone rongeur. The inferotemporal fossa is exposed. The top wing of the mandibular bone and muscles is retracted to provide more space. The mandibular nerve crosses the temporomandibular joint to the foramen ovale (Figure 12.6b).

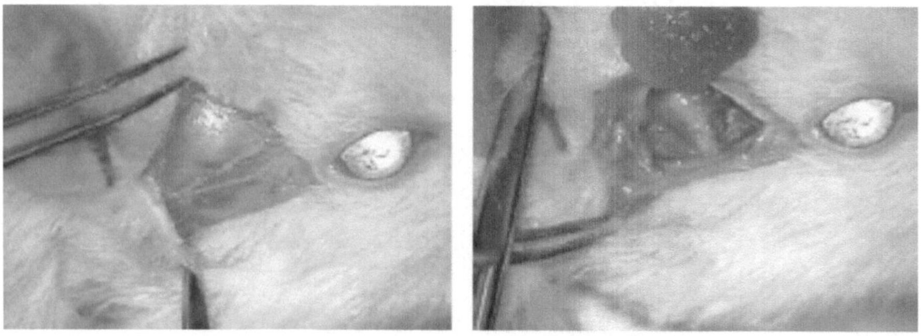

(a) (b)

FIGURE 12.5 An incision about 2 cm long is made between the lateral canthus of the right eye and the external auditory canal. The temporal muscle is exposed; there are some nerves and vessels overlying this muscle (a). This muscle is partially removed to expose the underlying skull (b).

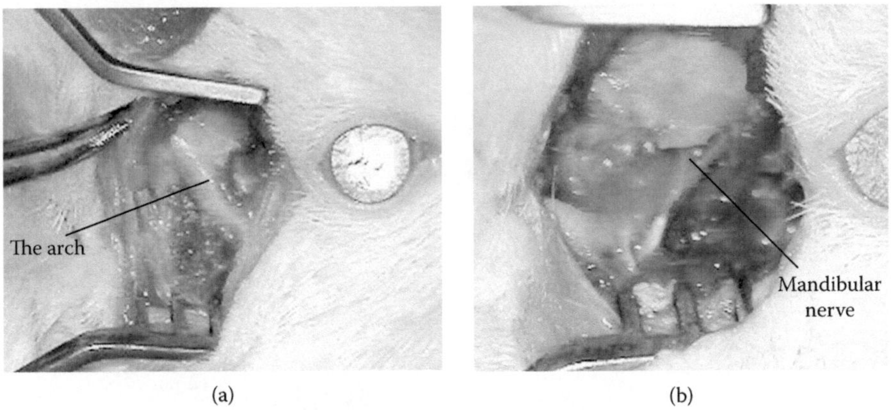

FIGURE 12.6 The zygomatic arch (a) and structures revealed after removing the zygomatic arch (b). The mandibular nerve crosses the temporomandibular joint to the foramen ovale (b). **(See color insert following page 146.)**

- A small hole (4 to 6 mm) is made by a drill at the squamosal bone and is centered at about 3 mm anterior and 1 mm lateral to the foramen ovale or the mandibular nerve, just near the arch rostrum (Figure 12.7a). The hole is drilled carefully, to a depth of 0.5 to 1.0 mm, while probing it frequently to determine its give or softness and its transparency by using saline in the cavity (Figure 12.7b). There is a shallow groove running diagonal vertically down to the base of the skull in this region.
- The dura is opened by a cruciate incision with a 23-gauge needle. The MCA runs directly from the bottom of the skull laterally to the top of the head (Figure 12.8). The MCA usually has several branches that vary among different individuals (Figure 12.9). Another major vein (inferior cerebral vein) crosses the MCA forward to its rostral end.
- Curved genuine Dumont tweezers are hooked around the right MCA and pulled 1 mm up just above or below the inferior cerebral vein; the MCA is then cauterized permanently by pinpoint electrocautery. Different branches are usually seen owing to anatomic variability. To limit the variability of infarction in this stroke model, only one main trunk should be occluded. If the main MCA trunk cannot be identified, the rat should be discarded. If bleeding is difficult to control, epinephrine injection solution (1:1000) is applied to the area to help control the bleeding.

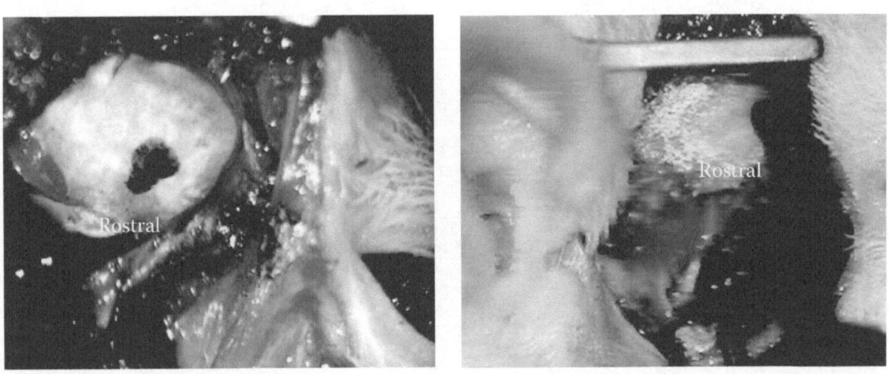

FIGURE 12.7 The drilling location at the squamosal bone is centered about 3 mm anterior and 1 mm lateral to the foramen ovale or the mandibular nerve, just near the arch rostrum (a). A small hole is made using a drill (b). **(See color insert.)**

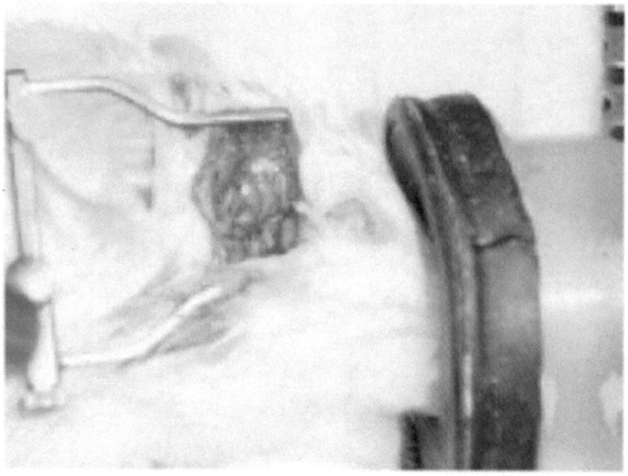

FIGURE 12.8 The dura opened with 23-guage needle by a cruciate incision. The middle cerebral artery runs directly from the bottom laterally to the parietal side. **(See color insert.)**

- The hole is sealed with Tissumend (2 mL/vial, 36 vials/case, Cat. No. 568-7612, Henry Schein) or Nexaband (formulated cyanoacrylate, Lot No. 087092, Cat. No. 92101, 2 mL/vial, Veterinary Products Laboratories, Phoenix, Arizona 85013, order information 888-241-9545, technical information 888-257-7633).
- 0.25% bupivicaine (lidocaine-epinephrine-tetracaine mixture) is given at the incision to reduce postoperative pain.
- The incision is sutured with wound clips.
- The left CCA is occluded by a small vessel clip (refer to Figure 10.10 in Chapter 10). About 60 minutes later, the clip on the left CCA is removed to allow reperfusion.
- The incision is closed with clips after giving 0.25% bupivicaine (0.1 mL).
- Spray furazolidone aerosol powder or other antibiotics on the wounds to protect from infection.
- The rat will wake up, usually in 10 to 20 minutes after discontinuing the isoflurane gas.

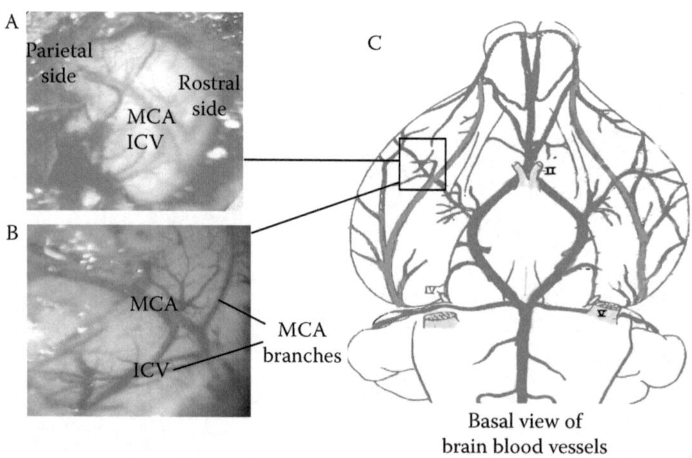

FIGURE 12.9 Individual variations in the branching pattern of the middle cerebral artery (MCA). (A) An MCA going straight up the parietal side without branching. (B) Two branches from the MCA. Another major vein (inferior cerebral vein, ICV) crosses the MCA forward to the rostral side (×25). (C) Basal view of the rat's brain vessel shows where the surgical window was opened to occlude the MCA. **(See color insert.)**

- Monitor the degree of paresis during the recovery period.
- Rats that do not demonstrate left arm paresis are excluded from further study.
- For the temporary MCA ligation model, the MCA is ligated with 10-0 suture tandem occlusion of right and left CCAs with small vessel clips for 1 hour, and then the suture and clips are released to allow reperfusion.[3,4]

PROBLEMS

The most common complication during rat MCAO is acute respiratory distress. The animals that experience this complication exhibit stereotypical symptoms, including rapid heartbeat, asthma-like difficult breathing, pink bubbly discharge from the nose or mouth, wheezing, and cyanosis. In the distal MCAO model with occlusion of both CCAs, the incidence of this problem is greater than 20% in other laboratories (unpublished data) and on average is 29% in our laboratory (50/169 rats from 11 shipments; 50% of stroked rats with this complication died). In the intraluminal suture model with temporary occlusion of one CCA, the incidence of respiratory distress was 4.9% (3/61 rats), lower than that in the distal MCAO model. Large differences were observed between shipments of animals, from 0% to 57% in 11 shipments (12 to 20 rats/shipment), suggesting that the susceptibility to this complication was animal dependent and not related to the skill of the surgeon. Although mortality is generally accepted as a potential consequence of MCAO, the possibility that a majority of deaths that occur during the procedure are not due to the severity of brain injury but to a preventable form of respiratory distress deserves to be addressed.

Electrocardiogram analysis and pathological examination of lungs, brain, and heart revealed that animals that showed signs of respiratory distress during surgical procedures had evidence of acute arrhythmia, myocardial ischemia, and left ventricular heart failure that resulted in lung congestion (see Chapter 3, Figures 3.1, 3.2, and 3.3). This was likely caused by occlusion of the CCAs, which increased cardiac afterload and stimulated the reflex arc of the carotid sinus and parasympathetic nerve. Local application of bupivacaine to block the reflex arc prevented heart failure and subsequent respiratory distress and decreased the mortality rate sixfold.[5]

ACKNOWLEDGMENT

We thank Francis Farrell from Johnson & Johnson for his support and wonderful leadership.

REFERENCES

1. Tamura, A., Graham, D.I., McCulloch, J., and Teasdale, G.M., Focal cerebral ischemia in the rat: Description of technique and early neuropathological consequences following middle cerebral artery occlusion, *J Cereb Blood Flow Metab,* 1, 53, 1981.
2. Brint, S. et al., Focal brain ischemia in the rat: methods for reproducible neocortical infarction using tandem occlusion of the distal middle cerebral and ipsilateral common carotid arteries, *J Cereb Blood Flow Metab*, 8(4), 474, 1988.
3. Chen, S.T. et al., A model of focal ischemic stroke in the rat: reproducible extensive cortical infarction, *Stroke*, 17, 738, 1986.
4. Wang, Y. et al., Bone morphogenetic protein-6 reduces ischemia-induced brain damage in rats, *Stroke,* 32, 2170, 2001.
5. Wang-Fischer, Y.L. et al., Refined technique for inducing and grading middle cerebral artery occlusion in rat stroke model, *American Association of Laboratory Animal Science 54th National Meeting*, Seattle, platform sessions speaker PS 43 2003, October 11–16, and *AALAS Tribranch Symposium, Biotechnology in the 21st Century and Beyond*, presenting poster 10, Philadelphia, June 8–10, 2003.

13 Surgical Models of Stroke Induced by Intraluminal Filament Implantation

Yanlin Wang-Fischer, Afshin A. Divani, Ricardo Prado, and Lee Koetzner

CONTENTS

SUTURES OR FILAMENTS: STUDIES USING DIFFERENT FILAMENTS OR SUTURES

The terms *monofilament* and *nylon suture* were used in the stroke model by different investigators. To distinguish from the general term *silk suture*, we use the term *monofilament* or *nylon suture* in this book. The intraluminal middle cerebral artery occlusion (MCAO) model via a monofilament has been widely used for studies of acute focal cerebral ischemia and edema since its first description in 1986 by Koizumi et al.[1] However, its relatively high variation and high mortality are of great concern. Besides the performer's surgical skill, which can be improved by practice, the nature of the monofilament is the most important factor that affects the model's outcome. Unqualified monofilaments may result in incomplete occlusion or intracranial bleeding. To overcome these problems, different preparations of the monofilaments have been developed for this model, including different sizes from 3-0 to 4-0, coatings with silicon, poly-L-lysine, or other material or without coating but blunting the filament tips. Koizumi et al. used a silicone-coated 4-0 nylon surgical filament (diameter 0.25 to 0.30 mm), while Zea Longa et al.[2] used a 4-0 uncoated nylon filament with a tip that was blunted by heating near a flame. Both groups showed that reperfusion occurred when the filament was removed.

Laing et al.[3] compared the effectiveness and reliability of the methods of Koizumi et al.[1] and Zea Longa et al.[2] By measuring cerebral blood flow (CBF) rate after MCAO, they showed a higher residual CBF in the area supplied by the occluded middle cerebral artery (MCA) with a relatively large variation (95% confidence interval, 28.3 to 33.7 mL/100 g per minute) in the Zea Longa et al. method compared with Koizumi et al.'s method (9.9 to 11.5 mL/100 g per minute). Subarachnoid hemorrhage occurred more frequently with the blunt tip (10/52) than with the silicone-coated filament. The authors concluded that the coated filament method (Koizumi et al.) was more reliable than the blunt tip method (Zea Longa et al.). Other investigators, however, criticized this conclusion.[4,5] Their discussions focused on several factors: (1) size (3-0 or 4-0) of the filament; (2) coating length; (3) length of filament insertion; (4) body weight of the animal; and (5) other endpoint measurements. In another study in which a silicon-coated filament was used, approximately 30% of experimental animals had to be excluded.[6] Unsuccessful outcomes included (1) animals with no or very mild neurological deficits (incomplete occlusion) and (2) animals with subarachnoid hemorrhage caused by rupture of the intracranial internal carotid artery (ICA).

In a study by Nagasawa and Kogure,[7] in which the MCA was occluded with a silicone rubber cylinder attached to a nylon surgical filament, 38 of 41 rats died within 48 hours after MCAO; the overall mortality rate was 92.7%. Kuge et al.[8] performed a study to compare the effects of two different types of 4-0 nylon monofilament on neurological and morphological outcomes in the filament stroke model. Nylon monofilaments (size 4-0, Ethilon, n = 15, and Nitcho, n = 15) were used to permanently occlude the origin of the right MCA intraluminally in male Sprague-Dawley rats weighing 275 to 345 g. The tips of the filament were rounded by flame heating and were not coated. Neurological outcome and lesion size were compared between the two groups 24 hours after MCAO. The diameter, rigidity, and extensibility of each filament were measured. Their data showed that the mortality within 24 hours was 13% (2/15) in the Ethilon group and 0% in the Nitcho group. Total lesion volume in the Ethilon filament group was 300 ± 100 mm^3 (mean ± standard deviation), significantly larger and more reproducible than that in the Nitcho filament group (200 ± 150 mm^3, $p = .02$). The Ethilon filament had a significantly larger diameter (0.19 ± 0.01 mm versus 0.18 ± 0.01 mm from Nitcho; $p < .01$). Ethilon filaments have better extensibility but poorer rigidity than Nitcho filaments. This study indicated that slight differences in filament characteristics significantly affected lesion volume and reproducibility. This result may explain some conflicting observations in this MCAO model.

Belayev et al.[9] modified the technique of filament preparation by blunting the tip of a 3-0 nylon monofilament and then coating it with poly-L-lysine. Poly-L-lysine has previously been shown to promote the adherence of cells and proteins to glass and plastic surfaces.[10] This substance spreads as a monolayer and is thought to change the negative charge of a filament surface to a positive one, which is then available to attract the anionic sites of endothelial cells, causing the filament to adhere to the endothelial surface[11] so that flow around the filament is prevented. They used Sprague-Dawley rats in their study, subjecting them to 60 minutes or 2 hours of temporary MCAO by a size 3-0 intraluminal filament. In one group of rats, the nylon monofilament was coated with poly-L-lysine, while in the second group a conventional uncoated nylon filament was used. Behavioral function was evaluated at 50 to 60 minutes after occlusion and during a 3-day period after MCAO. Brains were then perfusion fixed, and infarct volumes were measured. Their data showed, in rats with 60-minute MCAO, only three of seven animals with uncoated sutures had infarcts, whereas in the group with sutures coated with poly-L-lysine, all rats (n = 7) exhibited infarction ($p = .009$, Fisher exact test). With 2 hours of MCAO, total infarct volume (corrected for brain edema) was significantly larger in rats with sutures coated with poly-L-lysine than in the group with uncoated sutures (mean ± standard deviation, 122 ± 8 versus 70 ± 30 mm^3, respectively; $p = .03$; n = 4 in each group). In the 2-hour MCAO study, infarct volumes in the uncoated suture group tended to be variable and inconsistent (54% coefficient of variation) compared with the group in which sutures were coated with poly-L-lysine, in which a highly consistent infarct was produced (8% coefficient of variation of infarct volume). They concluded that a suture coated with poly-L-lysine provided a reliable and effective modification of this technique, yielding consistently larger infarcts and greatly reduced interanimal variability. A high rate of subarachnoid hemorrhage can, however, occur in this modified method owing to arterial damage resulting from suture withdrawal.

WHY DO SUTURES NEED TO BE PREPARED BEFORE USE?

The principle of the filament model is to induce brain ischemia by inserting a filament to completely block the blood flow at the origin of the MCA (see Figure 13.1).

Two possible sources of blood flow can reach the MCA in addition to the main blood supply from the ICA: one is from the anterior cerebral artery (ACA) (the backflow from the anterior

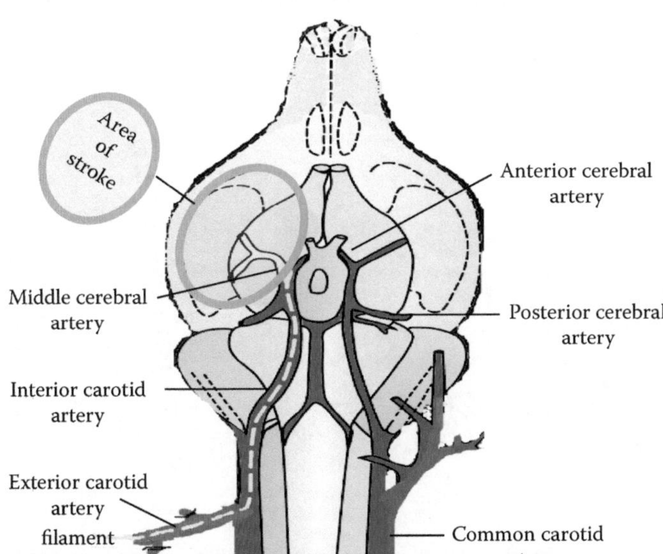

FIGURE 13.1 Schema of filament-induced focal ischemia stroke model.

communicating artery) and the other is from the posterior communicating artery (PCA); refer to Chapter 4 for a more complete discussion.

A properly prepared monofilament possesses enough length of consistent diameter or adhesiveness to increase the rate of successful occlusion by blocking backflow from both of these sources. By choosing the correct filament characteristics (diameter, coating length, and insertion length), the MCA can be occluded while still maintaining blood supply to the ACA and PCA. The between-animal variation may be dramatically reduced if the animal body weight, the length of monofilament inserted, and its location all have been controlled. The filament must be fitted very well to the subject artery to completely block the blood flow. The next sections describe variations in filaments and arteries and their interaction.

Variations in Monofilaments or Nylon Sutures

Monofilaments of the same nominal diameter are variable, which can cause blood flow leakage during attempted occlusion. The diameters of the filaments used in this model are smaller than the diameters of the vessels to be occluded. To overcome this problem, either the filament has to be coated to increase the diameter or the tip of the filament has to be rounded near a flame to a shape like a match tip (see Figure 13.2). This rounded-tip or silicone-coated body is big enough to block the blood flow near the MCA origin. The diameters of sizes 3-0 and 4-0 filaments are 0.220 mm and 0.190 mm on average, respectively. The exact diameter and quality of the filament, however, may not be completely the same even when a surgical nylon monofilament labeled 4-0 is used. The diameter of a 4-0 nylon monofilament ranges from 0.150 to 0.199 mm, according to the U.S. Pharmacopeia (USP) XXII standard. Quality may also vary from one 4-0 filament to another. This is why filaments must be prepared before use. The USP requirements for different sizes of sutures are listed in Table 13.1. (More information can be found at www.deknatel.com, www.supramid.com, and www.ethicon.com). The units are converted from pound force to kilogram force (kgf). The diameters of sutures labeled 3-0 can be 0.200 mm to 0.249 mm. The strength of the sutures varies from company to company (see Table 13.1). The characteristics of the sutures also vary from different companies; for example, size 3-0 nylon monofilament suture ordered from Henry Schein (Cat. No. 103-7025, blue color) is softer than the same size suture from Harvard Apparatus (Cat. No. AH51-7847, black color). The ability of the filament to withstand autoclaving also varies from company to company. In our experience, surgical filaments from Harvard Apparatus can withstand 15 minutes of autoclaving at 120°C (270°F).

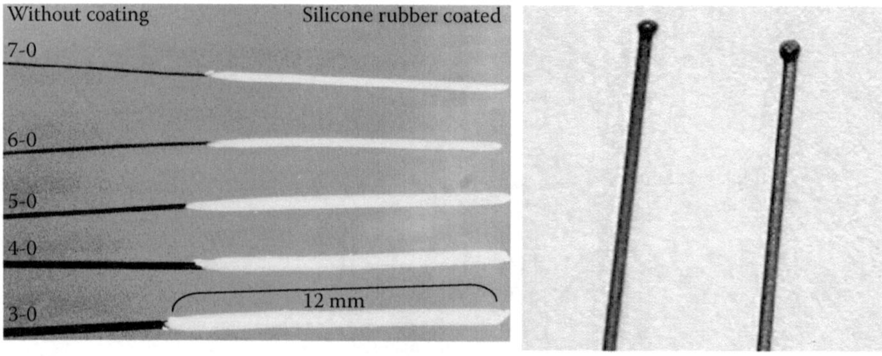

(a) (b)

FIGURE 13.2 The filaments were coated with silicone rubber to increase the diameter (a) or the filament tips (size 3-0) were rounded near a flame to be shaped like a match tip (b) (diameter of the tip is about 0.31 to 0.34 mm). The figures were taken under a surgical microscope (×5). (Figure 13.2a from Doccol. With permission.)

TABLE 13.1
Diameters and Strength of Nylon Monofilaments in Different Sizes

	Diameters (mm)		Knot Pull Strength Minimum (kgf)	
Suture Sizes, USP	**Min**	**Max**	**Supramid/Deknatel**	**Ethicon/Ethilon**
#3	0.6	0.699	3.68	7.29
#2	0.5	0.599	2.54	6.35
#1	0.4	0.499	1.81	5.08
1-0	0.35	0.399	1.45	3.9
2-0	0.3	0.339	1.02	2.68
3-0	0.2	0.249	0.66	1.77
4-0	0.15	0.199	0.458	0.95
5-0	0.1	0.149	0.231	0.68
6-0	0.07	0.099	0.2	0.25
7-0	0.051	0.069	0.109	0.14
8-0	0.041	0.048	0.059	0.07
9-0	0.031	0.038	0.043	0.05
10-0	0.02	0.028	0.019	0.025
11-0	0.01	0.019		
12-0	0.001	0.009		

Notes: kgf, kilogram force; Max, maximum; Min, minimum; USP, U.S. Pharmacopeia.

Variations in Diameters of Vessels to Be Occluded

The diameters of the ICAs vary among animal strains, animal body weight, age, and physiologic or pathological situations (refer to Chapter 4). Based on our experience, the elasticity of vessels in a young rat is superior to that in an older rat; when inserting a filament into the ICA to reach the origin of the MCA, a resistance may not be felt in young rats (<250 g) but can be felt clearly in older rats (>300 g). It is well known that physiologic and pathological changes affect vessel diameter; for example, a high blood CO_2 level dilates brain vessels.[12]

Divani's group at Zeenat Qureshi Stroke Research Center, University of Medicine and Dentistry of New Jersey, studied brain vessel anatomy in different rat strains. They injected resin under physiological pressure into the animals' vascular systems, harvested the vascular network, and then measured vessel diameters (Figure 13.3). Each animal was perfused with distilled water at 37°C prior to the injection of resin. The diameters of the arteries of interest were measured near their origin or proximal to the bifurcation or onto the next artery of interest. The aorta was measured in the ascending territory. There were up to 10 animals in each group with means plus or minus standard deviation. Table 13.2 shows the diameters (mm) of the intracranial section of ICA, MCA, and common carotid arteries (CCAs) in Sprague-Dawley rats with different body weights obtained from the resin injection method.

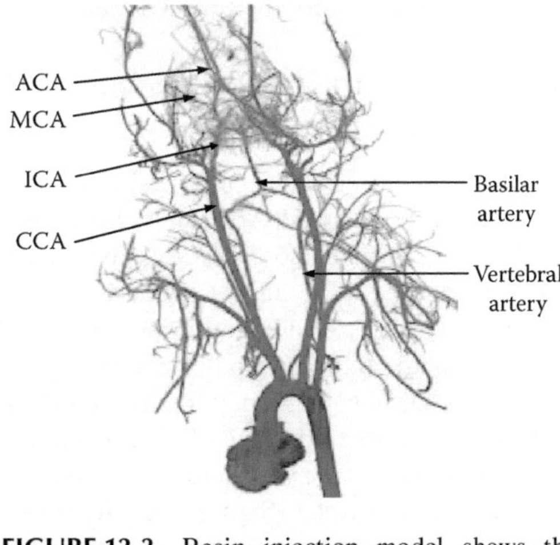

FIGURE 13.3 Resin injection model shows the detailed vessel structure. It is easy to measure vessel diameters in this model. **(See color insert following page 146.)**

TABLE 13.2
Size of Various Arteries in Sprague-Dawley Rats (Female and Male)

	200–249		250–299		300–349		350–399		400–449*		450–500*	
	L	R	L	R	L	R	L	R	L	R	L	R
CCA	0.78 ± 0.06	0.72 ± 0.05	0.71	0.64	0.77 ± 0.08	0.77 ± 0.07	0.77 ± 0.08	0.75 ± 0.03	0.84 ± 0.05	0.84 ± 0.02	0.75 ± 0.05	0.71 ± 0.04
ICA	0.37 ± 0.05	0.36 ± 0.02	0.32	0.31	0.41 ± 0.06	0.41 ± 0.08	0.46 ± 0.10	0.45 ± 0.02	0.48 ± 0.04	0.50 ± 0.04	0.50 ± 0.03	0.44 ± 0.05
MCA	NA	0.24	0.29	0.29	0.32	0.32	0.33	0.33	NA	NA	0.28 ± 0.03	0.29 ± 0.03

Notes: CCA, common carotid artery; ICA, internal carotid artery; L, left; MCA, middle cerebral artery; R, right.
* Male rats: Dr. Divani et al., unpublished data.

PREPARATION OF NYLON SUTURES OR MONOFILAMENTS

There are currently two categories of monofilaments (uncoated and coated) used in this MCAO model. The uncoated monofilaments include those that are heat blunted; the coated monofilaments include those treated with poly-L-lysine, silicone resin, or silicone rubber. The use of uncoated monofilaments frequently results in intracranial bleeding and incomplete occlusion because of their inflexible, irregular, and small tips. Poly-L-lysine–coated monofilaments result in improved occlusion but also increased bleeding because of their adhesive tips, which may tear the luminal surface of arteries when withdrawing the filament for reperfusion. Silicone resin coating does not adhere firmly enough to the monofilament surface and is frequently left behind after withdrawing the filament, which may affect reperfusion. Silicone rubber-coated monofilaments do not have these disadvantages, but they are difficult to prepare uniformly because of the "beading phenomenon," in which silicone sticks to the filament surface in the form of microdroplets ("beads") after it is applied. With this problem, a cylindrical shape cannot be achieved on a coated tip. The phenomenon is also called ball forming in some laboratories. The reason for the beading phenomenon is the high surface tension of the liquid silicone rubber before curing.

Here, we describe different methods for preparing filaments and introduce commercially available coated filaments for the reader's convenience.

Preparation of Size 3-0 Nylon Sutures by Rounding the Tip and Coating with Poly-L-Lysine

Nylon sutures coated with poly-L-lysine were first used in 1996 by Belayev.[9] Nylon sutures can be purchased from Harvard Apparatus, South Natick, Massachusetts (www.harvardapparatus.com). Size 3-0 sutures were chosen for this coating procedure (Cat. No. AH51-7847). These coated nylon sutures were used in Sprague-Dawley, Fisher 334, and Wistar rats with body weights between 250 and 350 g. The sutures are cut into 25-mm lengths, and then the tip is rounded by heating it near a flame. It is better to use a candle as a flame rather than a hot plate for this procedure because a hot plate makes the tips flat shaped like an umbrella. The diameter of the tip is rounded to about 0.31 to 0.34 mm; it is close to the size of the inner diameter of the ICA in Sprague-Dawley rats (250 to 320 g). These rounded-tip nylon sutures are then coated with poly-L-lysine solution (0.1% weight/volume) in deionized water (Sigma, Cat. No. P 8920). The nylon sutures are submerged into the poly-L-lysine solution for about 30 seconds and dried in a 60°C oven for a 1-hour period. This solution of poly-L-lysine from Sigma is ready for use (*do not dilute*). The bottle of poly-L-lysine must be shaken well before use. Prepared nylon sutures are stored at room temperature (18°C to 26°C) on a Petri dish before use. The coating process does not change the diameter of the nylon suture.[9]

Preparation of Different Sizes of Nylon Suture Coated with Silicone

Silicone elastomers have been employed for many years in the manufacture of medical devices, including device components and tubing. These elastomers are commercially available in two types: millable high-consistency silicone rubber and pumpable liquid silicone rubber. Low-viscosity silicone liquid is 100% polydimethylsiloxanes. It makes liquid silicone rubbers suitable for molding applications. (Silicone rubber can be purchased as Provil-L®, Bayer Dental D-5090, from Leverkusen, Germany, or Dow Corning, Midland, Michigan, or GE Sealant and Adhesives, Huntersville, North Carolina). Monofilament nylon sutures can be purchased from Ethicon, Somerville, New Jersey, or other companies. Here, we describe the method of Koizumi et al.[1] and Takano et al. (1997).[13]

Size 4-0 nylon suture (4-0 Ethilon) was chosen for this coating procedure. Sutures are cut into 25-mm lengths, and the tip is rounded by heating it near a flame to a diameter of 0.33 to 0.36 mm. The sutures are then coated over a 12-mm length with Provil-L silicone, which is characterized as a low-viscosity, type I silicone meeting the requirements for coating these sutures. The diameter of the tip and coated body of each suture (12 mm from the tip) is measured on a video screen calibrated with a stage micrometer (10-mm divisions) using a photomicroscope (2071, American Optical, Buffalo, New York) and a CCD color video camera (SSC-C374, SONY, Tokyo, Japan). The maximum diameter of the coated tip is 0.47 ± 0.01 mm, ranging from 0.44 to 0.49 mm with a 2.9% coefficient of variation. The mean diameter of the coated suture body was 0.36 ± 0.01 mm, ranging from 0.33 to 0.39 mm, with a coefficient of variation of 3.6%. This coated suture was used for modeling stroke on male Sprague-Dawley rats weighing about 350 g.

Monofilaments coated with silicone rubber are difficult to prepare because of the beading phenomenon owing to silicone having high surface tension before curing; further, the process is very time consuming. Doccol, a stroke research company, developed a method that is able to make high-quality silicone rubber-coated monofilaments. They provide a large variety of silicone rubber-coated nylon monofilaments for use in the intraluminal stroke model (see Figure 13.2a). The filaments are specially manufactured for the purposes of reducing both the risk of intracranial bleeding and the rate of incomplete occlusion. These coated filaments possess smooth and cylindrically shaped flexible tips that reduce the risk of vessel injury and intracranial bleeding, and increase the rate of successful occlusion. Moreover, they are autoclave safe, which reduces the chances of infection.

Doccol uses silicone rubber from Dow Corning (Cat. No. RTV Sealant 748) and GE (Silicone II). The molds are very good in controlling the coating thickness and smoothness, but extensive training is needed to yield perfect coatings. The coating process must be performed under a stereoscope. For more information, consult their Web site, www.doccol.com (7516 Treviso Court NE, Albuquerque, New Mexico 87113, 505-385-2084).

STEP-BY-STEP SURGICAL PROCEDURES FOR SUTURE-INDUCED ISCHEMIA MODEL IN RAT

Initial Steps

1. A rat is anesthetized with 2% isoflurane in a gas mixture consisting of 70% N_2O and 30% O_2 and is placed supinely on a table (see Chapter 8, Figure 8.11).
2. The fur on the ventral cervical area is shaved, and the skin is cleaned with 0.5% Betadine and 75% alcohol or 2% chlorhexidine.
3. Eye cream is placed on both eyes to protect the corneas from drying.
4. The neck is hyperextended by placing a 10-cc syringe under it to aid in the exposure of the CCAs.
5. The forelimbs are fixed on the table with tape.

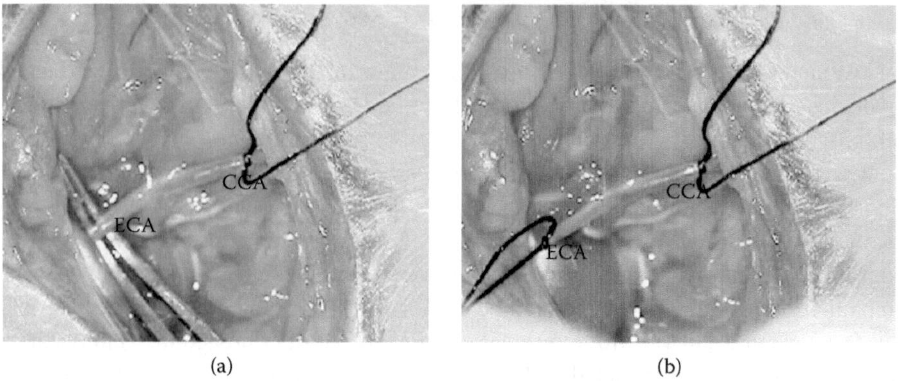

FIGURE 13.4 Dissection of the external carotid artery (ECA) and common carotid artery (CCA) (a). The CCA is temporarily ligated, and the ECA is permanently ligated with 4-0 silk sutures (b). **(See color insert.)**

Surgical Steps

1. Under an operating microscope or magnifier, a middle incision is made in the ventral aspect of the neck; the right sides of the sternohyoid and sternomastoid muscles are blunt dissected to expose the right CCA, which is temporarily ligated with a 4-0 silk suture (Figure 13.4) or temporarily blocked with a microvascular clip. A self-retaining retractor is positioned between the digastric and sternomastoid muscles, and the omohyoid muscle is divided.
2. The right sides of the external carotid artery (ECA) and ICA are isolated and mobilized (freed) from the surrounding tissues very near the skull base. The ECA is superior to the ICA with the rat in this position. The occipital artery (a branch of the ECA) is coagulated to prevent excessive bleeding. The ECA is dissected further distally, ligated with a 4-0 silk suture, and coagulated at the end. All other branches, including the superior thyroid, ascending pharyngeal, lingual, and maxillary arteries, are coagulated. The ECA terminal stump, about 0.5 cm long from the ECA-ICA bifurcation, is freed from surrounding tissue and cut at its distal end. This vascular segment is then slightly pulled back, kept under tension, and maintained straight with respect to the ICA (Figure 13.5).
3. Next, a 4-0 silk suture is tied loosely around the ECA stump near the bifurcation.
4. The ICA is isolated and carefully separated from the adjacent vagus nerve. Further dissection identifies the glossopharyngeal nerve near the origin of the pterygopalatine artery (PPA); this posteriorly directed extracranial branch of the ICA is ligated with a 6-0 suture close to its origin. At this point, the ICA is the only remaining branch of the CCA. Some investigators do not ligate the PPA owing to its deep location.
5. The ICA is temporarily occluded by a fine vessel clip. This blocks entry of blood from the CCA and ICA and prevents bleeding from the incision in the ECA stump (Figure 13.5).
6. A small incision is made in the ECA with spring scissors.
7. A 25-mm-long, size 3-0 nylon monofilament (AH51-7847 Harvard Apparatus; prepared as in the earlier section of this chapter, "Preparation of 3-D Nylon Sutures by Rounding the Tip and Coating with Poly-L-Lysine") is selected and is then gently inserted into the ICA through the ECA stump. Monofilaments can be prepared differently according to specific study paradigms. (Monofilament diameter is dependent on rat weight; 250- to 320-g rats require 3-0 sutures.)
8. The silk suture around the ECA stump is tightened around the intraluminal nylon filament to prevent bleeding, and the microvascular clip is removed from the ICA (Figure 13.6).

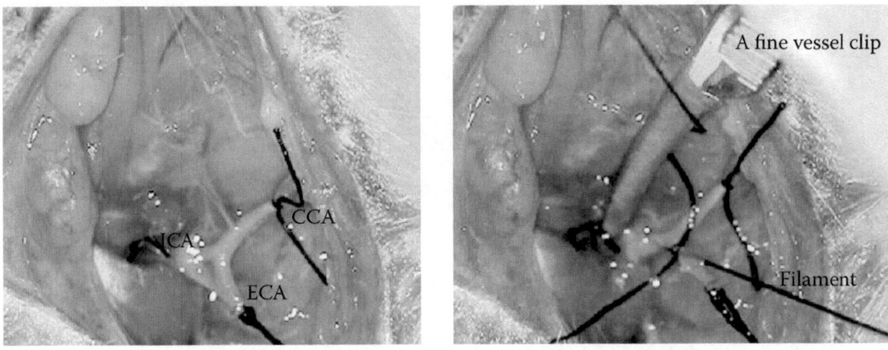

(a) (b)

FIGURE 13.5 The external carotid artery (ECA) terminal stump about 0.5 cm from the fork was separated from surrounding tissue and cut at the distal end. The ECA stump was pulled back under slight tension and then aligned with the internal carotid artery (ICA) (a). The ICA was temporarily blocked by a fine vessel clip before cutting a hole in the ECA stump. A 25-mm-long 3-0 monofilament nylon suture was gently inserted into the ICA through the ECA stump (b). A 4-0 silk suture was tied loosely around the ECA stump along with the filament near the bifurcation before releasing the clip. **(See color insert.)**

9. The nylon filament is then gently advanced from the ECA to the ICA lumen to within about 18 to 20 mm from the fork (dependent on the animal size; see Table 13.3, Figure 13.6a). To know the length of filament being inserted, the filament length from the bifurcation to the outside end is measured (Figure 13.6b). For example, if the length of suture from the bifurcation to outside ECA stump is 6 mm, a length of 19 mm is inserted into the ICA (25 mm − 6 mm = 19 mm). Beyond measuring the length of filament being inserted, other signs are also important, including feeling resistance or filament push back or observing a slight curving of the suture or stretching of the ICA, indicating that the blunted tip of the suture has passed the MCA origin and reached the proximal segment of the ACA, which has a smaller diameter. At this point, the intraluminal filament is blocking the origin of the MCA, occluding all sources of blood flow from the ICA, ACA, and posterior cerebral artery. About a 5- to 7-mm length of the filament remains outside, to be withdrawn to allow reperfusion. Below is a schema showing the inserted filament.

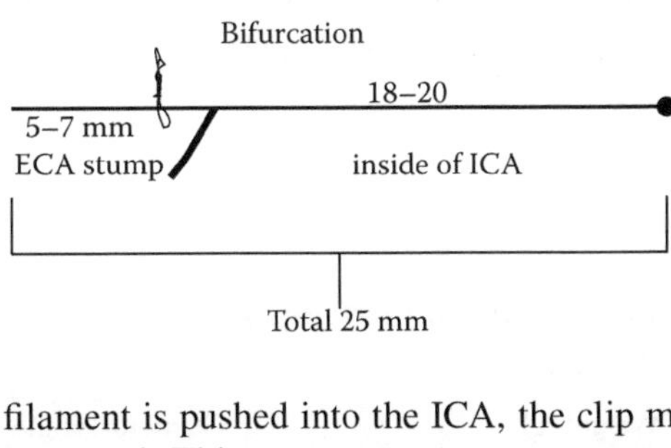

Note: Before the filament is pushed into the ICA, the clip must be released to permit blood flow into the vessel. This prevents spasm or contraction of the vessel, which would prevent successful insertion of the filament. If this happens, the filament should be retracted to the ECA stump and blood flashed into the vessel, after which another attempt to insert the filament can be made.

Never insert a filament length longer than 21 mm; otherwise, the MCA will be damaged, and bleeding will occur when the filament is withdrawn.

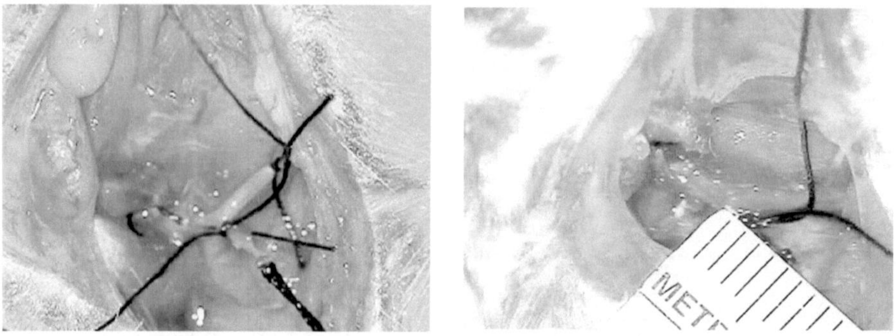

(a) (b)

FIGURE 13.6 The fine vessel clip was removed from the internal carotid artery (ICA). The nylon filament was then gently advanced from the external carotid artery (ECA) to the ICA lumen. It is about 18 to 20 mm from the fork (a). To know the length of filament being inserted, the filament length from the bifurcation to the proximal end of the filament was measured (b). **(See color insert.)**

10. The vessel clip or ligation on the CCA is released after tightening the silk suture on the ECA-intraluminal filament. Some investigators have occluded the CCA for 90 minutes to reduce the possibility of blood leakage from the ECA stump.
11. The incision is closed, and the animal is placed in a clean cage; recovery from the anesthesia should happen within 15 minutes. Some investigators have kept the animal under anesthesia until withdrawing the filament. In this case, the infarct size may be smaller than those in awake animals owing to anesthetic-induced neuroprotection.
12. Ischemia continues until the monofilament is withdrawn (from 60 to 120 minutes based on study design). Lesions resulting from 90 and 120 minutes of ischemia show good infarct size and edema (see Figure 13.7). To reperfuse the ischemic territory, the intraluminal filament is slowly retracted until resistance is felt, indicating that the tip has cleared the ACA-ICA lumen and is in the ECA stump. A small clip is placed on the stump to block blood flow, and then the entire filament is pulled out, followed by tightening the suture and releasing the clip. This procedure is done under light anesthesia.

Some investigators (for example, scientists from Skeletech, now a subsidiary of MDS) have used a longer filament (35 mm) and, after closing the incision, left about 10 mm of the nylon filament protruding outside so it could be withdrawn to allow reperfusion without anesthesia. The filament is pulled back until resistance is felt, indicating that the tip is in the ECA stump, and then trimmed. A small section of filament is left in the ECA stump.

TABLE 13.3
Filament Lengths Required for Different Body Weights in Sprague-Dawley Rats

Body Weight (g)	Filament Length (mm)
200 to 250	17.5 to 18.5
260 to 280	18 to 19
290 to 320	19 to 19.5
330 to 400	19.5 to 20

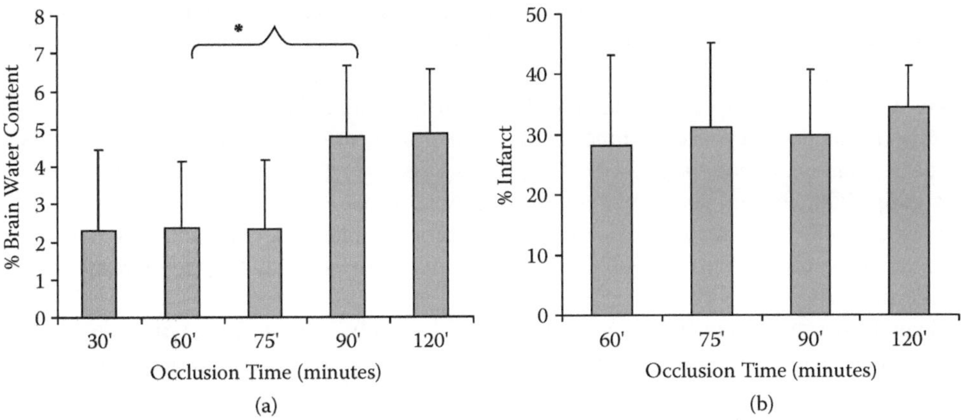

FIGURE 13.7 Brain percentage water content (dry/wet) and infarct volume (triphenyltetrazolium chloride stain) measured following different durations of filament occlusion in rats anesthetized with 2% isoflurane mixed with 100% O_2 and sacrificed 24 hours after reperfusion. Increase in brain water content (brain edema) was significantly different at 90 and 120 minutes of ischemia (a). Infarct size was not significantly different for occlusion times from 60 to 120 minutes (b) (mean ± standard deviation, n = 12, *$p < .05$).

Postsurgical Steps

1. The incision is closed with suture clips after locally giving 0.1 mL 0.25% bupivacaine to reduce postoperative pain.
2. Furazolidone (4%) aerosol powder is sprayed on the wounds to prevent infection.

Rats will wake up 10 to 20 minutes after the isoflurane gas face mask is removed.

The surgical procedures usually take about 10 to 15 minutes for a well-trained surgeon.

Dr. Shimin Liu, a neuroscientist from the University of New Mexico College of Pharmacy, made the following suggestions for this model:

1. Perform a pilot study to determine the optimal tip diameter and optimal inserted length for your animals. *This is critical to the success of your experiments.*
2. Measure the monofilament tip and coated body diameter as well as your animal body weight. Match the diameter with the body weight.
3. Mark the monofilament at the desired length using a Sharpie metallic silver marker pen.
4. Autoclave the monofilament. Loosen the container cap before autoclaving. Keep the filaments and container away from direct contact with metal while being autoclaved.
5. Use a 4-0 braided silk suture to make a loose knot above the position where the monofilament will be inserted.
   a. Knots tied with sutures smaller than 4-0 size tend to be tight and difficult to loosen. A tight knot may force the coating to be left behind as the filament is withdrawn. On the other hand, knots tied with sutures larger than 4-0 will be too clumsy for this fine surgery.
   b. The knot should be easy to loosen, or reopen, but also tight enough to stop bleeding from backflow.
6. Once the monofilament has been put in position, it is necessary to apply a microclip onto the ICA to immobilize the monofilament. *Dislodgement of filaments occurs frequently in freely moving animals.*

7. The operational sequence for doing reperfusion is to loosen the knot, temporarily open the microclip, withdraw the filament, reapply the microclip after the monofilament has passed through, retie the knot to stop bleeding, and then remove the microclip.
8. For a silicone rubber-coated filament, the recovered monofilament must be cleaned carefully if intended for reuse, and the surface of the filament must be intact. A poly-L-lysine–coated filament cannot be reused.
9. The tip diameter and surface smoothness may have been altered after being used in the MCAO model. Measuring the tip diameter is recommended before reusing a coated monofilament. Applying a little silicone oil to the coating surface may help to increase the surface smoothness.
10. Coated monofilaments should be used within 1 month because the coating texture may change after repeated use. Unused filaments may maintain their texture for 3 to 4 months.

PROBLEMS WITH INTRALUMINAL FILAMENT-INDUCED ISCHEMIA MODEL AND SOLUTIONS

Model Inconsistencies

Model consistency varies from laboratory to laboratory, from strain to strain, and from surgeon to surgeon. Table 13.4 shows the variations of infarct volume in rat strains of different body weights using uncoated monofilaments (refer to www.doccol.com). Occlusion time is important for model consistency. Figure 13.7 shows the result of edema (Figure 13.7a) and infarct volume (Figure 13.7b) at different occlusion times in our study. There were significant differences in brain edema between occlusion times greater than 90 minutes and less than 75 minutes.

These rats were anesthetized with isoflurane in 100% O_2 and sacrificed at 24 hours after occlusion.

Identifying the Types of Variation

1. Gases inhaled during surgery in rats interfere with the results: Smaller infarcts or edema result with inhalation of 100% O_2 compared to inhalation of 70% N_2O/30% O_2.
2. Body temperature during surgical and monitoring procedures: Lower temperatures are neuroprotective.
3. Incomplete occlusion:
   a. Mismatch of the filament diameter with animal body weight, either too small or large.
   b. Inserted length insufficient.
   c. Coating length insufficient.
   d. Filament tips are not in good condition; the ideal filament should have a cylindrical contour with a smooth surface.
4. Occlusion for a longer time than necessary (over occlusion):
   a. Occlusion of the ACA owing to the filament being inserted too far.
   b. Thrombosis without reperfusion.
   c. Hypotension during occlusion.

Preventing Incomplete Occlusion

1. A pilot study is required to deduce the optimal filament diameter, insertion length, and coating length for each body weight range.
2. Standardize the measurement methods for filament diameter and insertion length.
3. Deploy reusable monofilaments for the purpose of paired controls.

TABLE 13.4
Previously Published Infarct Volumes Using Uncoated Suture Middle Cerebral Artery Occlusion (MCAO) Methods among Untreated Rats

Author	Rat Strain	Body Weight (g)	Suture Size	MCAO Duration (hours)	Time (hours after MCAO) and Type of Histology	Infarct Volume (mean ± standard deviation) (mm³)	Coefficient of Variation (%)
Kawamura et al. (1991)[14]	SD	250 to 340	3-0	Perm	24, H-E, hemisph	166 ± 73	44
Minematsu et al. (1992a)[15]	SD	280 to 350	4-0	Perm	24, TTC, hemisph	330 ± 80[a]	24.1
Minematsu et al. (1992b)[16]	SD	280 to 370	4-0	2	24, TTC, hemisph	179 ± 100[a]	56
Minematsu et al. (1993a)[17]	SD	280 to 365	4-0	Perm	24, TTC, hemisph	301 ± 96	31.8
Minematsu et al. (1993b)[18]	SD	280 to 360	4-0	Perm	24, TTC, hemisph	225 ± 127	56.8
Warner et al. (1993)[19]	Wistar	310 to 320	b	1.5	96, NT, cortical	115 ± 104	90.4
Chopp et al. (1994)[20]	Wistar	270 to 300	4-0	2	48, H-E, hemisph	154 ± 31	20.3
Karibe et al. (1994)[21]	SD	280 to 340	3-0	2	24, CV, hemisph	160 ± 34[a,b]	21.4
Zhang et al. (1994)[22]	Wistar	270 to 290	4-0	2	48, H-E, hemisph	171 ± 38	22.5
Fisher et al. (1995)[23]	SD	300 to 360	4-0	Perm	24, TTC, hemisph	297 ± 65	21.9
Kuge et al. (1995)[8]	SD	276 to 346	4-0	Perm	24, TTC, hemisph	296 ± 97	32.8
	SD	276 to 346	4-0	Perm	24, TTC, hemisph	190 ± 145	76
Memezawa et al. (1995)[24]	Wistar	290 to 350	b	Perm	24, TTC, hemisph	262 ± 76	29
Warner et al. (1995)[25]	Wistar	275 to 330	b	1.5	96, NT, cortical	160 ± 123	76.9
	Wistar	275 to 330	b	1.5	96, NT, cortical	205 ± 137	66.8
Kim et al. (1996)[26]	SD	270 to 330	3-0	1	96, H-E, cortical	42 ± 19	45.3

Notes: CV, cresyl violet; H-E, hematoxylin and eosin; hemisph, cortical and subcortical infarct volume; NT, nitroblue tetrazolium; Perm, permanent occlusion; SD, Sprague-Dawley; TTC, 2,3,5-triphenyltetrazolium chloride; b, 0.25 in diameter.

[a] Originally expressed in mean ± standard error.

[b] Corrected for compensation of brain edema.

4. Measure the filament diameter and check the shape and surface of each filament with a microscope before use. An irregularly shaped filament tip will likely produce incomplete occlusion and thus an irregular infarct volume.
5. Avoid dislodgement by immobilizing the monofilament inside the ICA.

Preventing Overocclusion (Occlusion for an Excessive Time)

1. A pilot study must be conducted in the same manner as above.
2. Reduce thrombosis by preparing good filaments and reducing endothelial damage.
3. Clean and autoclave the filaments.
4. Use silicone oil to smooth the coating surface.
5. Avoid using larger-size filaments.
6. Choose the optimal coating material and filament material.
7. Avoid hypotension.
8. Avoid dislodgement of filaments by securing them with a clip or suture placed on the ICA to immobilize them.
9. Deploy reusable monofilament for paired control purposes.

Subarachnoid Hemorrhage

Intracranial bleeding (subarachnoid hemorrhage) occurs frequently after reperfusion, which causes high mortality. It occurs in about 20% of animals subjected to the poly-L-lysine–coated filament model in our laboratory. The common reason is artery perforation (see Figure 13.8).

Identifying the Causes

How can an artery be perforated? There are several causes: A filament can be inserted too far, the filament may be too rigid, the coating material may not be flexible enough, the filament diameter may be too large for the subject artery, or the coating may adhere to the artery wall.

Solutions for This Problem

Choosing the optimal inserted length, filament material, coating material, and filament size and smoothing the coating surface can help to avoid artery perforation.

The ECA approach may help to decrease subarachnoid hemorrhage.

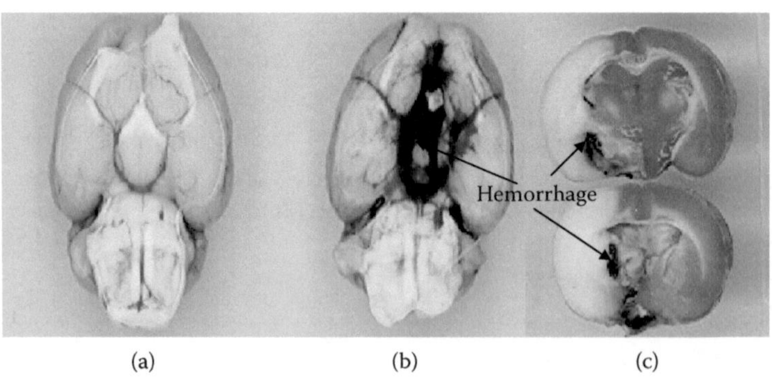

FIGURE 13.8 Basal view of subarachnoid hemorrhage in transient filament model (stained with triphenyltetrazolium chloride [TTC]). (a) A normal rat. (b) Subarachnoid hemorrhage from a stroked rat. The middle cerebral artery (MCA) was occluded for 2 hours by a 3-0 nylon monofilament coated with poly-L-lysine, and the rat was sacrificed at 24 hours. (c) Subarachnoid hemorrhage in the same rat (stained with TTC).

Incomplete Reperfusion

Identifying the Common Reasons

The common reasons for incomplete reperfusion are as follows: (1) thrombosis; (2) intracranial bleeding; (3) hypotension; (4) no-reflow phenomenon.

Reduce Incomplete Reperfusion

To reduce incomplete reperfusion:

1. Reduce thrombosis by selecting the best filaments and reducing excessive endothelial damage resulting from overocclusion.
2. Clean, smooth, and autoclave the filaments.
3. Optimize filament material, coating, length, and diameter.

Hyperthermia

Identifying the Causes

Postischemic spontaneous hyperthermia as a complication of occlusion of the MCA with an intraluminal filament has been observed by some authors.[27,28] This model induces hyperthermia up to 40°C early after MCAO. This pathological increase in body temperature is associated with hypothalamic injury, caused by occlusion of the hypothalamic artery (HTA) as it originates from the distal ICA (see Figure 4.5 in Chapter 4). The monofilament stroke model blocks not only the MCA, but also the entire ICA tree and thus should be considered a model of ICA territory ischemia. This includes the anterior choroidal artery (AChA), a branch of the ICA that arises just proximal to the MCA bifurcation, and also the HTA, thereby reducing flow within the arteries that supply the hypothalamus. The hypothalamus is the neural center for body temperature regulation. Ischemia in the hypothalamus causes the body temperature to change; most such situations result in increase of body temperature. Figure 13.9 shows body temperature changes in our 2-hour transient filament stroke model. The animals were under slight anesthesia with 2% isoflurane during the temperature measurement. The body temperature started increasing at 30 minutes and reached 39°C to 40°C at 2 hours after MCAO and slowly recovered at 24 hours after reperfusion. In this stroke model, the temperature did not increase in about 10% of the rats, whereas the body temperature increased in

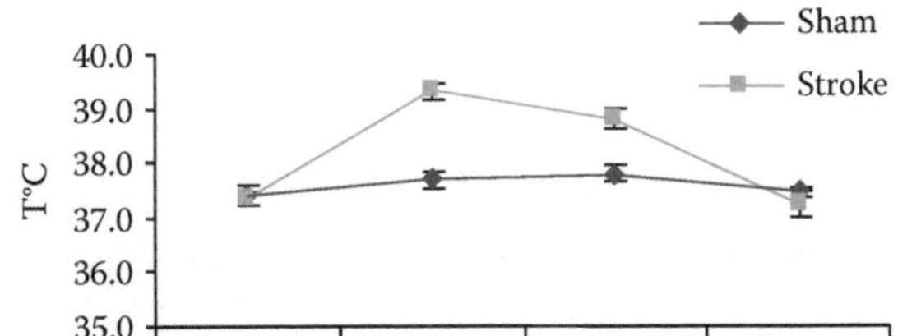

	0'	2 hrs	4–5 hrs	24 hrs
Sham	37.42	37.69	37.80	37.4
Stroke	37.34	39.33	38.79	37.2

Time

FIGURE 13.9 Temperature changes in monofilament stroke model. Sprague-Dawley male rats (280 to 350 g) were subjected to 2 hours of ischemia and sacrificed at 24 hours. Their body temperature started increasing at 30 minutes and reached over 39°C at 2 hours after middle cerebral artery (MCA) occlusion. The temperature recovered at 24 hours (n = 15 in each group).

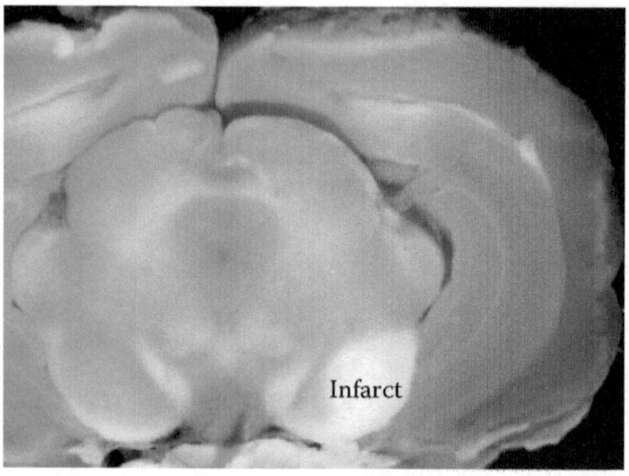

FIGURE 13.10 A very small infarct (3.37%) in the hypothalamus, stained with triphenyltetrazolium chloride at 24 hours after occlusion. This rat had a body temperature over 39°C after 2 hours of filament occlusion. **(See color insert.)**

90%; in the latter, 70% had a large infarct, and 20% had a smaller infarct with triphenyltetrazolium chloride (TTC) stain.

This temperature increase is closely related to the degree of injury in the hypothalamus. Figure 13.10 shows a very small infarct (3.37%) in the hypothalamus with TTC stain at 24 hours after occlusion; this rat had a temperature of over 39.0°C with a behavioral score of 2 after 2-hour occlusion. It is widely accepted from experimental and clinical studies that hyperthermia increases infarct volume and worsens clinical outcome[29,30] and therefore might influence the effects of neuroprotective treatments in animal studies.

Solutions for This Problem

1. Shave a large area of fur from the animal's back to help decrease the high temperature.
2. Place the animal in a cold room (10°C) if the animal's temperature starts to rise above 37°C. As soon as the temperature starts to fall below 37°C, animals are allowed to return to room temperature.
3. Alcohol (70%) application can be used in cases of a sudden rise in temperature.
4. Keep the animal under anesthesia without a heating pad.
5. We have tried to develop new methods to occlude only the MCA origin using a specially coated filament (the filament is coated about 1 mm at the distal end), which occludes the MCA without blocking the HTA.

Temporal Muscle Necrosis

Temporal muscle necrosis has also been identified as a complication of the intraluminal filament model of MCAO. These lesions resulted in impairment of body weight evolution and delayed restoration of neurological function in Wistar rats.[31] This is neither a laboratory-specific nor a rat strain-specific problem as it has been mentioned in several publications.[32,33] Magnetic resonance imaging in Sprague-Dawley rats on day 2 after temporary MCAO (suture technique) depicted extracranial lesions in the temporal muscle that were identical to those found in other laboratories. Unfortunately, the authors did not comment on this finding.[33]

Summary of Solutions

- Perform a pilot study to optimize critical factors.
- Standardize the measuring methods and surgical procedures, including

a. Use the ECA approach.
b. Use a gas mixture of 70% N_2O and 30% O_2.
c. Control the anesthesia time and method of application.
d. Control body temperature to avoid hypothermia or hyperthermia.
e. Avoid hypotension and hypoxia.
f. Design an appropriate time period of occlusion.
g. Pay close attention to filament diameter and coating method (measure and check the monofilament before using). Clean, smooth, and autoclave monofilaments before use. Immobilize the monofilament after occlusion. Use reusable monofilaments.

STEP-BY-STEP PROXIMAL MIDDLE CEREBRAL ARTERY OCCLUSION INTRALUMINAL FILAMENT MODEL IN MICE

In this book, we do not have a detailed discussion of the MCAO model in mice. There are some advantages to using mice in this model; here, we describe the surgical steps.

The MCAO model induced by intraluminal filament implantation in mice[21] is similar to the model in rats. The differences are as follows:

1. Mice have smaller arteries, so the filaments used in mice are smaller than those in rats.
2. Filaments can be coated with silicone or uncoated, but there is no need to blunt the tips with a flame.
3. The filament length inserted into the ICA in mice is shorter than that in rats (8 mm in mice versus 19 mm in rats).
4. The occlusion time in mice is shorter than that in rats (45 to 90 minutes in mice versus 60 to 120 minutes in rats).
5. The behavioral changes in mice are less distinguishable than in rats.
6. Postocclusion hyperthermia is very common in rats but is not common in mice.
7. The surgical procedures must be carried out under a microscope.

ANIMALS

Male C57 /BL6 (black mice), 20 to 25 g or 26 to 36 g, can be used for this model.

ANESTHESIA

Mice are anesthetized under 2.5% isoflurane in an induction chamber and are then transferred to a small anesthetic face mask and anesthesia under 1.5% to 2% isoflurane.

SURGICAL PROCEDURES

1. The mouse is placed in supine position. The fur on the ventral neck area is shaved; the skin is cleaned with 0.5% Betadine and 75% alcohol or 2% chlorhexidine.
2. Place eye cream on both eyes to protect the corneas from drying if the surgical procedure is longer than 20 minutes; a cotton applicator stick is placed under the neck to help expose the arteries. The forelimbs are fixed to the table by tape.
3. A middle incision is made in the ventral aspect of the neck; the sternohyoid muscle and sternomastoid muscles are blunt dissected to expose the CCAs.
4. The right CCA, the right ECA, and the ICA are isolated via a midline incision. The ECA is ligated with a 5-0 silk suture and cauterized to prevent bleeding.
5. The distal end of the right CCA is temporarily ligated with a 4-0 suture. The ICA is temporarily clamped using a microvascular clip (see Chapter 10 on surgical instruments).

6. A tiny hole is cut into the ECA stump (some investigators cut a hole in the CCA just before the bifurcation of ECA and ICA; in this case, the CCA is permanently occluded).
7. An 11-mm length of a 7-0 nylon intraluminal monofilament, of which about one-third is coated with silicone gel, is gently inserted into the ECA stump through the tiny hole, and a 5-0 silk suture is tied onto the ECA after the filament passes the fork and is placed in the ICA.
8. Remove the ICA clip while continuing to insert the filament into the ICA to block the origin of the MCA. It will be about 8 mm in length from the fork (dependent on animal size). Unlike in rats, there will be no feeling of resistance when the filament reaches the original MCA. The length of the inserted filament is the only way to judge if the MCA is occluded. About a 3-mm length of the filament is left outside for withdrawal to allow reperfusion.

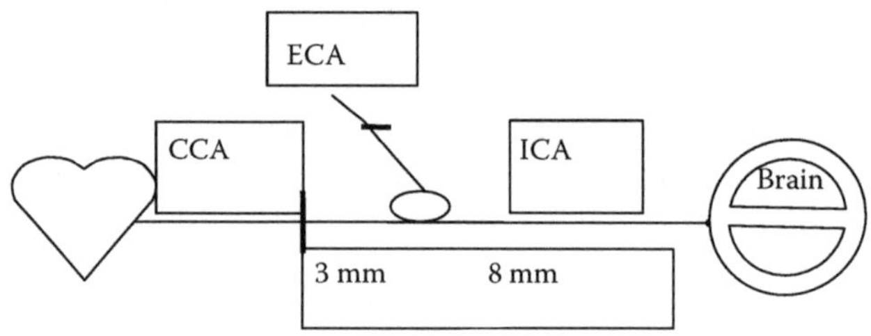

Note: Never insert a filament longer than 10 mm; otherwise, the MCA will be damaged, and bleeding will occur as the filament is withdrawn.

9. After 45 to 90 minutes, the intraluminal filament is slowly pulled out to allow reperfusion. The occlusion time is dependent on the needs of the individual study.
10. The incision is closed with suture clips after giving 0.1 mL 0.25% bupivacaine.
11. Furazolidone (4%) aerosol powder is sprayed on the wounds to prevent infection.
12. Discontinue isoflurane; recovery from anesthesia should take 10 to 20 minutes.

Preparation of Filaments

The filaments can be ordered from Johnson & Johnson Health Care System (800-255-2500, Cat. No. 1696G, Ethilon nylon monofilament suture 7-0, P-1) or from other companies. Some investigators used different sizes of nylon (polyamide) filaments, including uncoated 5-0 nylon filaments 13 mm long for 27- to 36-g mice and 6-0 nylon filaments 12 mm long for 22- to 26-g mice.[34]

The method to coat filaments with silicone is easily described but difficult to do. Much practice is needed to find the best proportion of hardener with silicone and best shape for the coated filaments. The diameter of the silicone-coated filament is about 0.11 mm for 23- to 27-g mice. If size 7-0 filaments are used for this procedure, the diameter of the coated filament is about double the size of 7-0 filaments (the diameter of a 7-0 filament is about 0.05 to 0.06 mm; see Table 13.1).

Low-viscosity silicone (Xantopren VL Plus) and silicone hardener (Universal Cutter Sil liquid) can be purchased from Heraeus Kulzer GmbH (Grüner Weg 11, D-63450, Hanau, Germany, Reorder No. 65647397) or Heraeus Kulzer (www.heraeus-kulzer-us.com, 99 Business Park Drive, Armonk, New York 10504, 800-343-5336).

Coated filaments for mice are commercially available (see www.doccol.com). Tabrizi and Wang modified the intravascular MCAO technique[35,36] with a nonsiliconized, uncoated, size 6-0, 8-mm-long prolene filament in mice weighing 22 to 27 g. This uncoated filament may have different occlusion and thrombogenic properties in comparison with the silicone-coated nylon filament (the coated filament has no thrombogenic properties).

Measuring Cerebral Blood Flow in Mice

Measurements of relative CBF were obtained as previously reported[37] using a straight laser Doppler flow probe placed 2 mm posterior to the bregma and 3 mm to each side of the midline using a stereotaxic micromanipulator, keeping the angle of the probe perpendicular to the cortical surface. These CBF measurements, expressed as the ratio of ipsilateral to contralateral blood flow, were continuously obtained at baseline, immediately before MCAO, during the occlusion, and at the time of withdrawal of the occluding suture.

ACKNOWLEDGMENTS

We thank Dr. Kenneth Rhodes for his support and wonderful leadership and Drs. Zhihong Haung and Jianya Ma from Johnson & Johnson, PRD, for their technical assistance and advice.

REFERENCES

1. Koizumi, J., Yoshida, Y., Nakazawa, T., and Ooneda, G., Experimental studies of ischemic brain edema, I: A new experimental model of cerebral embolism in rats in which recirculation can be introduced in the ischemic area, *Jpn J Stroke,* 8, 1, 1986.
2. Zea Longa, E.L., Weinstein, P.R., Carlson, S., and Cummins, R., Reversible middle cerebral artery occlusion without craniectomy in rats, *Stroke,* 20, 84, 1989.
3. Laing, R.J., Jakubowski, J., and Laing, R.W., Middle cerebral artery occlusion without craniectomy in rats: Which method works best? *Stroke,* 24, 294, 1993.
4. Garcia, J.H., A reliable method to occlude a middle cerebral artery in Wistar rats [letter], *Stroke,* 24, 1423, 1993.
5. Holland, J.P. et al., Rat models of middle cerebral artery ischemia [letter], *Stroke,* 24, 1423, 1993.
6. Matsuo, Y. et al., Role of cell adhesion molecules in brain injury after transient middle cerebral artery occlusion in the rat, *Brain Res,* 656, 344, 1994.
7. Nagasawa, H. and Kogure, K., Correlation between cerebral blood flow and histologic changes in a new rat model of middle cerebral artery occlusion, *Stroke,* 20, 1037, 1989.
8. Kuge, Y.J. et al., Nylon monofilament for intraluminal middle cerebral artery occlusion in rats, *Stroke,* 26, 1655, 1995.
9. Belayev, L. et al., Middle cerebral artery occlusion in the rat by intraluminal suture neurological and pathological evaluation of an improved model, *Stroke*, 27, 1616, 1996.
10. Huang, W.M. et al., Improved section adhesion for immunocytochemistry using high molecular weight polymers of l-lysine as a slide coating, *Histochemistry*, 77, 275, 1983.
11. Mazia, D., Schatten, G., and Sale, W., Adhesion of cells to surfaces coated with polylysine, *J Cell Biol*, 66, 198, 1975.
12. Eicke, B.M. et al., Influence of acetazolamide and CO_2 on extracranial flow volume and intracranial blood flow velocity, *Stroke*, 30(1), 76, 1999.
13. Takano, K. et al., Reproducibility and reliability of middle cerebral artery occlusion using a silicone-coated suture (Koizumi) in rats, *J Neurol Sci*, 153, 8, 1997.
14. Shingo Kawamura, S., Shirasawa, M., Fukasawa, H., and Yasui, N., Attenuated neuropathology by nilvadipine after middle cerebral artery occlusion in rats, *Stroke*, 22(1), 51, 1991.
15. Minematsu, K., Li, L., Fisher, M., Sotak, C.H., Davis, M.A., and Fiandaca, M.S., Diffusion-weighted magnetic resonance imaging: Rapid and quantitative detection of focal brain ischemia, *Neurology*, 42, 235, 1992a.
16. Minematsu, K., Li, L., Sotak, C.H., Davis, M.A., and Fisher, M., Reversible focal ischemic injury demonstrated by diffusion-weighted magnetic resonance imaging in rats, *Stroke*, 23(9), 1304, 1992b.
17. Minematsu, K., Fisher, M., Li, L., Davis, M.A., Knapp, A.G., Cotter, R.E., McBurney, R.N., and Sotak, C.H., Effects of novel NMDA antagonist on experimental stroke rapidly and quantitatively assessed by diffusion-weighted MRI, *Neurology* 43, 397, 1993a.
18. Minematsu, K., Fisher, M., Li, L., and Sotak, C.H., Diffusion and perfusion magnetic resonance imaging studies to evaluate a noncompetitive N-methyl-D-aspartate antagonist and reperfusion in experimental stroke in rats, *Stroke*, 24, 2074, 1993b.

19. Warner, D.S., McFarlane, C., Todd, M.M., Ludwig, P., and McAllister, A.M., Sevoflurane and halothane reduce focal ischemic brain damage in the rat: Possible influence on thermoregulation, *Anesthesiology*, 79, 985, 1993.
20. Chopp, M., Zhang, R.L., Chen, H., Li, Y., and Rusche, J.R., Postischemic administration of an anti-Mac-1 antibody reduces ischemic cell damage after transient middle cerebral artery occlusion in rats, *Stroke*, 25, 869, 1994.
21. Karibe, H., Zarow, G.J., Graham, S.H., and Weinstein, P.R., Mild intraischemic hypothermia reduces postischemic hyperperfusion, delayed postischemic hypoperfusion, blood–brain barrier disruption, brain edema, and neuronal damage volume after temporary focal cerebral ischemia in rats, *J Cereb Blood Flow Metab*, 14(4), 620, 1994.
22. Zhang, R.L., Chopp, M., Chen, H., and Garcia, J.H., Temporal profile of ischemic tissue damage, neutrophil response, and vascular plugging following permanent and transient (2H) middle cerebral artery occlusion in the rat, *J Neurol Sci*, 125(1), 3, 1994.
23. Fisher, M., Meadows, M.E., Do, T., Weise, J., Trubetskoy, V., Charette, M., and Finklestein S.P., Delayed treatment with intravenous basic fibroblast growth factor reduces infarct size following permanent focal cerebral ischemia in rats, *J Cereb Blood Flow Metab*, 15(6), 953, 1995.
24. Memezawa, H., Zhao, Q., Smith, M.L., and Siesjö, B.K., Hyperthermia nullifies the ameliorating effect of dizocilpine maleate (MK-801) in focal cerebral ischemia, *Brain Res*, 23, 670(1), 48, 1995.
25. Warner, D.S., Martin, H., Ludwig, P., McAllister, A., Keana, J.F., and Weber, E., *In vivo* models of cerebral ischemia: Effects of parenterally administered NMDA receptor glycine site antagonists, *J Cereb Blood Flow Metab*, 15(2), 188, 1995.
26. Kim, Y., Busto, R., Dietrich, W.D., Kraydieh, S., and Ginsberg, M.D., Delayed postischemic hyperthermia in awake rats worsens the histopathological outcome of transient focal cerebral ischemia, *Stroke*, 27(12), 2274, discussion 2281, 1996.
27. Gerriets, T. et al., The macrosphere model. Evaluation of a new stroke model for permanent middle cerebral artery occlusion in rats, *J Neurosci Meth,* 122, 201, 2003.
28. Li, F.H., Omae, T., and Fisher, M., Spontaneous hyperthermia and its mechanism in the intraluminal suture middle cerebral artery occlusion model of rats, *Stroke*, 30, 2464, 1999.
29. Reglodi, D. et al., Postischemic spontaneous hyperthermia and its effects in middle cerebral artery occlusion in the rat, *Exp Neurol,* 163(2), 399, 2000.
30. Abraham, H. et al., Filament size influences temperature changes and brain damage following middle cerebral artery occlusion in rats, *Exp Brain Res*, Epub 2001, 142(1), 131, 2002.
31. Dittmar, M. et al., External carotid artery territory ischemia impairs outcome in the endovascular filament model of middle cerebral artery occlusion in rats, *Stroke*, 34, 2252, 2003.
32. Palmer, G.C. et al., T2-weighted MRI correlates with long-term histopathology, neurology scores, and skilled motor behavior in a rat stroke model, *Ann N Y Acad Sci*, 939, 283, 2001.
33. Dittmar, M.S., Adverse effects of the intraluminal filament model of middle cerebral artery occlusion [letter to the editor], *Stroke*, 36, 530-a, 2005.
34. Choudhri, T.F. et al., Reduced microvascular thrombosis and improved outcome in acute murine stroke by inhibiting GP IIb/IIIa receptor-mediated platelet aggregation, *J Clin Invest,* 102(7), 1301, 1998.
35. Tabrizi, P. et al., Tissue plasminogen activator (tPA) deficiency exacerbates cerebrovascular fibrin deposition and brain injury in a murine stroke model. Studies in tPA-deficient mice and wild-type mice on a matched genetic background, *Arterioscler Thromb Vasc Biol*, 19, 2801, 1999.
36. Wang, L. et al., Chronic nicotine treatment enhances focal ischemic brain injury and depletes brain capillary tissue plasminogen activator in rats, *J Cereb Blood Flow Metab*, 17, 136, 1997.
37. Connolly, E.S., Jr. et al., Procedural and strain-related variables significantly affect outcome in a murine model of focal cerebral ischemia, *Neurosurgery*, 38, 523, 1996.

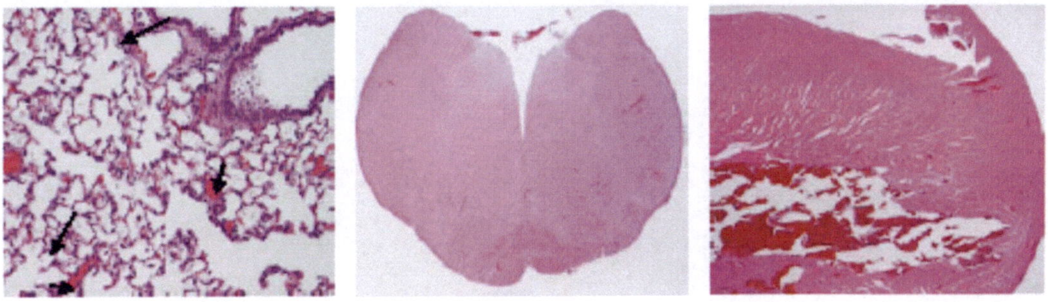

(a) (b) (c)

FIGURE 3.2 The pathological examination of the lungs (a) showed pulmonary edema (arrows, fluid in the alveoli) and congestion (arrows, blood in small arteries and capillaries, hematoxylin and eosin stain, ×20). No brainstem bleeding (b) or morphological myocardial damage (c) was observed.

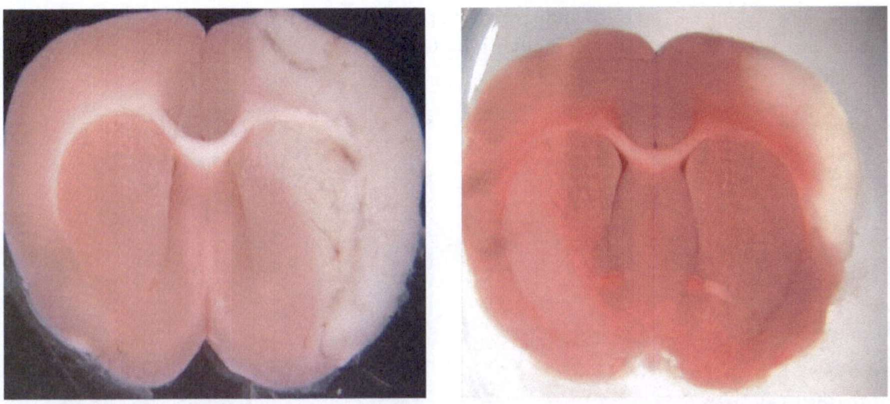

(a) (b)

FIGURE 6.1 Proximal middle cerebral artery (MCA) occlusion produces a big infarction that involves cortical and subcortical areas (a). Distal occlusion produces a smaller infarction that only involves the cortex (b).

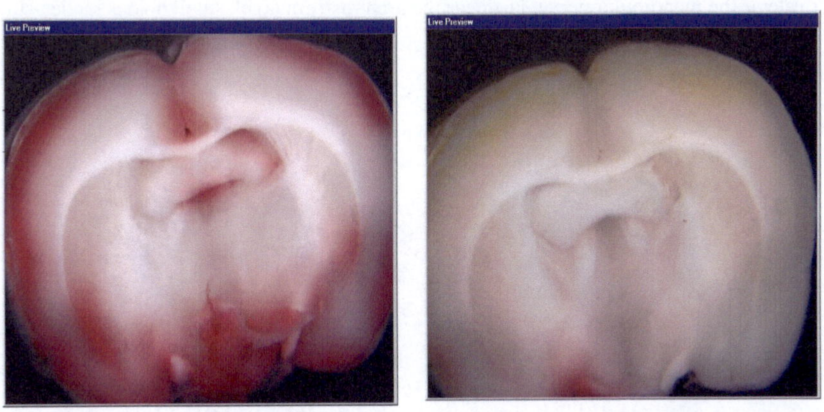

(a) (b)

FIGURE 8.1 Results from studies with rats perfused (35 minutes) with 2% triphenyltetrazolium chloride (TTC) showed that CO_2 anesthesia damaged the blood–brain barrier (BBB). Sprague-Dawley adult rats sacrificed under a mixture gases of 70% CO_2 and 30% O_2 (a) or pentobarbital (b). The red color in the brain tissue indicates the BBB leakages.

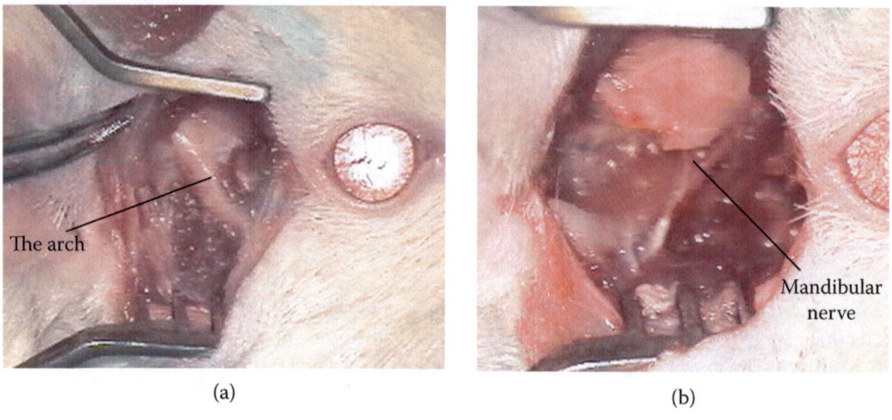

FIGURE 12.6 The zygomatic arch (a) and structures revealed after removing the zygomatic arch (b). The mandibular nerve crosses the temporomandibular joint to the foramen ovale (b).

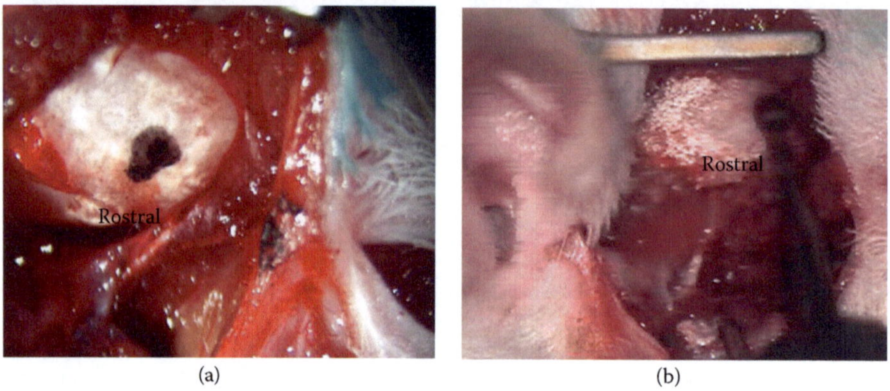

FIGURE 12.7 The drilling location at the squamosal bone is centered about 3 mm anterior and 1 mm lateral to the foramen ovale or the mandibular nerve, just near the arch rostrum (a). A small hole is made using a drill (b).

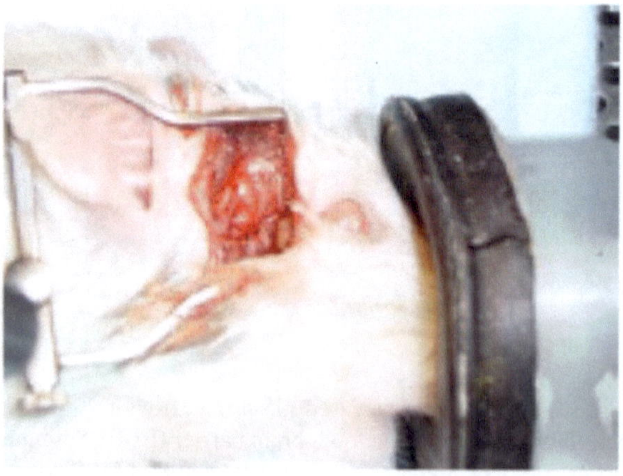

FIGURE 12.8 The dura opened with 23-guage needle by a cruciate incision. The middle cerebral artery runs directly from the bottom laterally to the parietal side.

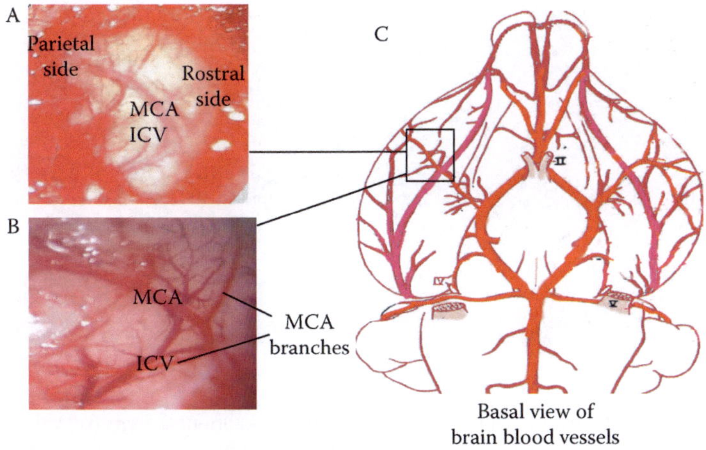

FIGURE 12.9 Individual variations in the branching pattern of the middle cerebral artery (MCA). (A) An MCA going straight up the parietal side without branching. (B) Two branches from the MCA. Another major vein (inferior cerebral vein, ICV) crosses the MCA forward to the rostral side (×25). (C) Basal view of the rat's brain vessel shows where the surgical window was opened to occlude the MCA.

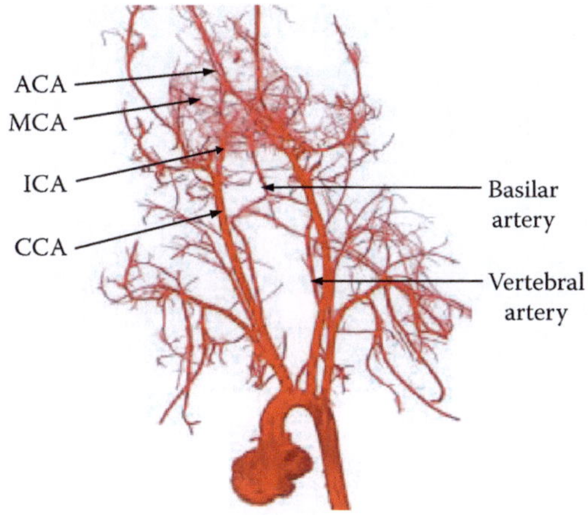

FIGURE 13.3 Resin injection model shows the detailed vessel structure. It is easy to measure vessel diameters in this model.

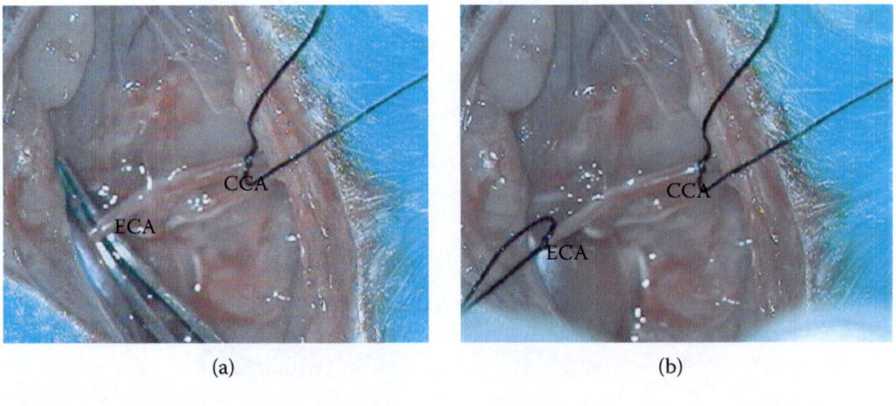

FIGURE 13.4 Dissection of the external carotid artery (ECA) and common carotid artery (CCA) (a). The CCA is temporarily ligated, and the ECA is permanently ligated with 4-0 silk sutures (b).

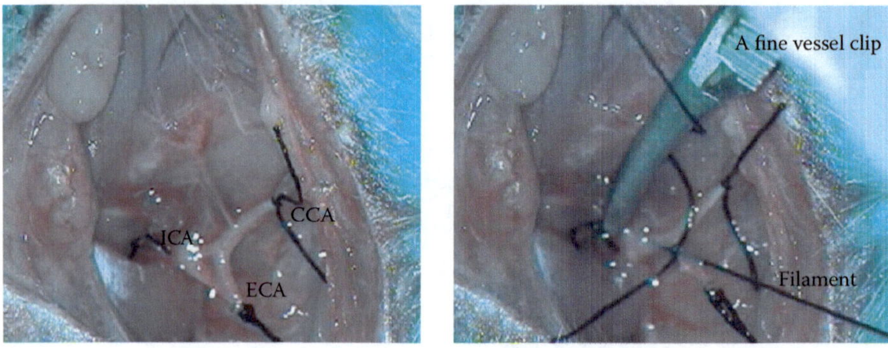

(a) (b)

FIGURE 13.5 The external carotid artery (ECA) terminal stump about 0.5 cm from the fork was separated from surrounding tissue and cut at the distal end. The ECA stump was pulled back under slight tension and then aligned with the internal carotid artery (ICA) (a). The ICA was temporarily blocked by a fine vessel clip before cutting a hole in the ECA stump. A 25-mm-long 3-0 monofilament nylon suture was gently inserted into the ICA through the ECA stump (b). A 4-0 silk suture was tied loosely around the ECA stump along with the filament near the bifurcation before releasing the clip.

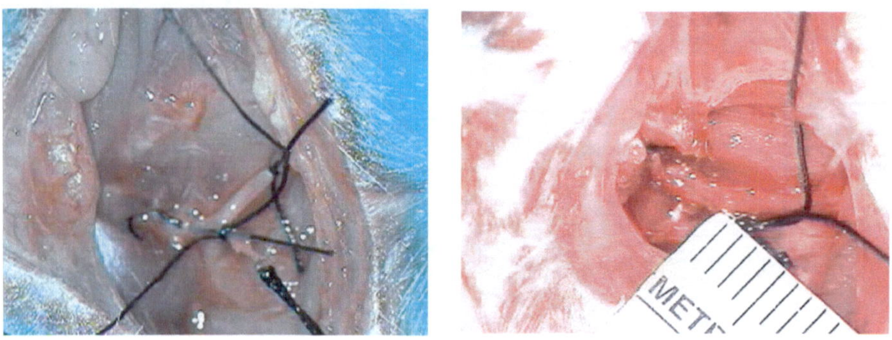

(a) (b)

FIGURE 13.6 The fine vessel clip was removed from the internal carotid artery (ICA). The nylon filament was then gently advanced from the external carotid artery (ECA) to the ICA lumen. It is about 18 to 20 mm from the fork (a). To know the length of filament being inserted, the filament length from the bifurcation to the proximal end of the filament was measured (b).

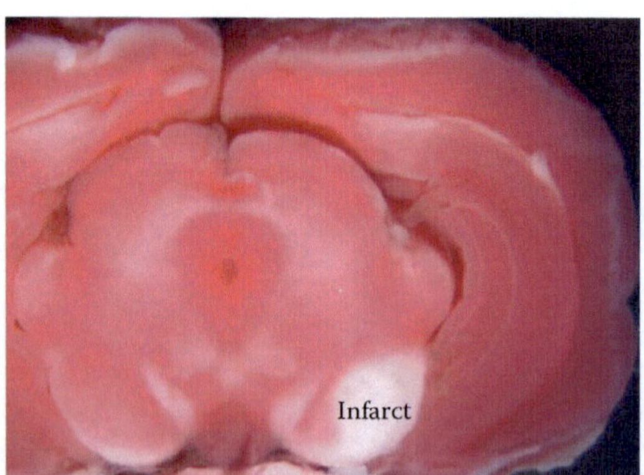

FIGURE 13.10 A very small infarct (3.37%) in the hypothalamus, stained with triphenyltetrazolium chloride at 24 hours after occlusion. This rat had a body temperature over 39°C after 2 hours of filament occlusion.

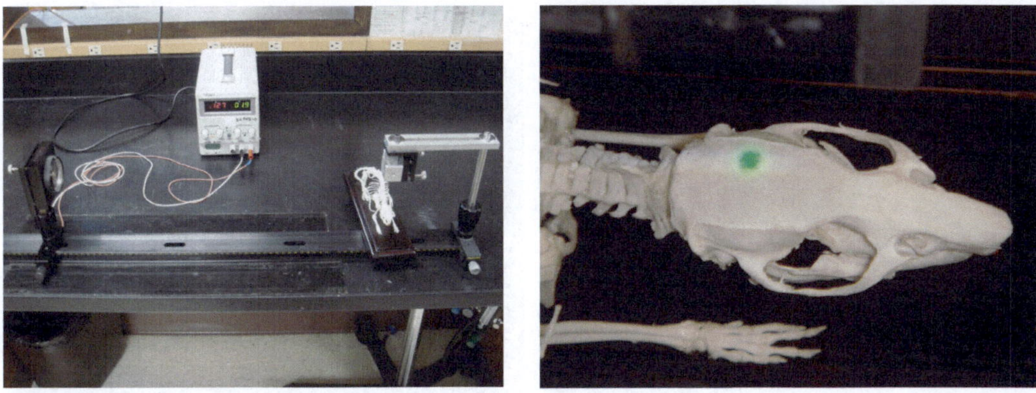

(a) (b)

FIGURE 15.5 Essential features of cortical spot lesioning apparatus. (a) A MicroGreen laser (Snake Creek) is shown illuminating a rat skull (b) after reflection from an elliptical mirror set at 5° from vertical incidence. The beam from this laser strongly diverges, so a spot 5 mm in diameter can be formed just by positioning the laser on the optical rail about 0.7 M away from the focus.

(a) (b)

FIGURE 15.6 Essential features of middle cerebral artery (MCA) occlusion apparatus. (a) The same MicroGreen laser as in Figure 15.5 is shown being magnified by an X5 beam telescope (Edmund Scientific, 101 East Gloucester Pike, Barrington, New Jersey 08007, 800-363-1992, www.edmundoptics.com) and focused with a 25-cm FL (focal length) planoconvex lens onto a deflecting mirror. The mirror directs the beam onto the skull region overlying the distal MCA territory (b). The diameter of the focused beam on the skull is about 200 μm.

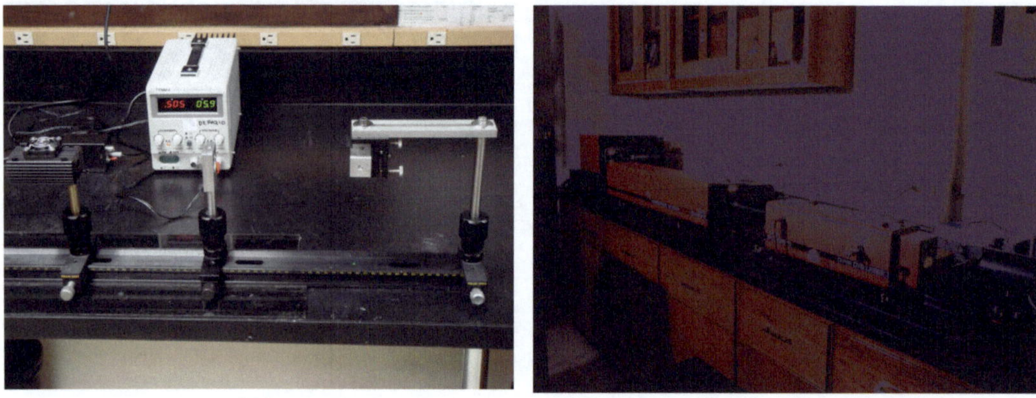

(a) (b)

FIGURE 15.7 Common carotid artery occlusion systems. (a) The beam from a compact 532-nm Nd:YAG (yttrium aluminum garnet) laser (at left; 100 mW, model LAGR100M, manufactured by Laserglow and sold by Information Unlimited, Amherst, New Hampshire) is shown traversing a 61-cm FL planoconvex lens (at center, in front of power supply) and internally reflecting from a right-angle prism (at right). The beam optics of the more powerful 1-W Laserglow model LLS-532 in Figure 15.4b are the same. The beam appears as a 1-mm diameter green spot on the optical rail below, on which the rat can be placed. (b) A model 70-4 argon ion laser (left) coupled to a CR599 dye laser (right) (Coherent, Fremont, California). This argon laser is rated at 4 W for all lines, but with a new plasma tube usually produces about 6 W. The dye laser (rhodamine 560) emits at least 1 W at 562 nm. These lasers are quite hardy but require external water cooling, regular tuning, and maintenance such as cleaning the dye jet and pumping system, replacing the dye solution periodically, and cleaning the many optical surfaces.

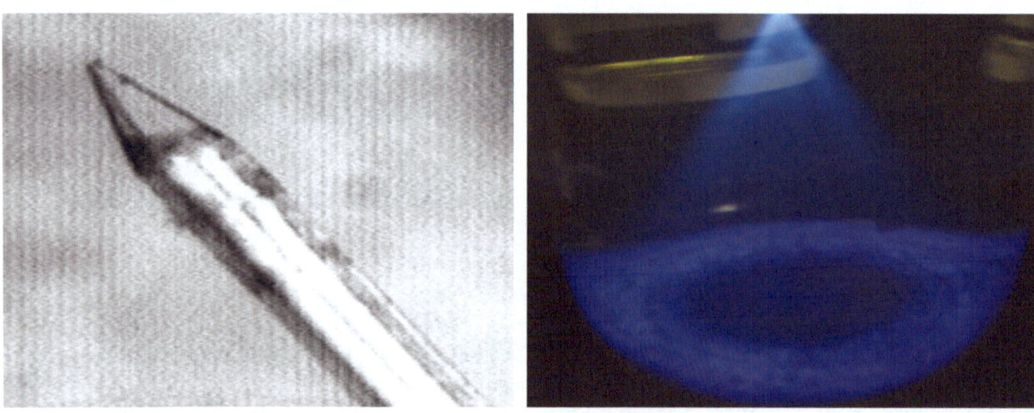

(a) (b)

FIGURE 15.13 Geometric properties of a conical-tip optical fiber. (a) Optical fiber with conical tip ground and polished at an apex angle of 35°. (b) A ring-shaped ultraviolet (UV) laser beam (351 nm argon) suitable for radially symmetric endovascular irradiation produced by this conical-tip fiber and projected into a beaker filled with water. The blue color is due to UV excitation of fluorescent impurities. A conical tip defect (a) permits an axial beam of negligible intensity to appear inside the ring beam in (b).

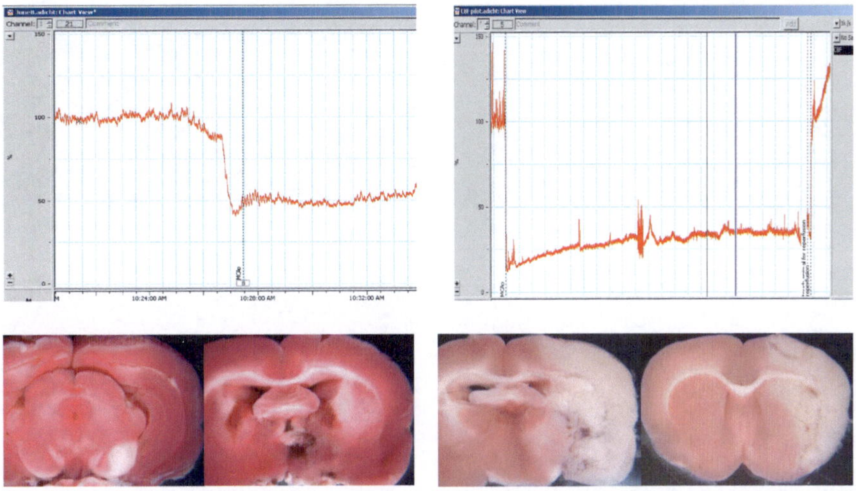

(a) (b)

FIGURE 19.7 (a) Incomplete ischemia: Cerebral blood flow (CBF) decreased only 50% after the suture was inserted; this rat had a temperature of 39.4°C and a behavioral score of 2 after 2-hours occlusion. Triphenyltetrazolium chloride (TTC) staining at 24 hours postocclusion showed a very small infarct (3.4%). This animal was excluded on the basis of CBF. (b) Full ischemia: CBF decreased to 20% to 30% of baseline after occlusion; the rat had a high temperature of 39.2°C and a behavioral score of 2. TTC staining showed a large infarct of 50% (white) on the ischemia side.

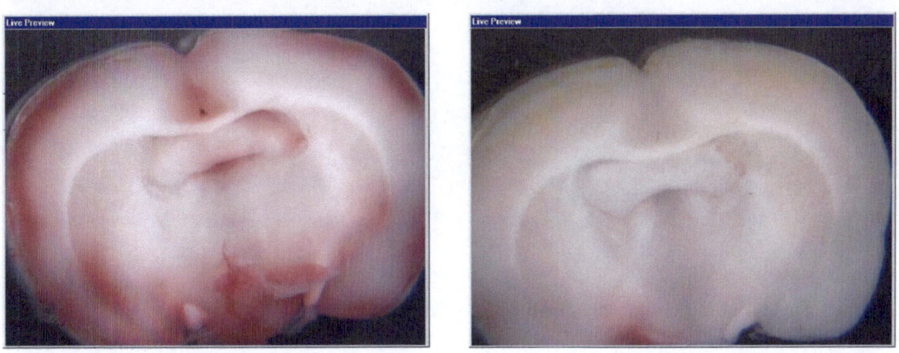

(a) 70%CO_2 + 30%O_2 (b) Pentobarbital

FIGURE 19.20 Triphenyltetrazolium chloride (TTC; 2%, 35 minutes) solution perfused into rats: (a) rats euthanized with CO_2 (5 minutes); (b) rats euthanized with pentobarbital.

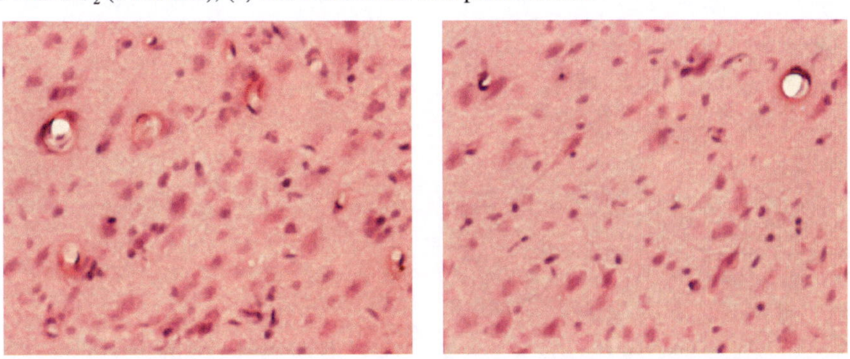

(a) Infarct side (b) Contralateral (normal)

FIGURE 20.7 Hematoxylin and eosin Y (H&E) stain on rat brain at 3 weeks after stroke, ipsilateral infarct side (a) and contralateral normal side (b); ×20.

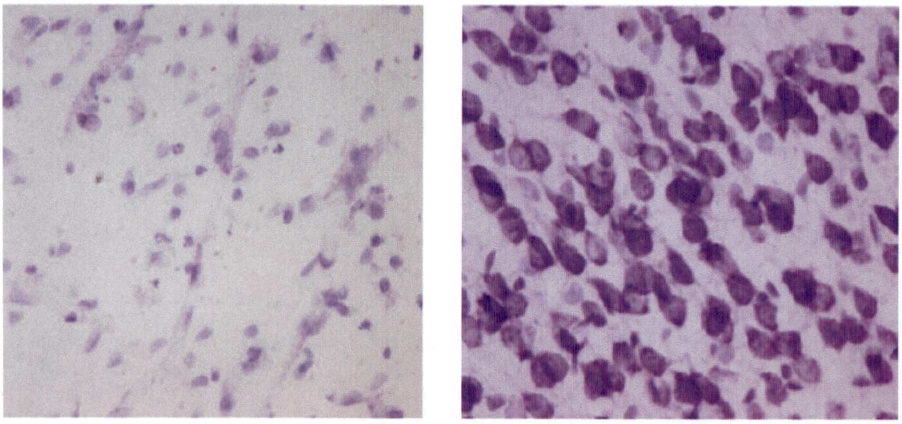

(a) Infarct area (b) Contralateral

FIGURE 20.8 Thionine stain on rat brain at day 2 after suture stroke (×40). (a) Infarct area shows that most neurons have disappeared; (b) the contralateral side shows normal neurons.

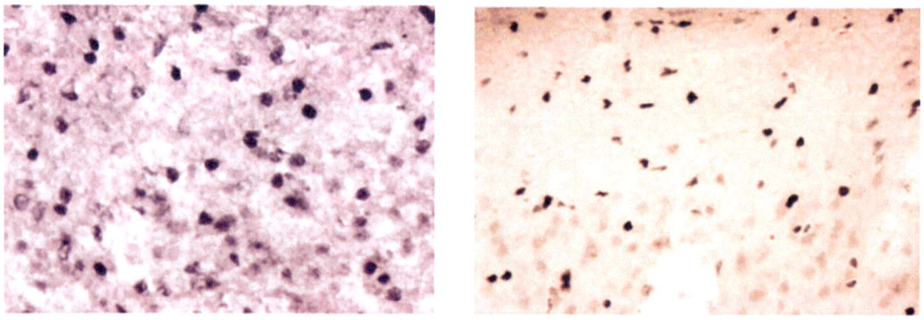

(a) Infarct side (b) Sham surgery

FIGURE 20.11 Immunohistochemical staining shows that caspase-3, a marker for apoptosis, is upregulated in a stroked rat. (a) Infarct side at 48 hours after stroke; (b) sham-operated rat (×40).

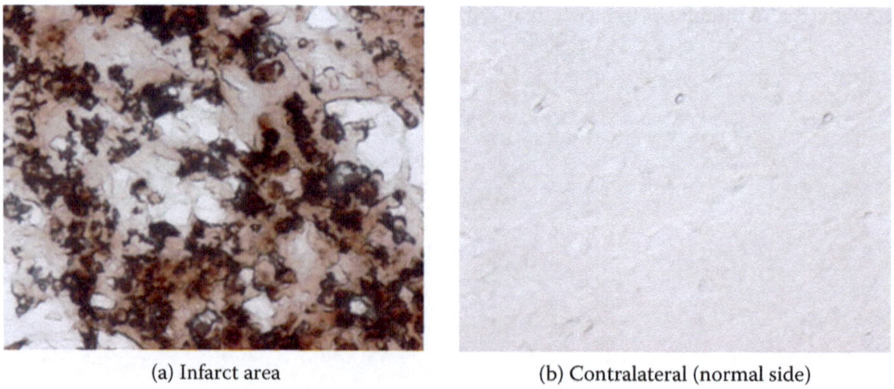

(a) Infarct area (b) Contralateral (normal side)

FIGURE 20.12 ED1 immunohistochemical stain shows macrophages in the infarct area (a) but not on the contralateral side (b) (×40x; 3 weeks postsuture stroke rats).

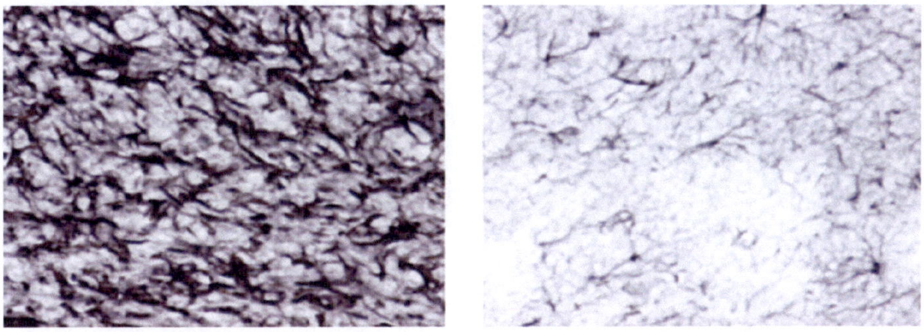

(a) Infarct side (b) Contralateral

FIGURE 20.13 Immunohistological staining shows increased GFAP (glial fibrillary acid protein) on the infarcted side of a stroked rat (a) but no increase on the contralateral side (b) (×40; 3 weeks postsuture stroke). Note that the morphology of astroglia changes from small bodies (b) to larger bodies and thicker processes (a).

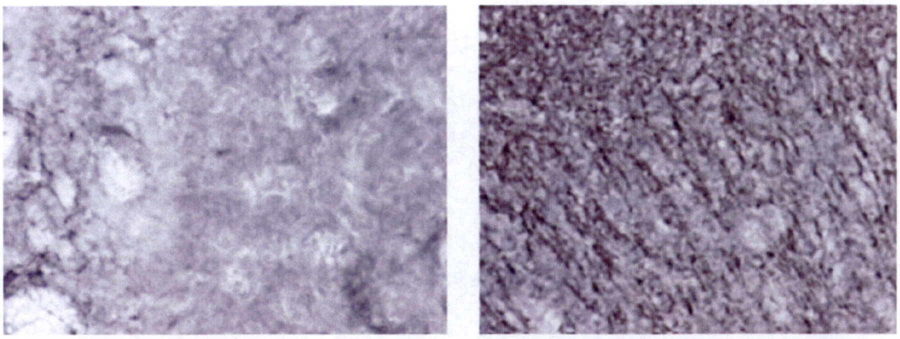

(a) Infarct core (b) Contralateral

FIGURE 20.14 Map2 (microtubule-associated protein) staining shows disappearance of neuronal dendrites in the infarct core (a) compared to the contralateral side of a stroked rat at 3 weeks poststroke (b); ×40.

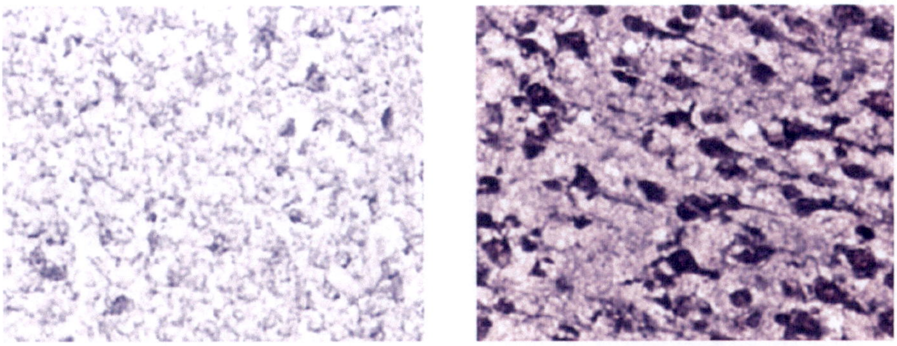

(a) Infarct core (b) Contralateral

FIGURE 20.15 Neuron-specific nuclear protein (NeuN). The stain shows disappearance of neuronal nuclei in the infarct core (a) compared to the contralateral side (b) (48 hours poststroke rat; ×40).

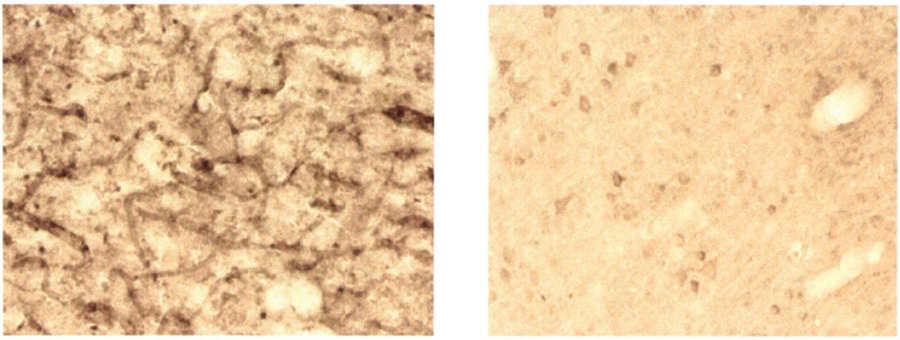

(a) Infarct side (b) Contralateral

FIGURE 20.16 Matrix metalloproteinase (MMP-9) staining of rat brain 24 hours poststroke. (a) Infarct side. (b) Contralateral normal brain. MMP-9 is upregulated in the vascular tree on the stroke side. MMP-9 rabbit antirat antibody 1:5000 diluted at ×20.

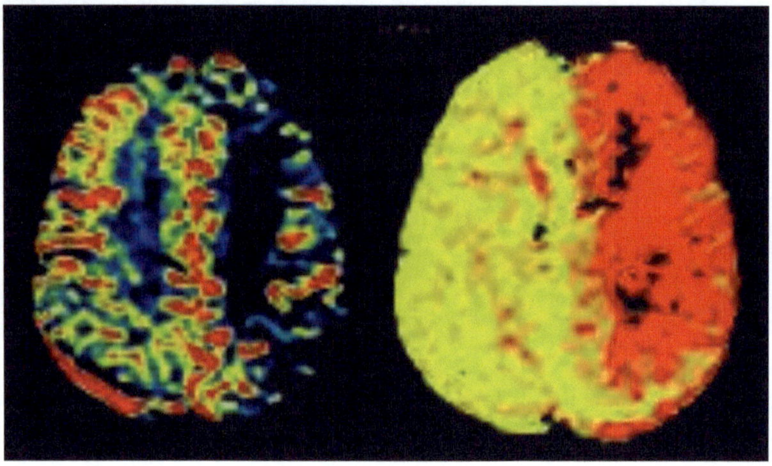

FIGURE 22.2 Magnetic resonance imaging (MRI) in acute stroke. Left: Perfusion-weighted MRI of a patient who presented 1 hour after onset of stroke symptoms. Right: Mean transfer time (MTT) map of the same patient.

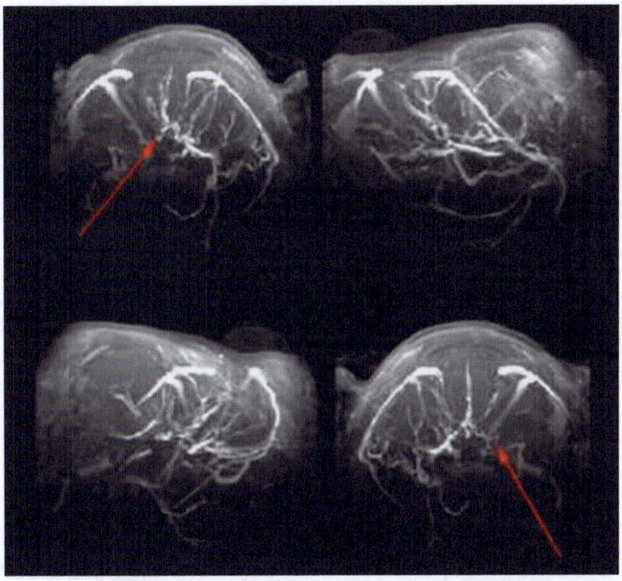

FIGURE 22.7 Four views of an MIP (maximum intensity projection) reconstruction of magnetic resonance imaging (MRI) data obtained with flow-compensated 3D FLASH sequence. Drastic reduction of flow is visible at the site of the suture-induced middle cerebral artery occlusion (MCAO) insult (red arrows). These images were obtained without contrast agent. BioSpec Applications—Stroke. (Courtesy of M. Neumaier, U. Pschorn et al., Boehringer Ingelheim Pharma KG, Department of CNS Research, Ingelheim, FRG.)

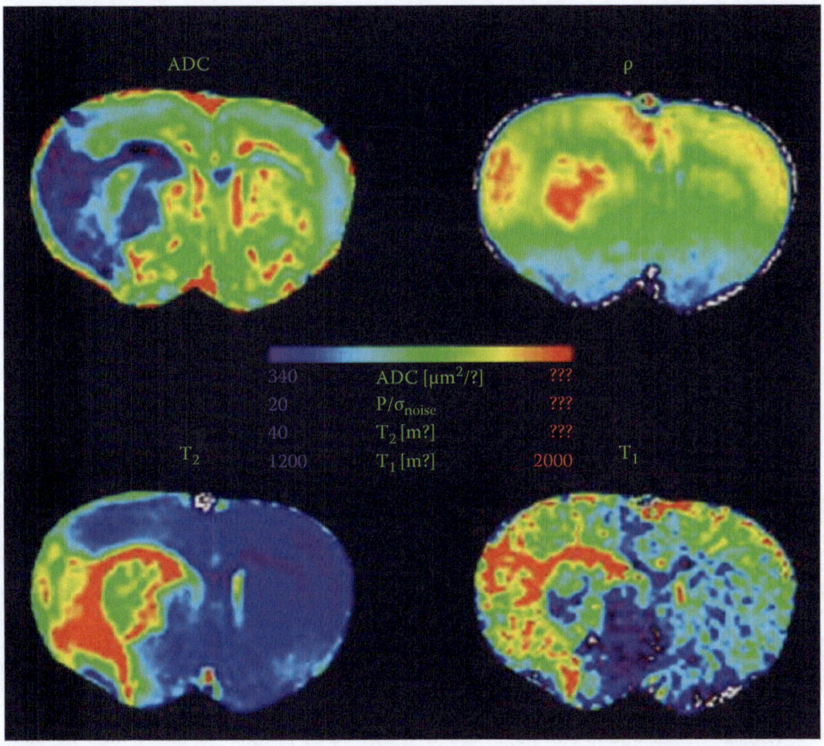

FIGURE 22.8 The damage resulting from a middle cerebral artery occlusion (MCAO) in rat brain can be investigated using several different magnetic resonance imaging (MRI) parameters. Each provides another insight into the changes in the physical environment of the ischemic region. The color-coded parameter maps presented here display the apparent diffusion coefficient (ADC), proton density (P), T2 and T1 relaxation times, all of which show a marked contrast between ischemic (left) and normal brain (right) as well as between the different parameter maps. BioSpec Applications—Stroke. (Courtesy of M. Eis, U. Pschorn, et al., Boehringer Ingelheim Pharma KG, Department of CNS Research, Ingelheim, FRG.)

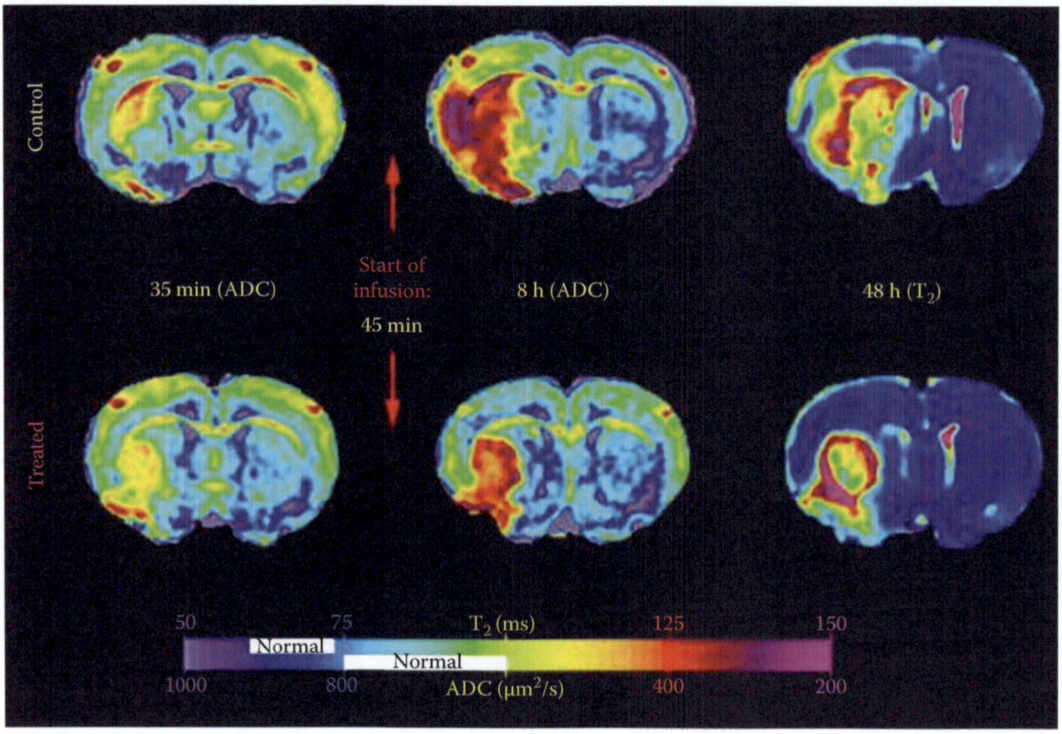

FIGURE 22.9 Rat brains at three time points after suture middle cerebral artery occlusion (MCAO) insult with (bottom row) and without (top row) cerebroprotective drug treatment. The compound reduces the growth of both lesion volume and its severity (note the very different lesion size at the outset). BioSpec Applications—Stroke. (Courtesy of M. Eis, M. Neumaier, and U. Pschorn, Boehringer Ingelheim Pharma KG, Department of CNS Research, Ingelheim, FRG.)

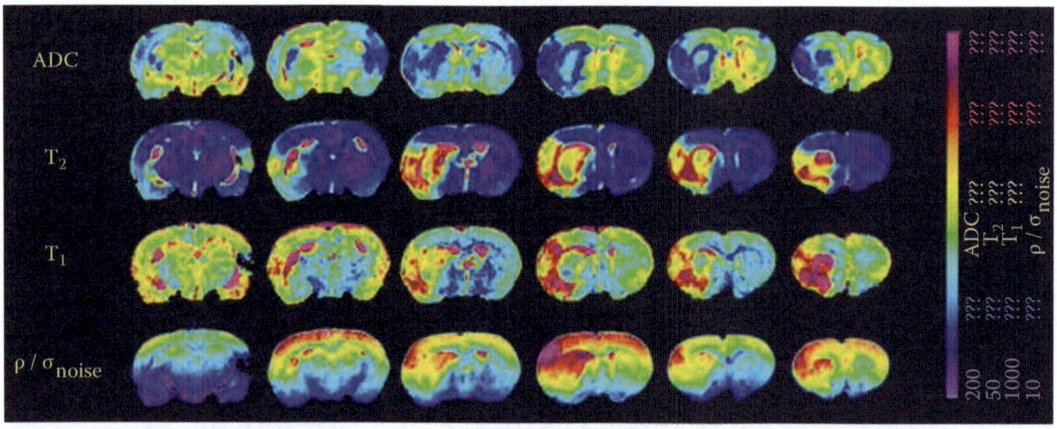

FIGURE 22.10 The extent of ischemic damage was investigated 48 hours after suture middle cerebral artery occlusion (MCAO) in the left hemisphere of a rat using calculated ADC (apparent diffusion coefficient), T2, T1, and proton density maps. The insulted area is divided into a core and an ischemic penumbra, the volumes of which can be assessed by contiguous multislice, full-brain coverage. BioSpec Applications—Stroke. (Courtesy of M. Eis, M. Neumaier, and U. Pschorn, Boehringer Ingelheim Pharma KG, Department of CNS Research, Ingelheim, FRG.)

14 Embolic Stroke Models

Yanlin Wang-Fischer, Afshin A. Divani, and Lee Koetzner

CONTENTS

EMBOLI: STUDIES BASED ON DIFFERENT EMBOLUS PREPARATIONS

There are three primary stroke subtypes by causes: thrombosis, embolism, and hemorrhage. Thrombosis and embolism are responsible for approximately 80% of human strokes.[1,2] In the past 20 years, emboli prepared in different ways have been used to model stroke in rodents (rats, mice, rabbits[3]), swine,[4] and cynomolgus monkeys.[5] Embolic materials included (1) blood thrombi (autologous or heterologous blood clots)[6]; (2) polyethylene microspheres (50 µm in diameter)[7]; (3) polyvinylsiloxane (PVS)[8]; (4) viscous silicone oil[9]; (5) latex microspheres; (6) polyvinyl acetate; and (7) ceramic macrospheres, which have been injected into the common carotid artery (CCA)[10,11] or into the middle cerebral artery (MCA).[12,13] Thrombotic cerebral ischemia has also been produced by photochemical irradiation of the CCA or MCA following the administration of a photosensitive dye.[14–16] The techniques of photochemically induced stroke, including embolic stroke, are described in Chapter 15.

CHARACTERISTICS AND PREPARATIONS OF EMBOLI AND SPECIAL PROCEDURES

Blood Thrombi

Models of focal cerebral ischemia induced by intraarterial autologous or heterologous blood clots provided great information on the safety and efficacy of antithrombotic therapies.[3] Kudo et al.[6] in 1982 and Kaneko et al.[17] in 1985 described thromboembolic stroke in rats by using a suspension of blood clot fragments to occlude the entire arterial tree. Later, Chopp's group and Hossmann's group made significant improvements to this technique. Their publications are the most widely cited; this chapter introduces their methods in detail.

Method of Chopp's Group

In 1997, Chopp's group[12,18] described MCA occlusion by injection of a single blood clot 25 mm long through an internal carotid artery (ICA) to near the origin of the ipsilateral MCA. The infarct was limited to the MCA distribution area in this technique. The infarct volume in rats was 32% ± 14% at 24 hours, n = 20; 33% ± 13% at 48 hours, n = 13; and 35% ± 16% at 168 hours, n = 12 (mean ± standard deviation; the original paper expressed mean ± standard error).

In this study,[18] animals (male Wistar rats weighing 320 to 400 g) were anesthetized with 3% to 4% halothane and maintained with 1.0% halothane in 70% N_2O and 30% O_2 through a face mask. Rectal temperature was maintained at 37°C ± 0.5°C throughout the surgical procedure using a feedback-regulated water heating system.

Preparation of the Emboli

1. Blood was withdrawn from the femoral artery in a donor rat and injected into 20 cm of PE-50 tubing and retained in the tube for 2 hours to clot at room temperature and subsequently retained for 22 hours at 4°C.
2. Of the PE-50 tube containing clot, 5 cm were cut and attached to two PE-10 tubes (20 cm) at each end. Two syringes filled with saline were connected to the PE-10 tubes. The clot was shifted by continuous alternating movement from one syringe to the other for 5 minutes (see Figure 14.1).
3. A single clot 25 mm long (3.14 × 0.01 mm², 0.8 μL) was transferred to a modified PE-50 catheter with a 0.3-mm outer diameter filled with saline. (The PE-50 catheter was modified to have a very small diameter, 0.3 mm, by stretching the catheter near a flame.)

Animal Model

1. Under an operating microscope (Carl Zeiss, Thornwood, New York) or a magnifier, the right CCA, external carotid artery (ECA), and ICA are exposed through a ventral midline incision in the neck.
2. The right ECA is isolated, and the superior thyroid and occipital arteries are ligated with a 5-0 suture and transected.
3. A 5-0 silk suture is loosely tied at the origin of the ECA and ligated at the distal end of the ECA. The ICA is further dissected distally, and the right pterygopalatine artery (PPA) is ligated with a 5-0 suture.
4. The right CCA and ICA are temporarily clamped using a curved microvascular clip or two separate small vessel clips (Codman and Shurtleff, Randolf, Massachusetts).
5. A modified PE-50 catheter with a 0.3-mm outer diameter filled with a 25-mm-long clot, attached to a 100-μL Hamilton syringe, is introduced into the ECA lumen through a small puncture.

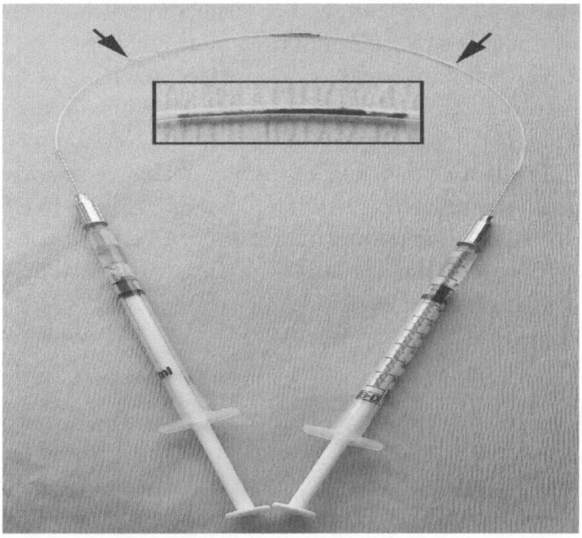

FIGURE 14.1 The connection between syringes and PE-10 tubes; arrows indicate the blood clot in a PE-50 tube.

6. The suture around the origin of the ECA is tightened around the intraluminal catheter to prevent bleeding, and the microvascular clip is removed.
7. A 15-mm length of catheter is gently advanced from the ECA into the lumen of the ICA. At this point, the intraluminal catheter is 2 to 3 mm from the origin of the MCA.
8. The clot along with 5 μL of saline in the catheter are injected into the ICA over 10 seconds (Figure 14.2a).
9. The catheter is withdrawn from the right ECA 5 minutes after injection. The right ECA is ligated. The duration of the entire surgical procedure is approximately 25 minutes.

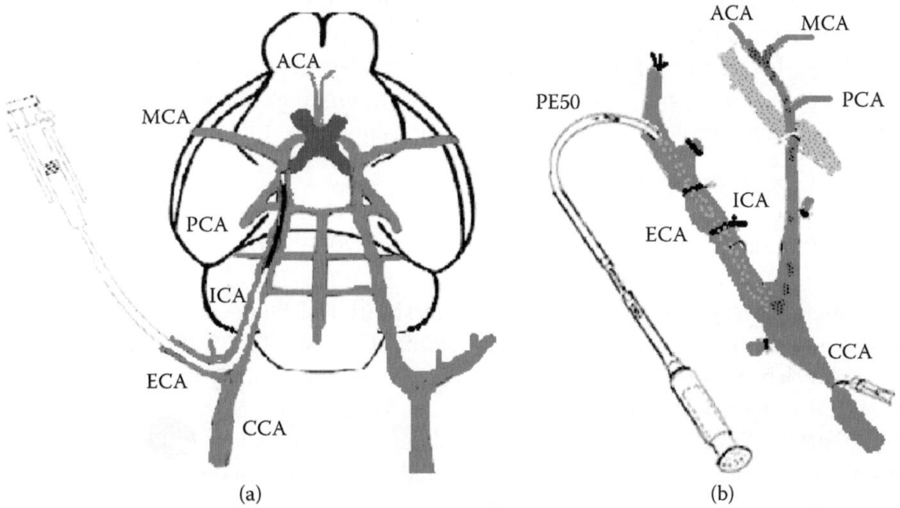

FIGURE 14.2 Schema shows the different catheter placements for two techniques. (a) A small-size catheter placed into the external carotid artery (ECA) and internal carotid artery (ICA) with its tip 2 mm from the origin of the middle cerebral artery (MCA). The catheter contains a clot (black) that will be injected into the MCA and anterior cerebral artery (ACA). (b) The catheter is placed into the ECA near the carotid bifurcation, and the clots are injected into the ICA and packed in toward the MCA origin and other ICA branches.

Heparin is not administered to any animal. For sham surgery, rats are subjected to the same surgical procedures but without intravascular insertion of an embolus.

To evaluate the optimal length of a clot for consistent lodgment at the origin of the MCA, Chopp's group tested various lengths (10 to 30 mm) in a preliminary study. They found that the probability of lodgment of a clot at the origin of the MCA was 40% in rats injected with a clot less than 25 mm long and 95% with a 25-mm-long clot. Of animals injected with a clot longer than 25 mm, 20% exhibited a hemispheric infarction. A clot less than 25 mm in length frequently lodged within a branch of the MCA or the anterior cerebral artery (ACA). In contrast, a clot longer than 30 mm blocked the ACA, the MCA, and the PCA. Therefore, in their study, a 25-mm-long clot was selected for intraluminal placement.

Method of Hossmann's Group

During 1997 and 2003, Hossmann's group[19,20] produced cerebral ischemia by injection of a small number (12) of medium-size cylindrical autologous clots into a rat ICA. The size of these cylindrical clots (1.5 mm long × 0.35 mm diameter) was small enough to reach the MCA origin, but large enough to be trapped in the proximal MCA and ACA portions (Figure 14.2b). The infarct involved the posterior, middle, and anterior cerebral artery territories, similar to the suture ischemia model.

They added thrombin to the blood to gain fibrin-rich blood clots. Thrombin is an endopeptidase that regulates blood coagulation.[21] It has profound effects on every aspect of vascular wall biology, including regulation of vessel tone, vascular permeability, smooth muscle cell proliferation, differentiation, migration, vascular development, atherogenesis, and angiogenesis. Thrombin is also neurotoxic, especially in combination with plasminogen activators.[21] The multiple biological effects of thrombin may be relevant for the outcome of experimental clot embolization studies. The clots were prepared by mixing freshly drawn arterial blood with a high concentration of thrombin. This process led to the formation of compact clots; the fibrin-rich parts of the clots were selected for intracarotid embolization. The thrombin coagulative response varies wildly among laboratories. Apparently, it is difficult to just copy anybody's method and expect to get the same results. Obviously, such clots differ from spontaneously developing clots in clinical patients, in whom thrombin generation is dependent on the plasma level of prothrombin; it varies considerably among different patients. One of the main reasons for using thrombin-induced clots instead of spontaneously forming clots is the higher density of the fibrin meshwork and, as a consequence, the higher reproducibility of infarct size. The other real reason they used thrombin to make fibrin-dense clots was to emphasize clot susceptibility to tissue plasminogen activator (tPA) treatment.

Preparation of Clots with Thrombin

1. Animals (for example, male Wistar rats weighing 320 to 400 g) are anesthetized. Body temperature is maintained at 37°C ± 0.5°C.
2. The femoral artery is cannulated; 0.6 mL fresh arterial blood is mixed with 0.15 mL thrombin (1 mg/mL, 30 U/mL in normal saline; Cat. No. T5772-100, from rat plasma, Sigma-Aldrich, St. Louis, Missouri). This mixture is immediately injected into 50 cm of PE-50 (inner diameter 0.58 mm). The blood clot forms in the catheter. The catheter in the femoral artery is sealed at the end (a stump about 2 cm long is left for future use) or withdrawn. The animals recover from anesthesia and are put back in their cages.
3. After 3 hours, the catheter containing the clot is cut into 5-cm pieces, and the clot material is removed by flushing with normal saline. The clot is washed with saline twice for 30 seconds to remove blood cells and put into a solution of rat albumin (1 mg/mL) and phosphate-buffered saline.
4. The clots are inspected under the microscope for selection of fibrin-rich segments. These segments are cut into small pieces, producing cylindrical clots about 0.35 × 1.5 mm.

Animal Model

1. Animals are anesthetized as above.
2. The CCA, ECA, and ICA are exposed as above.
3. A PE-50 catheter is implanted in the ECA and placed near the carotid bifurcation.
4. Twelve clots are put into a solution of albumin (1 mg/mL) and phosphate-buffered saline. After full clot retraction is completed, the clots together with the albumin solution are drawn up into a PE-50 catheter 80 to 100 cm long.
5. The catheter containing the clots is connected to the line already surgically inserted into the ECA using a 22-gauge needle.
6. The 12 clots are injected into the ECA over a period of 30 seconds, during which the CCA is temporarily closed.

The difference between these two techniques is shown in Figure 14.2.

Polyethylene Microspheres

The polyethylene microsphere-induced cerebral embolism model in rats induces widespread and irreversible formation of small emboli and multiple infarct areas in the brain.[22] It mimics multiinfarct dementia.[23]

Preparation of Polyethylene Microspheres

Polyethylene is a polymer ethylene and is resistant to water, acids, alkali, and most solvents. The diameter of the spheres for embolic stroke modeling is 50 μm (47.5 ± 0.5 μm); the number of spheres injected varies (from 500 to 2000) among different researchers.[7,22] The microspheres are suspended in 150 to 200 μL of 20% dextran solution. Microspheres can be purchased from Perkin-Elmer Life Science, Bangs Laboratory, or New England Nuclear (Boston, Massachusetts). A detailed description of the technique is provided next.[7]

Animal Model

1. The procedures for exposure of right CCA, ECA, and ICA are the same as described for the method of Hossmann's group.
2. The CCA is occluded temporarily by a vessel clip.
3. A PE-10 catheter is inserted into the ICA through the ECA. A 0.2-mL suspension of nonradioactive polymer microspheres (50.0 ± 0.5 μm diameter; 2000 spheres [10,000/mL of suspension]) in 20% dextran solution is injected into the ICA through the PE-10 catheter and is flushed with 0.2 mL saline.
4. The catheter is removed after the injection, and the proximal portion of the ECA is ligated with a 5-0 suture.
5. In the sham surgery group, the same volumes of vehicle (20% dextran without microspheres) and saline are injected.

Polyvinylsiloxane

Polyvinylsiloxane (hydrophilic vinyl polysiloxane [PVS], Reprosil®, Type 1, Dentsply International, Miford, Delaware) is a viscous, nontoxic substance that has been widely used for making dental impressions. When mixed with a catalyst, this agent sets within minutes to a rubbery consistency. In addition to its dental application, this material has been used for nerve cuff electrodes.[24] Injecting high-viscosity PVS into the CCAs to occlude the circle of Willis and all cerebral arteries induced an anemic decerebration in rats.[25] The low-viscosity formula allows easy introduction of PVS into the

MCA through a smaller catheter (0.28-mm internal diameter). PVS is x-ray transparent and therefore does not interfere with the angiographic assessment of the MCA. Microscopic observation suggested that injected PVS only entered the large arteries and did not pass into the capillary beds.[8]

Preparation of Polyvinylsiloxane

1. A 10:1 ratio (PVS base to catalyst) is mixed at room temperature; this ratio is sufficient to allow the mixture to be transferred to a syringe and injected through a microcatheter.
2. The mixture is prepared in a cell culture dish, transferred to a 1-mL syringe, and injected into the right MCA through a catheter. It is important to tightly hold the connection between the syringe and the needle connected to the catheter as the viscosity of the mixture creates substantial back pressure.

Animal Model

This animal model method is derived from the work of Yang and Shuaib.[8]

1. Animals (male Wistar rats weighing 400 to 450 g) are initially anesthetized with 3.0% isoflurane. Anesthesia is maintained with 2% isoflurane in a 70% N_2O and 30% O_2 (volume/volume) mixture via a face mask.
2. The surgical procedure is similar to that used by Chopp's group.
3. A PE-50 catheter (Becton Dickinson, Franklin Lakes, New Jersey) is modified to 0.3-mm outside diameter at the tip of the catheter (same as Chopp's group, 0.97-mm outside diameter in the catheter body).
4. The catheter is attached to a 1-mL syringe containing saline and is introduced into the right ICA from the ECA via a small cut. A silk suture is tied around the ECA to prevent bleeding and dislodgment.
5. The vascular clips are removed from the ICA and CCA.
6. The catheter (17 mm, calculated from the site of the bifurcation) is gently advanced through the ECA and extracranial portion of the ICA into the intracranial portion of the ICA; the target depth is 1 to 2 mm away from the origin of the MCA.[12]
7. The syringe with saline is replaced with a 1-mL syringe filled with PVS, and a total of 30 μL PVS (after calculating the volume of PVS left in the dead space of the catheter, about 30 μL) is injected at the point immediately proximal to the MCA bifurcation. The PVS is injected into the MCA in about 5 seconds.
8. The catheter is withdrawn 1 minute after injection to avoid removing the embolus from the artery.
9. The catheter is withdrawn, the ECA is ligated, and the incision is sutured.
10. The animals are allowed to recover from the anesthesia, and they are given free access to food and water.

The infarct created by this technique is limited to the distribution area of the MCA, similar to Chopp's method. Yang et al. compared three different methods of stroke models; they found PVS-induced ischemia had lower variability than that induced by filament and embolic blood clot methods.[8]

Viscous Silicone Oil

Preparation of Viscous Silicone Oil

The method of Lauer and Hudetz's group (2002) that used viscous silicone oil as emboli to induce stroke[9] is introduced next.

1. Viscous silicone was prepared by heating hydrogen functional polydimethyl siloxane (RT604A, Wacker Silicone, Adrian, Michigan) mixed in a 9:1 ratio with a platinum catalyst (X84, Wacker Silicone) in a 37°C water bath with continual manual agitation for 2 minutes.
2. This mixture cross-links to form a viscous agent that solidifies at body temperature in approximately 10 minutes.

Animal Model

1. Animals are anesthetized as above and laid in the supine position.
2. The left carotid artery bifurcation is exposed. The CCA is ligated with a 3-0 suture, and the ECA and PPA are ligated with small surgical clips.
3. The PE-50 catheter is filled with saline and cut to a standard length (26.5 cm) to ensure a constant dead space.
4. The CCA and ICA are cannulated with PE-50 tubing, which is advanced 4 to 5 mm.
5. The catheter is secured with 4-0 suture, the wound is closed, and the animal is turned prone. A schema of the preparation is shown in Figure 14.3.
6. Blood is allowed to flow back into the catheter, and then the silicone is injected until it fully replaces the blood, as confirmed visually. An 11-μL (small lesion) or 22-μL (large lesion) volume of silicone is manually injected over 30 seconds via the carotid catheter.
7. The syringe is left attached to a pressure micrometer for at least 1 hour. The success of embolization is confirmed by observing a drop in cerebral blood flow by Doppler flow meter. The size of brain lesion is dependent on the amount of silicone injected. (To support arterial blood pressure after embolization, methoxamine can be infused 0.5 mg/mL at 30 to 90 mg per minute; refer to Reference 9.)
8. The same procedure is carried out in a sham surgery group with the exception of silicone injection.

Ceramic Macrospheres

We mentioned in Chapter 13 that hypothalamic ischemia causes injury that results in hyperthermia and possibly worsened outcome. Gerriets et al.[26] in 2003 introduced a new middle cerebral artery occlusion (MCAO) model that avoids these disadvantages. In the filament model, a monofilament is advanced into the ICA until its tip is lodged in the ACA, where it blocks blood flow to the MCA as well as to the whole ICA arterial tree, including the hypothalamic artery (HTA) as it originates from the distal ICA (about 1.5 to 2 mm from the MCA in adult rats; refer to Chapters 4 and 13).

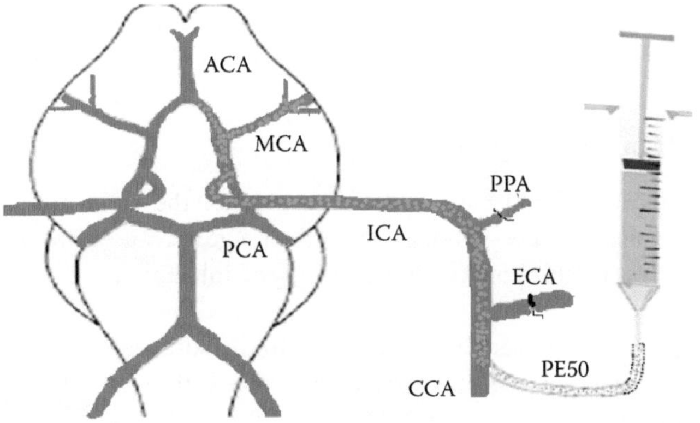

FIGURE 14.3 Schema of model preparation in the rat. A PE-50 catheter is advanced from the common carotid artery (CCA) 4 to 5 mm to the carotid bifurcation after ligation of the external carotid artery (ECA) and pterygopalatine artery (PPA). Viscous silicone is injected into the internal carotid artery (ICA).

The technique of using macrospheres is to obstruct the MCA main stem without occluding the HTA. Intraarterial embolization is induced with six ceramic macrospheres, 0.3 to 0.355 mm in diameter. Sprague-Dawley rats with body weight 290 to 350 g were used in this study.[26] Gerriets et al. compared this new model with the established filament occlusion model with respect to infarct size and localization, body temperature, and clinical outcome and described the natural course of lesion development over time on diffusion-weighted magnetic resonance imaging (DWI) and perfusion-weighted magnetic resonance imaging (PWI) in both models.

Preparation of TiO_2 Macrospheres

The TiO_2 macrospheres (0.3- to 0.4-mm diameter) are purchased from Brace GmbH, Alzenau, Germany.

Animal Model

1. The surgical procedure is similar to that of Hossmann group's blood thrombus model.
2. The distal portion of the ECA is ligated with a 3-0 suture and transected to create an ECA stump with a length of approximately 5 mm. The pterygopalatine branch of the ICA is also ligated with a 5-0 suture.
3. The CCA and ICA are temporarily ligated with two 3-0 sutures to avoid hemorrhage before and during the injection.
4. Nylon tubing (0.5-mm inside diameter, 0.63-mm outside diameter), filled with saline and six TiO_2 macrospheres 0.3 to 0.4 mm in diameter or with saline alone for sham procedure, is inserted into the ECA stump through a puncture.
5. The tip of the tubing is placed in the carotid bifurcation without affecting the blood flow from the CCA to the ICA and fixed with a 5-0 suture.
6. The CCA and ICA ligatures are released and the position of the tip of the tubing is corrected, if necessary, to avoid a restriction of the blood flow in the CCA/ICA.
7. The macrospheres are advanced separately into the ICA by slow injection of approximately 0.05 mL saline each until they move passively into the cerebral circulation.
8. Following the macrosphere injection, the CCA is again temporarily ligated, and the ICA is carefully flushed with 0.5 mL saline; the carotid artery is observed. Excessive dilation of the ICA, indicating high flushing pressure, is to be avoided.
9. The tubing is removed, and the ECA stump is occluded with 5-0 suture.

This technique also can be done using PE-50 tubing, filled with saline and six macrospheres, placed into the ICA. Then, the macrospheres are injected into the internal artery to occlude the MCA.

STEP-BY-STEP SURGICAL PROCEDURES

Rats weighing between 270 and 400 g are chosen depending on the experiment at hand. It is recommended that the body weight be kept within a 15- to 25-g range per study. Rats are anesthetized with 2.5% to 3% isoflurane in a mixture of 70% N_2O/30% O_2 by inhalation.

1. The animal is placed on its back with head toward the surgeon's left side. The fur on the ventral neck area is shaved; the skin is cleaned locally with 0.5% Betadine and 75% alcohol or 2% chlorhexidine.
2. Eye cream is placed on both eyes to prevent drying. The neck is slightly hyperextended by placing a 10-cc syringe beneath the neck to aid in the exposure of the CCAs. The forelimbs are fixed on the table with sticky tape.

3. An incision is made in the ventral aspect of the neck. The sternohyoid muscle and sternomastoid muscles are blunt dissected to expose the underlying CCAs.
4. The right CCA, the right ECA, and the ICA are isolated via a midline incision. A 4-0 silk suture is loosely tied at the origin of the ECA and ligated at the distal end of the ECA. The ECA is superior with respect to the ICA with the animal in the supine position (see Figure 13.4 in Chapter 13).
5. Other distal branches from the ECA are the occipital artery and thyroid artery, which bifurcate at this level. In rats, the ICA distal to its origin from the CCA bifurcates near the skull base into the PPA, which irrigates the ear artery and the ICA proper, which traverses the skull base irrigating the brain and giving origin to the MCA.
6. The right CCA and ICA are temporarily clamped using a curved microvascular clip. The distal ECA at the level of the occipital and thyroid artery bifurcation is permanently ligated and divided (cut). The remaining ECA stump (0.5 cm long) originating from the CCA bifurcation is freed from surrounding tissues.
7. The ECA stump is retracted in the posterior direction and toward the surgeon's side to reduce its angle with respect to the ICA. Slight tension is exerted by pulling the suture at the ECA end (refer to Figures 13.4 and 13.5 in Chapter 13).
8. A small hole is cut in the ECA stump at the distal end.
9. A PE-50 tube filled with normal saline is inserted into the ECA stump and advanced into the ICA. The clip on the ICA is released while the tube is continuously inserted close to the carotid canal.
10. Embolic materials are injected while the CCA is temporarily occluded.
11. The PE-50 tube is withdrawn after injection, and the ECA stump is ligated as close as possible to the CCA bifurcation.
12. The incision is infiltrated with 0.1 mL 0.25% bupivacaine and closed with suture clips.

Antibiotic cream or 4% furazolidone aerosol powder is applied to the wounds to prevent infection. Anesthetics are discontinued, and the animals usually regain consciousness in 10 to 20 minutes.

PROBLEMS AND SOLUTIONS

Model Inconsistencies

Data inconsistency is the major problem in all embolic stroke models. However, with the advent of noninvasive imaging techniques, variability is less of a concern because pretreatment control recordings are obtained in each individual animal.

Incomplete Occlusion

Incomplete occlusion results in model inconsistency. A pilot study to test for suitable size and numbers of emboli helps to mitigate this problem.

Hyperthermia

Hyperthermia, the pathological increase in body temperature, is associated with hypothalamic injury caused by occlusion of the HTA as it originates from the distal ICA (refer to Chapter 13).

Subarachnoid Hemorrhage

Subarachnoid hemorrhage occurs in embolic stroke models as well as in the filament stroke model (refer to Chapter 13).

SURGICAL PROCEDURES IN MOUSE EMBOLIC MODELS

Here, we introduce the method of Zhang and Chopp's group (2005).[27]

1. Animals (male C57BL/6J mice, 2 to 3 months old) are anesthetized with 3% isoflurane and maintained with 1.0% isoflurane in 70% N_2O and 30% O_2 using a face mask.
2. Rectal temperature is maintained at 37°C throughout the surgical procedure by means of a feedback-regulated water heating system.
3. Under an operating microscope (Carl Zeiss), a midline incision is made in the neck, and the right CCA, the right ECA, and the right ICA are isolated and carefully separated from the adjacent vagus nerve.
4. A 6-0 silk suture is loosely tied at the origin of the ECA, and another suture is ligated at the distal end of the ECA.
5. The right CCA and ICA are temporarily clamped using a curved microvascular clip (Codman and Shurtleff).
6. The PE-50 catheter is modified to an outside diameter of 0.15 to 0.18 mm by stretching the catheter near a flame.
7. The catheter containing a single intact fibrin-rich clot 8 mm long is attached to a 100-µL Hamilton syringe and introduced into the ECA lumen through a small puncture.
8. The suture around the origin of the ECA is tightened around the intraluminal catheter to prevent bleeding, and the microvascular clip is removed. The catheter (8 mm long) is gently advanced from the ECA into the lumen of the ICA as far as the origin of the MCA.
9. The clot is gently injected with 100 µL of saline through the catheter.
10. The catheter is removed immediately after injection of the embolus.

The present model of embolic stroke using a clot 8 mm long is modified from that used previously (clot 20 mm long) to induce infarction encompassing the territory of the MCA.[28] To identify localization of an embolus after injection, the fibrin-rich clot is labeled by Evans blue before injection. Briefly, the embolus is placed into 2% Evans blue solution (Sigma Chemicals) for 10 minutes and then washed twice with saline (for detailed information, refer to Reference 18).

REFERENCES

1. Albers, G.W., Antithrombotic agents in cerebral ischemia, *Am J Cardiol*, 75, 348, 1995.
2. Sloan, M.A., Thrombolysis and stroke: Past and future, *Arch Neurol*, 44, 748, 1987.
3. Hamilton, M.G. et al., A comparison of intra-arterial and intravenous tissue-type plasminogen activator on autologous arterial emboli in the cerebral circulation of rabbits, *Stroke,* 25, 651, 1994.
4. Ringer, A.J., Guterman, L.R., and Hopkins, L.N., Site-specific thromboembolism: a novel animal model for stroke, *Am J Neuroradiol,* 25, 329, 2004.
5. Kito, G. et al., Experimental thromboembolic stroke in cynomolgus monkey, *J Neurosci Meth,* 105, 45, 2001.
6. Kudo, M. et al., An animal model of cerebral infarction, *Stroke,* 13, 505, 1982.
7. Omae, T. et al., Inapparent hemodynamic insufficiency exacerbates ischemic damage in a rat microembolic stroke model editorial comment, *Stroke*, 31, 2494, 2000.
8. Yang, Y. et al., A new reproducible focal cerebral ischemia model by introduction of polyvinylsiloxane into the middle cerebral artery: A comparison study, *J Neurosci Meth*, 118(2), 199, 2002.
9. Lauer, K.K. et al., Focal cerebral ischemia in rats produced by intracarotid embolization with viscous silicone, *Neurol Res*, 24, 181, 2002.
10. Zivin, J.A., A model for quantitative evaluation of embolic stroke, *Brain Res*, 435, 305, 1987.
11. Uchiyama, T. et al., Prostacyclin analogue TTC-909 reduces memory impairment in rats with cerebral embolism, *Pharmacol Biochem Behav,* 52, 555, 1995.
12. Zhang, R.L. et al., A new rat model of thrombotic focal cerebral ischemia, *J Cereb Blood Flow Metab*, 17, 123, 1997.

13. Yang, Y. et al., Lamotrigine reduces the level of neuroactive substances in rat middle cerebral artery permanent occlusion model of focal ischemia, *Can J Neurol Sci,* 24 Suppl. 1, S50, 1997.
14. Markgraf, C.G. et al., Comparative histopathologic consequences of photothrombotic occlusion of the distal middle cerebral artery in Sprague-Dawley and Wistar rats, *Stroke,* 24, 286, 1993.
15. Matsuno, H. et al., A simple and reproducible cerebral thrombosis model in rats induced by a photochemical reaction and the effects of a plasminogen activator chimera in this model, *J Pharmacol Toxicol Methods,* 29, 165, 1993.
16. Dietrich, W.D. et al., Microvascular and neuronal consequences of common carotid artery thrombosis and platelet embolization in rats, *J Neuropathol Exp Neurol,* 53, 351, 1993.
17. Kaneko, D., Nakamura, N., and Ogawa, T., Cerebral infarction in rats using homologous blood emboli: Development of a new experimental model, *Stroke,* 16, 76, 1985.
18. Zhang, R.L. et al., A rat model of focal embolic cerebral ischemia, *Brain Res,* 766, 83, 1997.
19. Busch, E., Kruger, K., and Hossmann, K.A., Improved model of thromboembolic stroke and rt-PA induced reperfusion in the rat, *Brain Res,* 778, 16, 1997.
20. Niessen, F. et al., Differences in clot preparation determine outcome of recombinant tissue plasminogen activator treatment in experimental thromboembolic stroke, *Stroke*, 34, 2019, 2003.
21. Figueroa, B.E. et al., Plasminogen activators potentiate thrombin-induced brain injury, *Stroke*, 29, 1202, 1998.
22. Miyake, K., Takeo, S., and Kaijihara, H., Sustained decrease in brain regional blood flow after microsphere embolism in rats, *Stroke,* 24, 415, 1993.
23. Naritomi, H., Experimental basis of multi-infarct dementia: memory impairments in rodent models of ischemia, *Alzheimer Dis Assoc Disord*, 5, 103, 1991.
24. Pearson, K.G., Ramirez, J.M., and Jiang, W., Entrainment of the locomotor rhythm by group Ib afferents from ankle extensor muscles in spinal cats, *Exp Brain Res*, 90, 557, 1992.
25. Fouad, K. and Bennett, D.J., Decerebration by global ischemic stroke in rats, *J Neurosci Meth,* 84, 131, 1998.
26. Gerriets, T. et al., The macrosphere model. Evaluation of a new stroke model for permanent middle cerebral artery occlusion in rats, *J Neurosci Meth*, 122, 201, 2003.
27. Zhang, Z.G. et al., Model of mini-embolic stroke offers measurements of the neurovascular unit response in the living mouse, *Stroke*, 36, 2701, 2005.
28. Zhang, Z. et al., A mouse model of embolic focal cerebral ischemia, *J Cereb Blood Flow Metab*, 17, 1081, 1997.

15 Photochemically Based Models of Focal Experimental Thrombotic Stroke in Rodents

Brant D. Watson and Ricardo Prado

CONTENTS

INTRODUCTION

More Realistic Models of Stroke Are Needed

Stroke in its hemorrhagic or more common thromboembolic form continues to be a vexing chronic disease for its survivors and their caregivers. For treatment of acute thromboembolic stroke in particular, only the recombinant tissue plasminogen activators (rt-PAs) have been approved by the U.S. Food and Drug Administration (FDA). These are known as thrombolytic (clot-dissolving) agents, but in practice, thrombolysis means fibrinolysis. However, the appropriateness of their use is compromised by fear of complications, the most significant being conversion of a thromboembolic stroke into a hemorrhagic one.[1] This can occur if the arterial occlusion that induces the stroke is more resistant to rt-PA-mediated lysis than are existing, fibrin-stabilized hemostatic plugs normally produced during hemostasis. Thus, in practice, the original intention of the rt-PAs to bind specifically to clot-bound plasminogen to avoid induction of a lytic state in the blood can be subverted. Further, the embolization of platelet-rich thrombi sequestered by fibrin is unavoidable with rt-PA use because the platelets are tightly bound to each other by rt-PA-resistant fibrinogen bridges. Finally, bound thrombin is exposed during embolus release, which can trigger rethrombosis via platelet deposition *in situ*. In compensation, inhibitors of thrombin and platelet activity have been utilized in clinical trials but have yet to demonstrate major benefit.

In addition to fundamental studies of cerebral ischemia, much effort in experimental stroke research has been expended on the development and testing of "neuroprotective" drugs, usually in rodents, to mitigate stroke damage. The underlying rationale is that the partially compromised region (penumbra) surrounding the irreversibly damaged "core" is amenable to treatment, but again, no drug that has been shown to reduce infarct volume in experimental animals has demonstrated efficacy in humans (at least 49 consecutive failures).[2] Several fundamental reasons accounting for this mismatch of drug to disease[2] are discussed extensively in Chapter 7, but others are proposed here:

1. Experimental "stroke," as simulated by the methods described elsewhere in this book, is usually induced by mechanical occlusion of a major cerebral artery by ligature or by filament, often in concert with extracranial arterial occlusion. These methods produce cerebral ischemia, but not stroke *per se*, because they circumvent active thrombotic processes such as release of platelet secretions in response to atheromatous plaque rupture. Such secretions can damage the blood–brain barrier at sites remote from the thrombosis[3] and enhance the severity of reperfusion injury compared to the mechanical approach.[4]
2. The ligature or filament models, if allowed to generate ischemia in the absence of recirculation, also generally produce large brain infarcts to ensure reproducibility and thus render them "suitable" for drug testing in surviving animals. However, this feature comes at the cost of suppressing the natural protective response of collateral channel activation and thus conduction of the drug into the compromised tissue. In effect, these models are intended to simulate stroke when the offending "thrombus" cannot be dissolved, but drugs have often been administered before the onset of ischemia to ensure some effect.
3. To resolve this conundrum, ligature and filament models are now used to emphasize neuroprotection in the context of reperfusion, as induced by removing the mechanical obstruction. A drug introduced in this context is expected to reduce the amount of ischemic damage developing in the metabolically compromised (penumbral) region surrounding the core of the developing infarct. Direct histopathological evidence for suppression of such "reperfusion injury" in which reversibly injured tissue is converted into a state of irreversible injury is still very difficult to obtain, however.[5] Instead, drug-mediated reduction of factors that are *believed* to contribute to reperfusion injury (e.g., oxygen radicals, lipid or protein peroxidation) is usually measured. Such indirect evidence muddles accurate assessment of the potential effectiveness of radical inhibitor drugs (and many others) in human stroke. Further, reperfusion injury by this pathway and related ones (e.g., glutamate toxicity) is of little active concern to neurointerventionalists, whose overriding mission is to restore, by any available means, even a small amount of cerebral circulation.
4. The structure of the human brain cortex is convoluted by gyri, so human strokes almost always involve gray matter intertwined with its white matter projections. But, the rodent brain cortex is smooth, and white matter does not intercalate with gray matter. Thus, most, if not all, neuroprotective drugs are (by definition) designed and tested to mitigate ischemic damage only in the gray matter of rodent brains. The gray/white matter volume ratios in rodent brains are much higher than that in humans (mice 90/10; rats 88/12; humans 60/40).[6] For this reason, any clinically positive effects of a rodent-certified, neuroprotective gray matter drug might be masked by nonresponsive white matter damage in humans.
5. Other experimental work is often based on embolic stroke models (see Chapter 14). The composition of the homologous blood clots used in these models is determined by the classical coagulation cascade, in which clotting of blood *ex vivo* often leads to a mixed-composition thrombus in which are distributed platelets, red blood cells, and fibrin. Their composition can be manipulated by incubation with agents such as thrombin; in some coagulation protocols, fibrin-rich nets can even be selected. Many studies have been conducted with such model emboli, but conclusions drawn from these studies regarding the efficacy of rt-PA may be unsuitable for clinical stroke, because the cellular structure and composition of thromboemboli extracted from cerebral arteries in living humans are *completely different*. For decades, such clots were believed with pontifical certainty to be "red" (cardiogenic, arising from atrial fibrillation) or "white" (atherosclerotic, from atheromatous

plaque rupture). But in fact, after their extraction from live patients and analysis, these clots were found recently to be *indistinguishable*.[7] They consist of alternating layers of platelets and fibrin (classically, the lines of Zahn), which often entrap pools of red cells.[7,8] Because the layered platelets are bound chiefly by intraplatelet *fibrinogen* bridges between GPIIb-IIIa receptors, the effectiveness of *fibrinolytic* agents in human stroke should be diminished. This is consistent with clinical observations inasmuch as intravenous injection of rt-PA in stroke is still not dependably efficacious.[1] To compensate, low-frequency ultrasound is being used to enhance the penetrability of intravenous rt-PA into occlusions,[9] with equivocal-to-undesirable results (hemorrhage). Frank emission of platelet emboli during reperfusion can be expected from clinical clots in light of their now-known structure but is usually not considered worrisome compared to the risk of no reperfusion at all.

6. Finally, in the likely interest of throughput, the vast majority of studies in animal models of focal cerebral ischemia or stroke are performed on nonrespirated animals, despite the fact that respirated animals yield more accurate and reproducible data.[10] Neuroprotective drugs shown to mitigate penumbral damage in animals have nonetheless been discovered, although without human benefit as yet.[2] The facts that (1) stroke victims are rarely ventilated and (2) penumbra-specific neuroprotective drugs derived from nonventilated animal studies are ineffective, appear congruent with the recent notion that the penumbra in humans does *not* contribute significantly to reperfusion injury.[11] In this light, recent efforts to improve perfusion *per se* (e.g., with albumin[12]) by directly treating the vascular compartment may translate to the clinic more successfully than will current efforts to instigate neuroprotection in the presence of deteriorating perfusion leading to delayed neuronal death.

Given what is now known about clinical clot composition, our aim here is to describe experimental stroke models that are intended to simulate, more accurately, the occlusion-initiating event in *human* stroke. This chapter describes methods and procedures (in sometimes excruciating detail) to address that purpose. Most of the studies done by us and mentioned here have utilized invasive surgical techniques and thus have been conducted in anesthetized, artificially ventilated rats. In addition, we describe the process of ultraviolet (UV) laser-facilitated dethrombosis,[5] a relatively unknown method of recanalizing occluded arteries, in which intraplatelet fibrinogen bonds are destabilized followed by platelet release. This process is complementary to thrombolysis.

Photothrombotic Approach to Stroke Induction in Rodents

To mimic active clot formation in human stroke, in 1985 we developed a rodent model of stroke based on a thrombus-producing photochemical method for occlusion of cerebral microvasculature.[13] This approach was pioneered by Rosenblum and el-Sabban,[14] who used the interaction of sodium fluorescein dye and blue light in mice to occlude isolated pial vessels by platelet microthrombus formation. The platelets were believed to aggregate in response to endothelial injury induced by heat transmitted to endothelium by the light-absorbing dye. But, this is unlikely because fluorescein very efficiently reemits the absorbed light energy by fluorescence, its namesake process. In contrast, we recognized that rose bengal (disodium tetrachlorotetraiodofluorescein) or erythrosin B (disodium tetraiodofluorescein) should be much more efficient in damaging endothelium than heat transfer because (1) although all these dyes bind to endothelium, much of the light energy absorbed by rose bengal or erythrosin B is transferred to its relatively long-lived lowest triplet state[15] owing to the internal heavy atom effect induced by the iodine atoms in their structures; (2) oxygen (as O_2) efficiently interacts with this metastable state and accepts its *electronic-state* energy to become excited-state (singlet) oxygen ($O_2{}^1\Delta_g$); (3) singlet oxygen directly interacts with endothelial lipids and proteins, peroxidizing them; and (4) the resultant endothelial damage specifically attracts and stimulates platelets to aggregate.[16,17] Such structural and functional damage to molecules describes

Photosensitizing dye (rose bengal, erythrosin B), injected i.v.

+

"Green" arc lamp or laser irradiation with 514 nm argon, 532 nm Nd:YAG, 543 nm HeNe, 562 nm argon/dye, 568 nm krypton

▼

Singlet oxygen ($O_2{}^1\Delta_g$) generation via dye triplet state energy transfer

▼

Peroxidative damage to endothelium and vasoconstriction

▼

Platelet adhesion and aggregation

▼

White thrombus formation and vascular occlusion

▼

Distal-territory ischemia (resulting in stroke)

FIGURE 15.1 Outline of the process of intravascular type II photothrombosis, a photochemical reaction for producing stable platelet thrombi to the point of occlusion. Only erythrosin B and rose bengal are capable of eliciting pure type II photothrombosis, that is, the formation of purely platelet thrombi in response to endothelial peroxidation by their sole reaction product, singlet molecular oxygen ($O_2{}^1\Delta_g$).

a photochemical (*photodynamic*) reaction, but for this platelet-stimulating reaction we invented the specific adjective *photothrombotic* or the noun, *photothrombosis* (Figure 15.1).

In our initial effort, we found that the photochemical interaction of intravenously injected rose bengal dye with filtered arc lamp light (530 to 590 nm) was able to facilitate platelet occlusion of small cortical vessels *en masse,* resulting in the first model of focal cortical thrombotic stroke.[13] Rose bengal-induced photothrombosis utilized 35 times less blood concentration interacting with 7 times less irradiation intensity compared to fluorescein.[13,14] The high concentration of fluorescein (ca. 8 m*M*) likely generated singlet oxygen also, but less efficiently.[15] Further, irradiation could be done through the intact skull overlying the cortical region of interest because the skull is translucent to visible light. This feature renders the model minimally invasive.

Soon thereafter, we found that 514.5 nm argon laser light could be used with rose bengal to occlude the rat middle cerebral artery (MCA) with a platelet-only thrombus,[18] yielding the first model of arterial thrombotic stroke. No intercalated fibrin is present,[4,5,13,16–18] only fibrinogen (Figure 15.2), which renders this thrombus resistant to tissue plasminogen activators (tPAs). Soon afterward, a model of thromboembolic stroke was characterized. By 514.5 nm argon laser irradiation of a carotid artery in rose bengal-injected rats, a consistent degree of stenosis could be produced in the form of a mural, continuously embolizing platelet thrombus.[19] A similar photochemical reaction mediated by erythrosin B dye was used later to create the ring model of stroke in evolution, in which a ring-shaped laser beam (514.5 nm argon) produced a congruent locus of ischemia engendered by small-vessel occlusion.[20] The ring lesion physically sequesters and gradually compromises the tissue *inside* this zone metabolically, thus reproducibly simulating the apparent dynamics of an evolving penumbra, in contrast to standard rodent models of focal cerebral ischemia in which a penumbra of uncertain extent lies outside the ischemic core.

It must be emphasized that photothrombosis of actively conducting blood vessels is a threshold effect that is driven by the rates of endothelial damage and the platelet response to it, and is thus dependent on the laser irradiation intensity interacting with the photosensitizing dye. The irradiation dose is relevant as a measure of the time taken to occlude the subject vasculature with platelets, but only if the beam intensity is sufficient to stimulate a rate of platelet adherence and degranulation sufficient to overcome dilution of proaggregant platelet secretions by flowing blood. Therein arises a caveat: High-intensity pulsed lasers based on Q-switching (532 nm Nd:YAG; YAG stands for yttrium aluminum garnet) or superradiance (337 nm nitrogen or nitrogen-pumped dye lasers) may be set at average intensities equivalent to continuous-wave lasers, but cannot occlude arteries

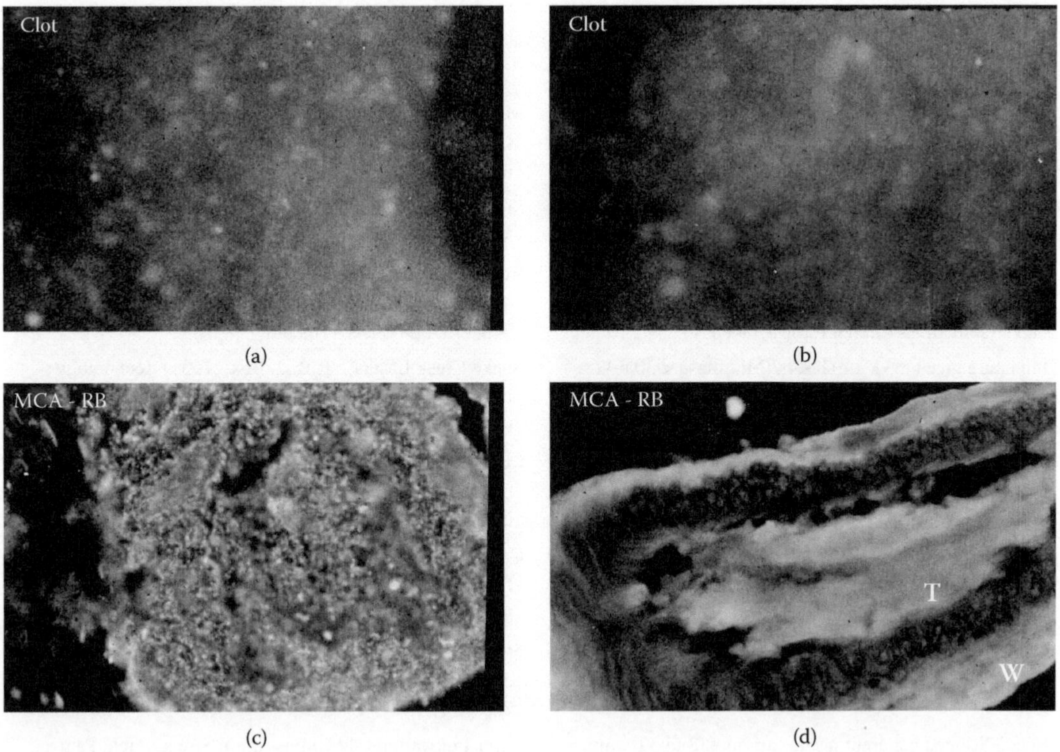

FIGURE 15.2 Immunofluorescent histochemical responses to blood clotted in air (a and b) and of platelet thrombi (c and d) formed in rat middle cerebral artery (MCA) in response to rose bengal dye-mediated photochemical damage instigated by argon 514.5 nm laser irradiation. Human fluorescein isothiocyanate (FITC)-conjugated antibodies to fibrinogen (a and c) and factor VIII (b and d) were used. No fibrinogen (A) or factor VIII (B) staining was seen for blood clotted in air, but both antibodies reported fluorescence far above this coagulation background when reacted with the respective MCA platelet thrombi (c and d), indicating little or no fibrin content. *Note:* T = thrombus inside arterial wall (W).

or arterioles because their 3- to 5-nanosecond pulse widths are much too short to sustain synergistic accumulation of platelet aggregation responses leading to occlusion. (However, if the focused laser intensity approaches the tissue breakdown threshold, coagulum can be formed.)

Just as important as simulation of thrombotic/embolic stroke is the development of dependable and specific methods for removing, with minimal side effects, cerebral arterial occlusions. While neurologists use mainly tPA or platelet inhibitors and anticoagulants, interventional neuroradiologists employ a wide array of mechanical devices in addition to tPA or other drugs. But, whether mechanically or pharmacologically based, nearly every current recanalization method will generate (or induce the generation of) clot fragments of different sizes. The mechanical methods usually do this on purpose by direct disruption, but so do the rt-PAs by dissolving fibrin concomitant with release of indissoluble platelet emboli, as does ultrasound either by itself or in combination with rt-PA.

To circumvent this inherently undesirable aspect, we have developed a unique photophysical method that specifically dissociates the platelet component of any thrombus to a degree sufficient to recanalize an occluded cerebral artery. This method, called UV laser-facilitated dethrombosis,[5] is administered by either direct (exovascular) irradiation, or nominal endovascular intervention (catheterization/optical fiber insertion) followed by UV laser irradiation of the arterial wall via the fiber. The negative consequences of the now-discredited technique of laser thrombolysis induced by photoacoustic shock, which depended on the presence of light-absorbing material in the clot (red blood cells), are avoided. In our method, only the vascular wall segment proximal to the thrombus is irradiated at relatively low intensity, and the platelet matrix in the nonirradiated adjacent thrombus

TABLE 15.1
Product and Manufacturer List (PML)

Erythrosin B (Cat. No. 190449), MW 880	MP Biomedicals (formerly ICN), 15 Morgan, Irvine, California 92618, 440-337-1200, 800-854-0530, www.mpbio.com
Rose bengal (Cat. No. 33,000-0), MW 1018 Flavin mononuclotide (Cat. No. 83810), MW 514	Sigma-Aldrich, Box 14508, St. Louis, Missouri 63178, 314-771-5765, www.sigmaaldrich.com
Laser goggles: E-G-LOTG-DYE II/CM for wavelengths 355 to 568 nm; for 532 nm YAG, E-G-LOTG-ARGON/KTP (sometimes offered for less on ebay)	Rockwell Laser Industries, 7754 Camargo Road, Box 43010, Cincinnati, Ohio 45243-0010, 513-271-1568, 800-945-2737, products@rli.com
Miniature green (532 nm) lasers (Microlase < 30 mW; Minilase < 150 mW)	Snake Creek Lasers™, R.R. 2, Box 2753, 1 Technology Drive, Hallstead, Pennsylvania 18822, 570-879-4992, www.snakecreeklasers.com
Compact green (532 nm) and yellow (556 nm) lasers	Laserglow.com, 5 Adrian Avenue, Toronto, Ontario, Canada M6N 5G4, 416-729-7976, http://laserglow.com/index.php?laboem
Compact (561 nm) yellow lasers; Compass®, 40 mW	Coherent, 5100 Patrick Henry Drive, Santa Clara, California 95054, 800-527-3786, 408-764-4983, tech.sales@Coherent.com
Compact yellow (561 nm) lasers (Model 85 YCA 025-115, 25 mW)	Melles-Griot Laser Group, 2051 Palomar Airport Road, 200, Carlsbad, California 92011, 800-645-2737, 760-438-2131
Reconditioned coherent lasers (argon 458 to 529 nm; krypton 568 nm; dye, tunable from visible to infrared, 351 or 364 nm UV with special mirrors)	Laser Innovations, 1150 East Main Street, Santa Paula, California 93060, 805-933-0015, www.laserinnovations.com
Optical components (lenses, optical rails, positioners)	Melles-Griot Optics Group, 55 Science Parkway, Rochester, New York 14620, 800-775-7558, 1-585-244-7220
Coherent LaserCheck power meter (Model U54-018, lowest price)	Edmund Optics, 101 East Gloucester Pike, Barrington, New Jersey 08007-1380, 800-363-1992

Note: MW, molecular weight.

dissolves, in the complete absence of forced mechanical effects (explosions). Our observations thus mimic the disaggregatory platelet response to thrombin inhibitors reported by the discoverers of dethrombosis, but we caution that their results were obtained in a crush-injury model of arterial thrombosis.[21] Such clots contain seepage channels that facilitate drug delivery,[22] unlike the tightly packed platelet thrombi formed photochemically.

In this chapter, we describe the basic equipment (listed in Table 15.1) and experimental techniques needed to incorporate these methods and models into the laboratory, often in terms of modern, less-expensive systems equivalent to ones we have used. Typical concentrations of dye used to effect photothrombosis with lasers of the appropriate wavelength are given in Table 15.2. Several quite common pitfalls that have arisen in applying these techniques are also described. Complications such as the ancillary production of heat during irradiation of target tissues are not widely appreciated but unnecessarily compromise the rigor of these models. When simulating a complex condition such as stroke, it is important to anticipate and minimize technical oversights that needlessly confuse the basic concepts or bias analysis of the results.

PRINCIPLES OF PHOTOTHROMBOSIS WITH LASERS

The following descriptions of our photochemically mediated techniques are in prose, not step-by-step as in other chapters of this book, with emphasis on the concepts utilized. Regardless of

TABLE 15.2
Summary of Laser Irradiation Parameters*

Vessel Diameter (μm)	Laser Wavelength (nm)	Laser Intensity (W/cm²)	Compatible Dye Concentration (mg/kg)
10 to 40	458 to 488	0.2	FMN (37)
180 to 240	458	15	FMN (>37)
10 to 40 (ring)	514.5	1	ErB (35)
180 to 240	514.5	16	RB (20)
700 to 1,400	514.5	160	RB (40)
700 to 1,400	514.5	50	ErB (35)
10 to 40	532	0.20 to 0.25	RB (10), ErB (12.5)
50 to 150	532	0.5 to 5.0 (N/A)	RB (10), ErB (12.5)
180 to 240	532	10 to 20 (N/A)	RB (20), ErB (25)
700 to 1,400	532	130 to 180 (N/A)	RB (40), ErB (N/A)
10 to 40	562	0.25	RB (10)
180 to 240	562	13	RB (20)
700 to 1,400	562	22	RB (40)

Notes: ErB, erythrosin B; FMN, flavin mononucleotide; N/A, not available; RB, rose bengal; W, watt.
* If 532 nm is replaced by 562 nm, then only RB can be used efficiently.

specificity or rationale, in our experience many users would rather develop their own procedures. A main reason seems the desire to use available equipment (such as arc lamps) as sources of excitation light rather than obtain newer, simpler, inexpensive, and far more efficient devices (such as lasers). Nonetheless, results have been obtained and published (but not without heat-related complications), likely owing to the intrinsic power of the photochemical approach for generating thrombosis *in vivo*. In contrast, lasers with a nonpulsed (continuous-wave) output are very much superior at stimulating reproducible photothrombosis while minimizing complications due to artifact, so we *emphatically* recommend their use in stroke models based on photothrombotic occlusion of cerebral vessels. Two of the five models involve occlusion of cortical microvessels (the "spot" focal model and the "ring" model); the third uses occlusion of an MCA; the fourth involves continuous formation of mural thrombi that embolize from a common carotid artery (CCA); and the fifth uses occlusion of a CCA by exo- or endovascular photosensitization.[16]

Photothrombosis is not confined to the production of platelet thrombi described previously, however. In the complementary process of type I photochemistry, the triplet state of the sensitizing dye acts like a free radical and thus can stimulate a very complicated array of reactions based on electron or hydrogen atom transfer.[16] We have used flavin mononucleotide (FMN) as a type I dye sensitized by blue argon laser light (458 to 488 nm) to obtain a mixed-composition, reddish, and tPA-sensitive thrombus in the rat MCA.[17] The principal drawback is that red mural thrombi are readily formed but then washed away by arterial pressure because fibrin cannot form quickly enough to stabilize them, so a stable occlusion can be obtained only at reduced blood pressure. Further, FMN operationally displays a minor type II component, which at higher laser intensities is sufficiently expressed to induce occlusion at the distal end of a red thrombus with a platelet-rich portion.[17] Unfortunately, no purely type I, biologically nontoxic photosensitizing substance seems to be available for photoproduction of red thrombi in arteries.

An interesting aspect of FMN photochemistry is that irradiation of rat cortex with an unfocussed blue (458 to 488 nm) laser beam specifically occludes venous-side microvessels, resulting in multiple petechial hemorrhages (that is, small-vessel hemorrhagic stroke).[17] This is the exact opposite of rose bengal- or erythrosin B-sensitized small cortical vessel occlusion, which occurs on the arteriolar side and does not result in hemorrhage. Another significant type I/type II difference is arterial dilation instead of constriction.[16] It remains unknown why peroxidative chain processes putatively sensitized by the type I photoreaction elicit such opposite vascular responses from those sensitized by direct type II photoperoxidation. However, superoxide anion and hydrogen peroxide are vasodilators[23] produced during a type I reaction,[16] which also produces tissue factor.[24] Photosensitizers used in photodynamic therapy for tumors, such as hematoporhyrin derivative (Visudyne) and chloroaluminum-sulfonated phthalocyanine, are claimed to produce singlet oxygen, but these result in unstable red thrombi, indicating that the type II pathway is *not* predominant. The *sine qua non* of a pure platelet thrombotic response is pure type II photochemistry, a condition satisfied only if mediated by erythrosin B or rose bengal. Purported susceptibility of a type II photothrombus to rt-PA is not possible unless fibrin is generated during irradiation, likely by excessive heating.[25]

Each of the five stroke models is readily expressed with a laser of suitable laser wavelength and irradiation intensity interacting with an appropriate intravenously injected type II photosensitizing dye. The type I process can be used to occlude small vessels and arteries less than 200 µm in diameter. A list of basic laser equipment and associated devices appears with manufacturers in Table 15.1.

The type II models are most often used to specifically generate platelet thrombi in capillaries, arterioles, and arteries, as desired, and if the technique is done correctly (without ancillary heat generation), the thrombi are fibrin free. These thrombi are extremely stable owing to strong intraplatelet fibrinogen bridges between GPIIb-IIIa receptors expressed on tightly intertwined platelet pseudopodia. This specific platelet response is activated by a type II photochemical reaction mediated by the sole photoproduct, singlet molecular oxygen,[26] which directly peroxidizes endothelial components and stimulates concomitant vasoconstriction.[4,5,15–18] The fluorescein dyes rose bengal and erythrosin B are the most efficient photochemical generators of singlet oxygen[15] and thus of type II photothrombosis. All the laser wavelengths listed in Figure 15.1 can excite rose bengal, but the most efficient wavelength is 562 nm, the absorption maximum of rose bengal in tissue.[27] A new laser operating at 556 nm (model LRS-556 from Laserglow, Table 15.1) should also work well with rose bengal but has not been tested. For excitation of erythrosin B (Figure 15.3), all the lasers listed can be used, but those at 556, 562, and 568 nm will be much less efficient.

Near-resonance matching of laser wavelength to dye absorption peak is very important to ensure maximal reaction efficiency and consistency of thrombus composition. To illustrate, after rose bengal injection, a 200-µm diameter MCA in rats can be occluded with platelets during low-power (several milliwatts) focused irradiation with either a 514.5 nm argon laser[18] or with a 562 nm argon/dye laser.[5] But, photothrombosis of a 1-mm diameter CCA with the same 514.5 nm beam at 1 W produces an occlusion containing some red blood cells and platelets in a milieu of heat-denatured protein.[17] Such a thrombus is inappropriate for modeling stroke. In stark contrast, photothrombosis with 562 nm laser irradiation produces CCA occlusion by a large platelet aggregate in much less time and with no unphysiological temperature increase.[17] Recovery, in a large-diameter artery, of this fundamental platelet response to type II peroxidative endothelial injury illustrates the importance of resonance matching of laser wavelength to dye absorption maximum. At first glance, this condition would not seem to be so important because the absorbance of rose bengal at 514.5 nm is at least moderate (52%) compared to that at 562 nm. But, the first response of platelets is to adhere to the damaged endothelium and then to rapidly degranulate, secreting proaggregatory chemicals such as thromboxanes and serotonin. While the first response is a linear function of focused laser beam intensity, the second is certainly nonlinear, in the manner of an exponentiating chain reaction. Obviously, the rate of the aggregatory response must be maximized to induce occlusion of free-flowing arteries just by platelets. Platelets entering the irradiation field can respond to aggregatory stimuli if they are not photooxidized.[28] Type II photoperoxidative damage thus converts a nonthrombogenic surface (endothelium) into a thrombogenic one and achieves the reverse for platelets.

Note that 532 nm, the wavelength of the frequency-doubled Nd:YAG laser, is very close to the protein-bound erythrosin B absorption maximum at 537 nm (Figure 15.3). Owing to the wide availability of such DPSS (diode-pumped solid-state) lasers now, we predict that this laser/dye

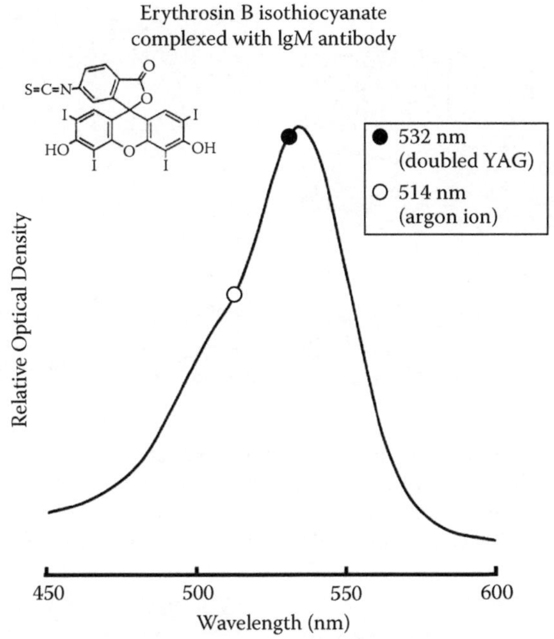

FIGURE 15.3 Red-shifted absorption (excitation) spectrum of protein-conjugated erythrosin B in water at two common laser wavelengths. While the efficiency of platelet adhesion response to endothelial injury is apparently a linear function of endothelial-bound dye absorbance, the self-sensitized aggregation response is a highly nonlinear process (in three dimensions). Accordingly, the efficiency of the platelet response increases rapidly toward the dye maximum at 537 nm, and thus 532 nm will elicit a far more avid platelet response with erythrosin B than will 514 nm.

couple affords the least-expensive and most effective way to achieve photothrombosis of any cerebral artery in the rat. Relatively low-power 532 nm lasers (Figure 15.4) can easily produce the photothrombotic cortical spot (Figure 15.5) and ring models and occlude rat arteries larger in diameter than the MCA (Figure 15.6). However, more focused power is needed to embolize or occlude

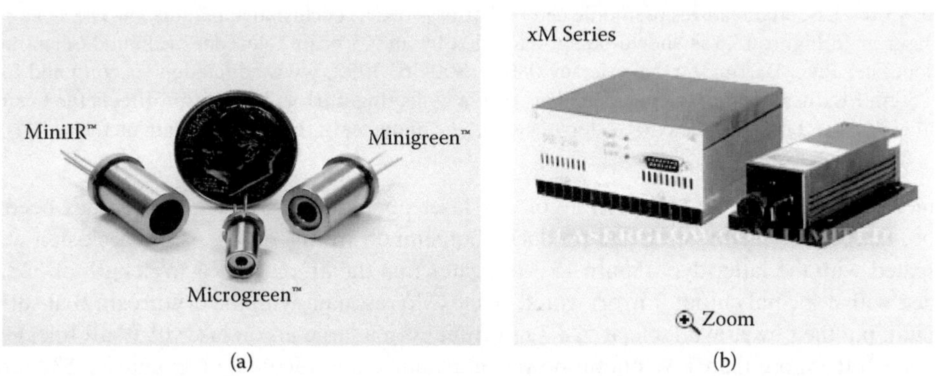

FIGURE 15.4 Examples of diode-pumped solid-state (DPSS) lasers suitable for photothrombosis. (a) Small, frequency-doubled Nd:YAG (yttrium aluminum garnet) continuous-wave lasers operating at 532 nm (MicroGreen, up to 30 mW; MiniGreen, up to 200 mW; the 1064 nm MiniIR is not intended for this purpose). Courtesy of Snake Creek Lasers. The output powers can be varied by a current-controlled power supply. (b) Continuous-wave, compact, 532 nm Nd:YAG (model LLS-532, Laserglow) operating at up to 8 W, which can be used for photothrombotic occlusion of any cerebral artery in the rat (the LCS-532 is suitable for middle cerebral artery [MCA] occlusion or possibly common carotid artery [CCA] embolization).

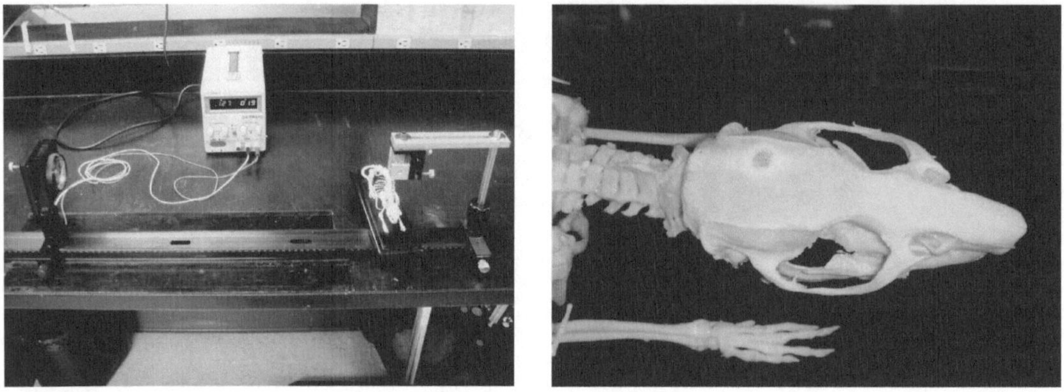

(a) (b)

FIGURE 15.5 Essential features of cortical spot lesioning apparatus. (a) A MicroGreen laser (Snake Creek) is shown illuminating a rat skull (b) after reflection from an elliptical mirror set at 5° from vertical incidence. The beam from this laser strongly diverges, so a spot 5 mm in diameter can be formed just by positioning the laser on the optical rail about 0.7 meters away from the focus. **(See color insert following page 146.)**

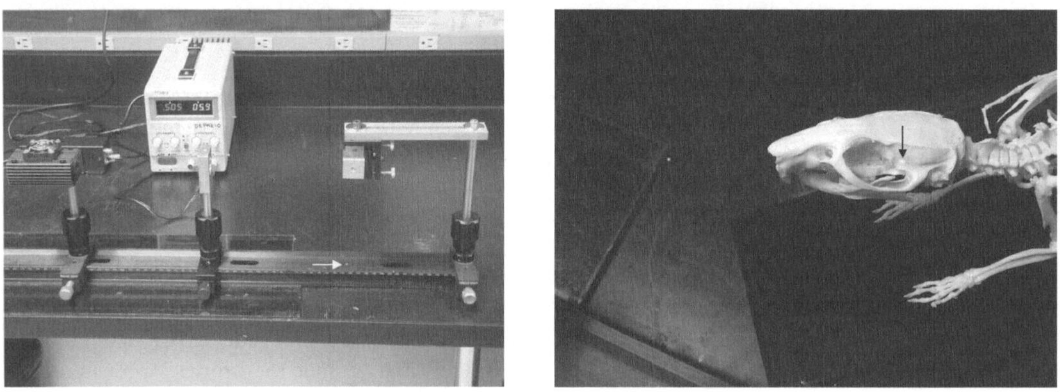

(a) (b)

FIGURE 15.6 Essential features of middle cerebral artery (MCA) occlusion apparatus. (a) The same MicroGreen laser as in Figure 15.5 is shown being magnified by an X5 beam telescope (Edmund Scientific, 101 East Gloucester Pike, Barrington, New Jersey 08007, 800-363-1992, www.edmundoptics.com) and focused with a 25-cm FL (focal length) planoconvex lens onto a deflecting mirror. The mirror directs the beam onto the skull region overlying the distal MCA territory (b). The diameter of the focused beam on the skull is about 200 μm. **(See color insert.)**

a 1-mm CCA in rat (Figure 15.7). An argon ion laser operating at 514.5 nm often has been used with rose bengal[19] or erythrosin B[29] to produce photoemboli from a rat CCA, but occlusion was not investigated with the latter dye. Figure 15.3 indicates that the argon laser wavelength of 528.7 nm (obtained with a special output mirror) would be close to resonant with the erythrosin B absorbance maximum, but the power available at 528.7 nm from even a large argon laser (9 W all lines) would not approach the more than 1 W output power of available and relatively inexpensive 532 nm systems (Table 15.1).

Most of our arterial occlusion research over the past 20 years has utilized an argon-pumped/dye laser operating at 562 nm (Figure 15.7) in conjunction with rose bengal. A krypton laser operating at 568 nm has been used to photothrombose rat arteries efficiently with rose bengal (the 90-K version of the argon laser in Figure 15.7 produces 150 mW maximum at 568 nm).[30,31] A frequency-doubled DPSS Nd:VO4 at 556 nm and Nd:YAG lasers at 561 nm are also available but are moderately expensive compared to a 532 nm Nd:YAG at equivalent powers (Table 15.1).

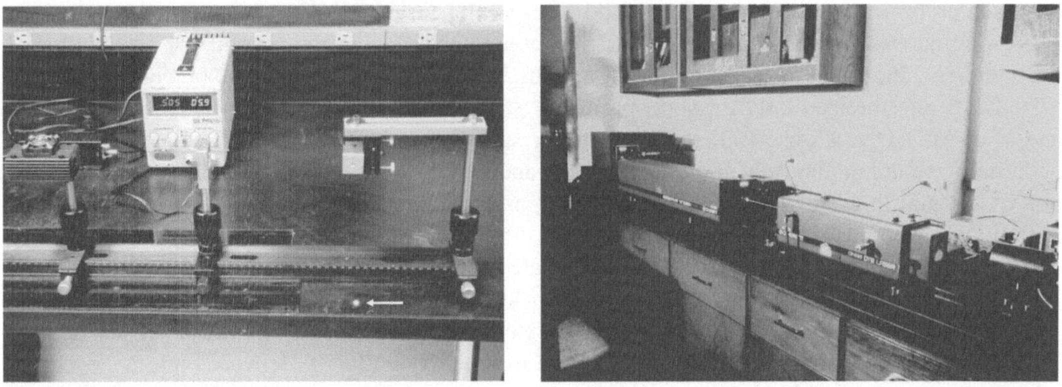

(a) (b)

FIGURE 15.7 Common carotid artery occlusion systems. (a) The beam from a compact 532-nm Nd:YAG (yttrium aluminum garnet) laser (at left; 100 mW, model LAGR100M, manufactured by Laserglow and sold by Information Unlimited, Amherst, New Hampshire) is shown traversing a 61-cm FL planoconvex lens (at center, in front of power supply) and internally reflecting from a right-angle prism (at right). The beam optics of the more powerful 1-W Laserglow model LLS-532 in Figure 15.4b are the same. The beam appears as a 1-mm diameter green spot on the optical rail below, on which the rat can be placed. (b) A model 70-4 argon ion laser (left) coupled to a CR599 dye laser (right) (Coherent, Fremont, California). This argon laser is rated at 4 W for all lines, but with a new plasma tube usually produces about 6 W. The dye laser (rhodamine 560) emits at least 1 W at 562 nm. These lasers are quite hardy but require external water cooling, regular tuning, and maintenance such as cleaning the dye jet and pumping system, replacing the dye solution periodically, and cleaning the many optical surfaces. **(See color insert.)**

Eliminating thrombi by means of dethrombosis requires UV lasers, which are also expensive (>$15,000) at this time. If a suitable argon laser is available (Figure 15.7 and Table 15.1), special mirrors can be obtained to operate it at 351 nm at a power above 200 mW. The beam can be focused exovascularly directly onto the occluded artery or be conducted endovascularly via a fused silica optical fiber ensheathed by a microcatheter (see sections on CCA photothrombotic embolization and occlusion and recanalization by UV laser-facilitated dethrombosis). Otherwise, a frequency-tripled Q-switched Nd:YAG laser operating at 355 nm can be obtained for exovascular use but cannot be readily used endovascularly except under special conditions (as discussed in the same sections).

GENERAL PREPARATION OF ANESTHETIZED, INTUBATED, AND ARTIFICIALLY VENTILATED ANIMALS

To obtain consistent results from surgically invasive models of cerebral ischemia or stroke, blood gas levels must be maintained by means of endotracheal intubation and artificial respiration during the required anesthesia because general anesthesia depresses the respiratory drive. The primary emphasis of artificial respiration is thus to keep blood gases within normal limits and constant across animal groups. Otherwise, CO_2 will accumulate, artifactually dilating cerebral arteries, while the O_2 level will decrease, compromising the animal metabolically. In the extreme case of cerebral ischemia induced by temporary cardiac arrest, it is obvious that the animal, if not artificially respirated, will be severely if not lethally affected by this imposition of essentially whole body ischemia (see Chapter 16). But, gas anesthetics can themselves confound outcome because they are not only neuroprotective but also might lower blood pressure (notoriously lethal in guinea pigs). The most realistic and accurate models would utilize minimal to no anesthesia during ischemia, but this really cannot be done in the context of invasive surgery.

The effects of uncontrolled blood gas levels on physiologic homeostasis are extensive and, in a given context, could ameliorate or exacerbate outcome. For example, in early investigations, nitric oxide (NO) synthase inhibition by L-nitroarginine was reported to reduce infarct volume in a model of focal cerebral ischemia induced by middle cerebral artery occlusion (MCAO) in anesthetized nonrespirated rats.[32] In contrast, L-NAME, a water-soluble nitroarginine inhibitor of endothelial NO synthase, increased infarct volume in a suture model of MCAO in spontaneously breathing Wistar rats anesthetized only for short periods.[33] A severe degree of blood flow reduction during and after MCAO was observed, involving frank stasis of collateral vessels. Later work in mice undergoing suture model ischemia indicated that endothelial NO synthesis was critical to preservation of vascular integrity and cerebral blood flow, and that *neuronal* NO synthesis was toxic.[34] The initially reported positive effect of NO synthase inhibition in nonventilated rats was likely due to constriction of cerebral vessels that had been dilated by retained CO_2, resulting in less edema and infarct volume, thus accounting for the observed pseudoprotection.[35]

To prepare animals for artificial ventilation (requiring intubation), rats are first anesthetized with 4% isoflurane and a mixture of oxygen and nitrous oxide (30/70) delivered into a closed jar. After several minutes, the subject rat should be sufficiently unconscious to be withdrawn and placed supine with its head over the table edge. The tongue is gently withdrawn with the fingers, and then intubation is begun. A guide wire made from a blunt stylet inserted into a close-fitting plastic tube (Figure 15.8) is gently introduced over the tongue into the trachea; the device is curved to slide into the trachea smoothly over the esophagus.[36] An otoscope may be used to visualize the procedure directly if the model involves acute respiratory distress, as induced, for example, by temporary cardiac arrest (Chapter 16). The stylet must be kept aligned with the axis of the rat as much as possible and inserted (quickly) during the inspiration cycle to avoid laryngospasm (closure of the vocal cords). If this is not timed correctly and the device is misaligned, resistance due to stiffened throat muscles will be felt. Successful insertion is indicated by sensing the tracheal rings and can be confirmed by observing breath condensation on a shiny metal surface (for example, a forceps handle) held at the hub of the tube. If the rat is *not* sufficiently anesthetized, introduction of the

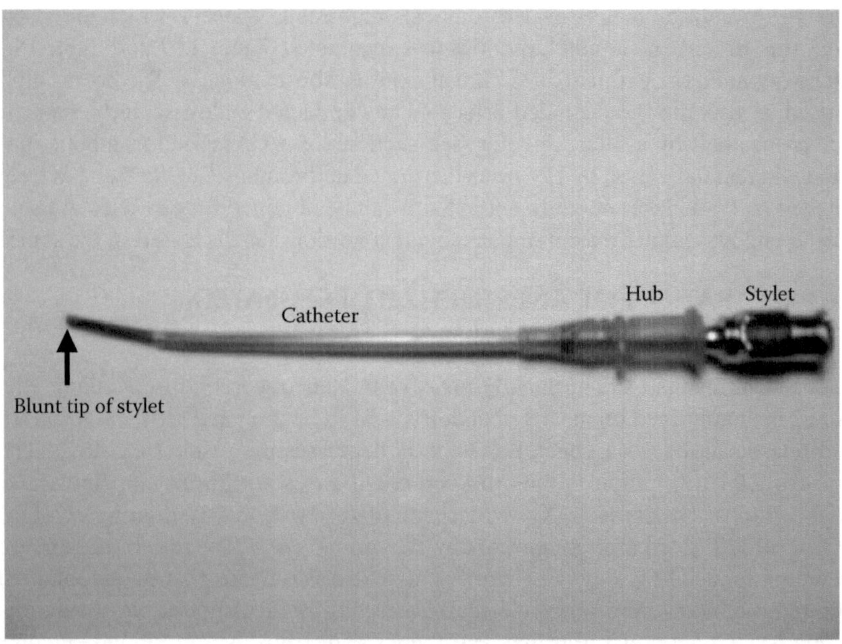

FIGURE 15.8 Blunt stylet acting as a guide wire for insertion of tracheotomy tube. The blunt end of the stylet (Popper and Sons, New Hyde Park, New York) is bent smoothly at about 10° and enveloped with a 14-gauge catheter (2.1 × 45 mm) capped with a plastic hub (extracted from a BD Insyte Autoguard intravenous catheter (REF 381467, Becton Dickinson Infusion Therapy Systems, Sandy, Utah). With the rat supine, the stylet/tube unit is inserted in the depicted orientation until the hub of the inserted catheter is flush with the mouth, and the stylet is then withdrawn.

stylet into the back of the throat, which is encountered first, will induce a gag reflex, so it is best to make certain the rat is deeply anesthetized before advancing the stylet farther. The insertion procedure is usually accomplished "blind" as described and takes some practice to acquire the necessary skill. Abortive attempts are limited to one or two before the onset of laryngospasm, which would make intubation difficult and traumatic to the rat. If this happens, the animal should be deeply anesthetized before attempting to reintubate. If the intubation is done but is traumatic nonetheless, supplemental humidified oxygen, atropine to minimize secretions, and mild suctioning must be administered in the postinsertion period.

After removing the stylet, the hub of the intubation tube is attached to a T- or Y-tube by Tygon tubing connected to the output and input ports of a small animal respirator (Chapter 16). The rat is artificially ventilated at a respiration rate of 60 breaths per minute with 1.5% ± 0.5% isoflurane and 70% NO_2/O_2. Pancuronium bromide is given intravenously at an hourly rate of 0.35 mg/kg to ensure that the ventilator completely controls respiration. To avoid atelectasis, a slightly lower-than-normal respiratory rate (60 breaths per minute) and higher-than-normal tidal volume are selected. Rectal temperature, maintained at 37°C, is measured with a thermocouple connected to a temperature-regulating device (Model 150, CMA), which actuates a heating pad beneath the animal. Head temperature is monitored and regulated by placing a needle probe into the temporalis muscle to maintain temperature between 36°C and 37°C via a servo-controlled, filtered, high-intensity lamp placed 15 cm from the rat's head. We have previously established that temporalis muscle temperature is an adequate indirect indicator of brain temperature and its effects on ischemic damage.[37]

Femoral venous and tail artery catheters (PE-50) are prepared for fluid administration and monitoring of physiological parameters. With the animal supine, an oblique skin incision is made in the right inguinal region and the underlying fascia dissected between the inguinal ligament and the superficial epigastric artery to expose the right femoral vein, artery, and nerve, which are then carefully dissected free from each other. The right femoral artery and vein are enveloped with 3-0 silk sutures, and a 2-cm segment is identified in each for catheter placement. Starting with the venous side, the distal (proximal) end of the venous (arterial) segment is temporarily occluded by gently suspending needle holders from the clamped suture ends, while the respective proximal and distal ends are suture occluded permanently. A small transverse incision is then made in the middle of each segment by means of microscissors and the analysis catheters inserted. The temporary sutures are then loosened to accept the respective catheters, which are secured inside the vessels by tying the sutures around them. Arterial blood samples are drawn at frequent intervals and analyzed for PCO_2 and PO_2 in a Radiometer acid-base analyzer (ABL-50). Readings are corrected automatically for any deviation of body temperature from 37.0°C. Arterial PCO_2 is maintained at 35 to 40 mm Hg. Arterial PO_2 is maintained at or above 110 mm Hg. Arterial blood pressure is monitored with a pressure transducer (Statham) via an arterial catheter and interfaced to a personal computer through an analog-to-digital converter, with a polygraphic recording system as backup.

For a given laser, a stock solution of the compatible photosensitizing dye is prepared (Table 15.2). If the solution has been made up as *X* mg/mL, and if the solution is injected at 1 μL/g rat weight, the dose will conveniently appear as *X* mg/kg. These dyes (Tables 15.1 and 15.2) are highly soluble in doubly distilled water but not necessarily in saline owing to precipitation of variable amounts of salts along with the dye during its manufacture. Salt (as sodium chloride) must be added afterward, if desired. Rose bengal and erythrosin B (but not FMN) are stable for months in ambient light, but to ensure uniformity, shielding by brown bottle or aluminum foil is recommended. To prevent thrombosis in an infusion line due to retrograde blood stagnation, the lines should be monitored visually and flushed with saline as needed. Dyes are best administered into a femoral vein catheter (tail vein optional) by an infusion pump set at a constant rate; in particular, rose bengal cannot be delivered manually without decreasing blood pressure. The plasma half-lives of erythrosin B and rose bengal are only a few minutes,[38] so the irradiation period usually need not be more than 10 minutes per single injection. Depending on the diameter of vessels to be occluded, the dye solutions range from 10 to 40 mg/mL for rose bengal, from 12.5 to 35 mg/mL for erythrosin B, and 37 mg/mL for FMN, all infused as discussed (Table 15.2). The resultant doses in milligrams per kilogram are somewhat arbitrary in that all will produce the effect desired if the irradiation intensity is sufficient but not needlessly excessive. While FMN is quickly eliminated through the urine, the fluorescein dyes are taken up by the liver and eliminated through the feces.

Finally, suitable laser goggles must be on hand. Just a few types will serve for all the lasers described here (Table 15.1). One must be aware, however, that if visible laser power approaches the watt level and the irradiation is being observed through an operating microscope (for example, during irradiation of a CCA), the laser light will be strongly scattered by the forming platelet thrombus. Specular reflections from metal or water surfaces are also dangerous, and this possibility must be checked before beginning irradiation at high power.

CORTICAL STROKE IN RAT BY PHOTOTHROMBOTIC OCCLUSION OF MICROVESSELS

This essentially noninvasive method of cortical stroke by photothrombotic occlusion of microvessels generates a discrete, reproducible zone of evolving infarction within any preselected region of the rat (or mouse) neocortical convexity. The results of histopathological, morphological, and rheological studies based on this model of cortical microvascular occlusion have been described in hundreds of publications from many laboratories[16] since our initial study.[13] Although this method generates a thrombotic stroke, it has no direct clinical counterpart because occlusion is observed mainly in small cortical vessels (smaller than ca. 40 μm), not in a major artery or branch. Nonetheless, in this Type II photochemically induced focal cerebral ischemia model, "MRI changes quantitatively reflect histopathology, revealing reproducible primary and secondary damage characteristics noninvasively, which essentially replicate those reported for other animal stroke models and clinically."[39]

After initial preparation, the animal is placed prone in a metal frame (stereotaxic if such precision is required) and its head secured, usually by ear bars. The scalp is reflected and the skull surface exposed and positioned for laser irradiation at perpendicular incidence to the skull (Figure 15.5). The skull surface over the cortical region selected for lesioning may be fitted with a piece of brass shim to outline the desired cortical region (or just delimit a "spot" beam). This is attached to the overlying calvarium, on which a drop of mineral oil is placed to present a smoother optical surface and thus reduce surface scattering; the lesion edges will nonetheless appear quite well defined. Because the skull is translucent (50% visible transmission per millimeter thickness), photochemically effective light intensities are conducted to the cortical surface, rendering craniotomy unnecessary.

A distillation of the many published studies leads us to suggest the following steps as suitable for beginning this procedure. An aliquot of a 10 (12.5) mg/mL solution of rose bengal (erythrosin B) is then infused over an interval of 1.5 (1.0) minute at 1 μL/g rat weight via a femoral vein. After 30 seconds of infusion, irradiation with the proper laser may commence at an intensity of 200 mW/cm^2. At this time, the most effective but least expensive lasers operate at 532 nm and are power tunable, such as a 30-mW MicroGreen or 100-mW MiniGreen from Snake Creek Lasers (Table 15.1). A 532 nm laser can be used with either dye; a 556 nm or 561 nm yellow laser (Table 15.1) can be obtained for resonant excitation of rose bengal but at considerably more expense. Irradiation sufficient to obtain a complete cortical lesion may require as little as 3 minutes.[40] Assuming that animal preparation is properly done, establishment of irradiation conditions leading to reproducible lesions should be attainable with five to seven animals. The advantage of using a laser wavelength close to the dye absorption maximum is that no external cooling of the irradiated skull is needed at the recommended intensity. In contrast, irradiation with even an optically filtered arc lamp does require cooling because its light on average is less photochemically effective, and thus more beam intensity is required.[13,16,17]

Although platelet occlusion of the microvasculature in this model is usually obvious, it is actually an epiphenomenon. The lesion is mainly induced by endothelial leakage (that is, blood–brain barrier breakdown), from which arises vasogenic edema and microvascular compression.[41,42] This fundamental aspect can be appreciated quite directly if the brain is cooled to 34°C during irradiation; an identical (at first glance) lesion will form in the complete absence of platelet occlusions.[41] On the other hand, we found that edema could be minimized by allowing the laser beam to graze the skull surface at 13° incidence from horizontal. A cylindrical lens focused the beam into a horizontal line parallel to the plane of the skull and then "skimmed" the cortical surface. At 37°C, this resulted in occlusion of pial vasculature, stasis of blood in noncompressed pial projections throughout the cortical thickness, and formation of a static (nonswollen) lesion.[43]

An alternative to the cortical spot model is the ring model, in which a ring-shaped laser beam photochemically creates an annular cortical lesion, and the region inside the thin annulus reproducibly undergoes stroke in evolution in similar fashion, it is presumed, to a classical penumbra (as described in the section, "Photothrombotic Approach to Stroke Induction in Rodents").[20] The ring beam was produced by introducing a 514.5-nm argon laser beam into an optical fiber at 10.8° off axis (which was also the beam's radial exit angle), where it was focused onto the skull of rats injected with erythrosin B (17 mg/kg) and irradiated at an intensity of about 0.9 W/cm^2. Vasogenic edema, radially propagating from the ischemic locus, evidently was responsible for lesion development. Later, Gu, Wester, and colleagues showed that a thinner ring beam (entering same fiber at 29° and exiting it radially at the same angle)[44] could display the same initial hypoperfusion in the cortical region at risk either to the point of histologically verified ischemic death or, with half the irradiation intensity, a remarkable spontaneous recovery of reperfusion after 2 days, along with near-complete morphologic reversal of tissue injury, including the growth of new neurons.[45] Such rapid progression and then reversal of edema is quite unusual and suggests that a new type of preconditioning has been evinced in this case of sublethal photochemically facilitated injury.

Both the spot and ring cortical models can be produced by photosensitization of FMN (see the discussion of "Principles of Photothrombosis with Lasers" and Reference 17) to produce petechiae on the cortical surface following venular occlusion and hemorrhage. However, the composition of the putative venular thrombi and the effect of temperature on lesion development (that is, whether venular edema exists and whether it determines lesion extent at lower temperatures) are not known. These topics remain open for investigation.

MIDDLE CEREBRAL ARTERY (MCA) PHOTOTHROMBOSIS AS MEDIATED BY INTRAVASCULAR PHOTOCHEMISTRY *IN SITU*

Occlusion of the MCA is widely regarded as the most clinically relevant animal model of stroke, but if instigated mechanically (by filament or ligation) it more accurately induces cerebral ischemia instead because occlusive thrombi are not produced intentionally. (Platelets can accumulate, however, if the endothelium is damaged on suture withdrawal.) The alternative technique of embolus implantation also induces downstream ischemia, but none of the several choices of experimental mixed-composition clot structure reproduces the layered structure of clots extracted from stroke patients (see points 4 and 5 in the "Introduction"). Clots extracted from living patients contain a significant fraction (20% to 40%) of cross-linked platelets,[8] which in their alternating sheeted configuration act as barriers to rt-PA transport. This situation cannot be mimicked accurately with the mechanically induced ischemia models or by injection of externally coagulated thrombi. Alternatively, we maintain that producing an occlusive mass of fibrinogen cross-linked platelets in extra- or intracerebral arteries by means of type II photothrombosis fairly models the most rt-PA-refractory component of a stroke-inducing occlusion. If the platelet sheets can be removed selectively (see the next section), perfusion of any thrombus will be facilitated, and the efficacy of rt-PA, if needed, will be much enhanced.

Animals are initially prepared as discussed and then the distal MCA territory is surgically exposed, preferably under magnification with an operating microscope. A vertical skin incision is made between the lateral canthus of the eye and the ear, the skin is retracted with silk sutures, and the underlying temporal muscle is incised and retracted ventrally. The distal MCA territory (left or right side) is exposed by means of a small craniectomy (ca. 3 mm^2) in the frontal-squamosal bone ventral to the coronal suture and above the zygomatic arch. For a period of 30 minutes to 1 hour before laser irradiation, the animal is monitored physiologically until stabilized, and all wounds are infiltrated with 1% xylocaine on closure.

Photothrombotic occlusion of the distal MCA is then instigated. Based on our knowledge of this process, it is no longer necessary to utilize an argon laser or an argon-pumped dye laser (Figure 15.7) for this purpose because a 15-mW MicroGreen or Laserglow model LCS-532 (Figure 15.4) should be quite sufficient in conjunction with erythrosin B (or rose bengal) dye injected at 25 (or 20) mg/kg. The beam is first focused at very low power through a 25-cm focal length (f.l.) BK7 antireflection-coated spherical glass lens and then reflected onto the chosen distal MCA segment of diameter D at perpendicular incidence to the arterial segment (Figure 15.6). This low-intensity beam is positioned to just cover the segment diameter (it will be necessary to remove the laser goggles to see the beam on the artery). The beam power P is then calculated to provide an intensity I of about 13 W/cm^2 from the simple formula I (W/cm^2) = 12.73 P (mW) /D (100 μm)^2, where D is in multiples of 100 μm (usually between 1.8 and 2.4 for the typical distal MCA diameter range of 180 to 240 μm). For a 200-μm diameter segment ($D = 2$), we calculate that $P = 4$ mW.

As the laser power is increased to the desired level, an intense straw yellow (orange) fluorescence owing to excitation of the circulating erythrosin B (rose bengal) is visible through the corresponding laser goggles. This fluorescence indicates the true position of the beam on the arterial segment, and the micrometer-controlled deflecting mirror must be adjusted to ensure that the segment fluoresces symmetrically across its diameter. The onset of thrombosis is presaged by vasoconstriction (Figure 15.9). Fluorescence from the circulating dye decreases during irradiation owing to the formation of a white, platelet-specific thrombus concomitant with exclusion of dye-bound plasma from the irradiated arterial segment. Stable occlusion is marked by loss of dye fluorescence and its replacement by backscattered laser light and should occur within 4 minutes, although the onset of

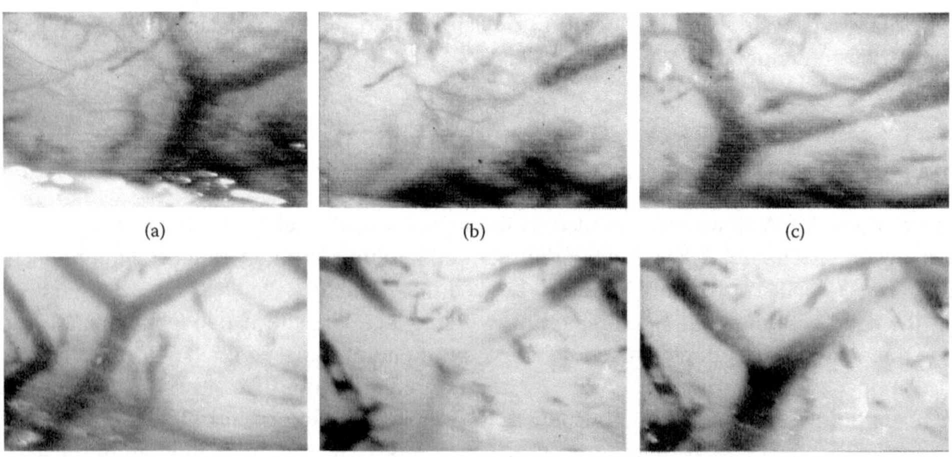

(a) (b) (c) (d) (e) (f)

FIGURE 15.9 Photothrombotic occlusion of rat distal middle cerebral artery (MCA) and its recanalization by dethrombosis (cf. p. 156). (a) Baseline appearance of rat MCA; (b) 120 minutes after rose bengal-mediated photothrombosis with 562 nm argon/dye laser irradiation at 13 W/cm^2; (c) recanalized MCA at 150 minutes after treatment with 10 W/cm^2 of ultraviolet (UV) laser irradiation (355 nm Nd:YAG [yttrium aluminum garnet]); (d) baseline appearance of MCA in another rat; (e) 180 minutes after photothrombosis; (f) 210 minutes after treatment with 10 W/cm^2 of UV laser irradiation (355 nm Nd:YAG).

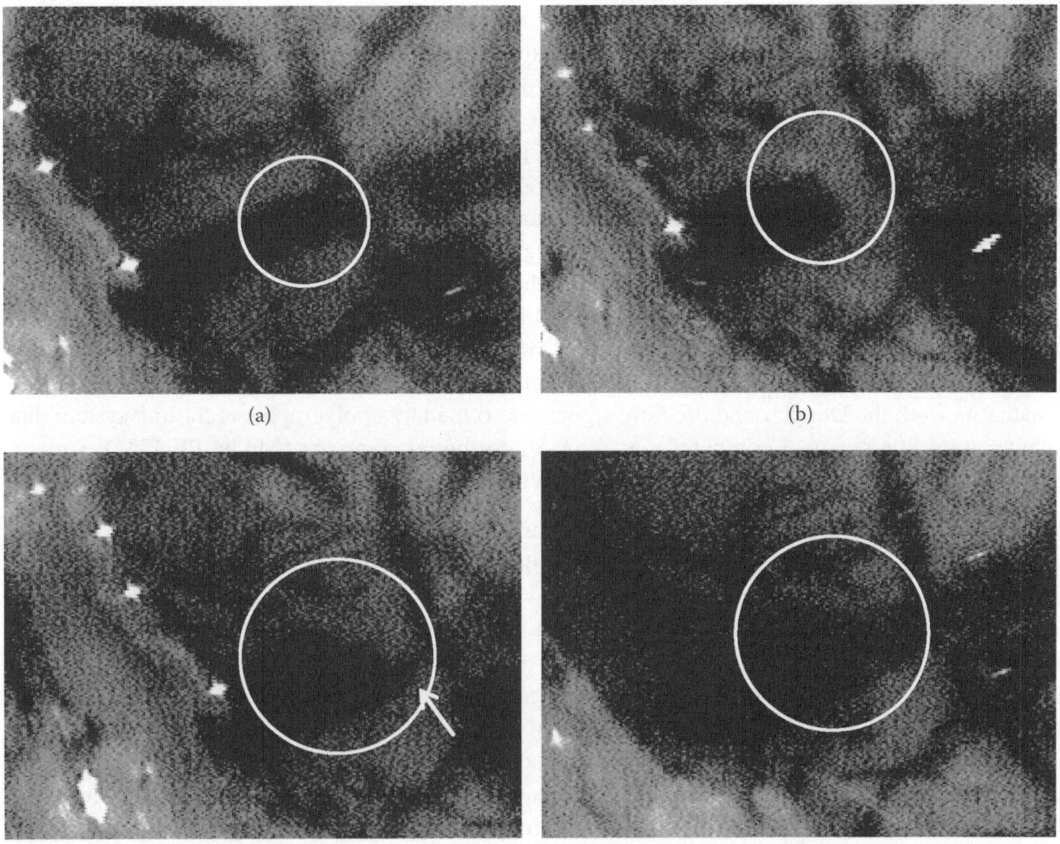

FIGURE 15.10 Type I photochemical occlusion of rat middle cerebral artery (MCA) and its recanalization by ultraviolet (UV) laser-facilitated dethrombosis (cf. p. 156). (a) Rat MCA preirradiation. The circle is the region of externally focused irradiation. (b) At 1 hour postirradiation with 458 nm argon at 21 W/cm^2 interacting with 1.3 m*M* flavin mononucleotide (FMN) delivered intravenously.[55] The circle encloses the region of occlusion. Note the lack of a blood column. (c) At 30 minutes posttreatment with 5 W/cm^2 of 355-nm Nd:YAG (yttrium aluminum garnet) laser light. The circle represents the area of treatment. Arrow indicates a small portion of the blood column starting to extend into the thrombosed region. (d) At 1 hour posttreatment with 355 nm Nd:YAG laser light. Note the reestablished blood column in the occluded region (circle).[55]

occlusion usually occurs within about 2 minutes, with limited proximal extension. Irradiation should be continued for 30 seconds afterward to ensure stability. The final thrombus consists of two parts. The primary thrombus at the point of irradiation is composed of very tightly aggregated platelets within a highly constricted segment. The secondary thrombus is induced distally by primary platelet thrombus secretions and extends for about 10 times that length in a less-constricted segment.[5] Our colleague H. Yao simplified this procedure by focusing the beam as a line along the MCA by means of a 30-cm f.l. cylindrical lens, resulting in formation of a long, stable thrombus directly.[30,31]

Alternatively, the MCA can be occluded with a mixed-composition, rt-PA-susceptible thrombus by means of an FMN-mediated photochemical reaction (Figure 15.10).[17] The procedures given are followed, but the FMN is injected to a body concentration of 37 mg/kg and is excited by means of a blue YAG laser (Table 15.2) operating at 473 nm (or an argon laser at 458 or 488 nm) at a focused intensity of about 15 W/cm^2 (ca. 5 mW of laser power).[17] During irradiation, unstable red mural thrombi are formed, which are often dislodged by flowing blood unless the blood pressure is reduced to about 70 mm Hg. Eventually, a fibrin-containing, mottled red thrombus will result; its formation must be completed before the FMN disappears from the bloodstream (ca. 6 minutes), as indicated by disappearance of its yellow fluorescence, and into the urine. The dye may be reinjected if the fluorescence begins to disappear while the thrombus is still forming.

It is often desired to monitor cortical blood perfusion in the MCA territory. To use the laser Doppler technique, a 2-mm^2 burr hole is made over the left frontoparietal cortex approximately 1.3 mm posterior and 5.5 mm lateral to bregma, just above the area of maximal histologic injury induced by occlusion of the distal MCA. Under a Zeiss operating microscope, the bone is drilled to a thin layer with a cutting burr under saline irrigation, a cortical area with blood vessels less than 50-μm diameter is selected by visualization through the thin bone layer, and a fiber-optic probe (0.8 × 30 mm) is placed on it. The fiber-optic probe, when coupled to a PeriFlux 4001 Master laser Doppler blood perfusion monitor (Perimed, 6785 Wallings Road, Suite 3A, North Royalton, Ohio 44133, 440-877-0537, Perimed@aol.com), measures cerebral blood perfusion in a 1-mm^3 tissue region. Arterial blood pressure is monitored via a femoral artery catheter coupled to a pressure transducer (Statham). Both the Doppler and pressure signals are routed to a polygraphic recording system that is interfaced to a personal computer via an analog-to-digital converter (Model PF 472, Perimed), utilizing data acquisition software (Perisoft for Windows).

Middle Cerebral Artery Recanalization by Ultraviolet Laser-Facilitated Dethrombosis

Recanalization of occluded cerebral arteries is the minimum step required for acute therapy of stroke, and here we describe a technique for achieving this goal. In contrast to tPA treatment, which is focused on thrombolysis via dissolution of fibrin, the relatively unknown process of dethrombosis dissolves platelet aggregates by specifically disrupting intraplatelet fibrinogen cross-links.[21] The significance of platelet barriers to thrombolysis in clinical clots is described in the "Introduction" (point 5). We first reported that UV light emitted at 355 nm from a Q-switched frequency-tripled Nd:YAG laser could very effectively dilate the distal MCA in normal[46] or platelet-thrombosed rats[5] and facilitated reflow in the latter by means of concurrently dilating the thrombus, as evidenced by continuous formation of blood-permeant microchannels in the platelet matrix. This was a consequence of destabilizing intraplatelet fibrinogen-GPIIb-IIIa cross-links, resulting in disintegration of the thrombus via departure of individual platelets. We later found that the 351-nm line from a continuous-wave argon ion laser is also suitable for this purpose (cf. p. 154). Dethrombosis was originally observed as a consequence of thrombin inhibition in platelet-rich clots containing seepage channels,[21,22] but in our case dethrombosis is apparently mediated by nitric oxide released by UV laser-induced scission from adducts in smooth muscle cells.[47]

As first used by us, the 355 nm beam from a Nd:YAG UV laser (Minilase II, New Wave Research, Fremont, California) was expanded with a high-quality X10 beam telescope (Model BXUV-4.0-10x-354.7, CVI Laser, Albuquerque, New Mexico), and then the beam was refocused onto the distal MCA with a 25-cm spherical lens (as above) to encompass the arterial diameter. The telescope improves the sharpness of focus, producing a well-demarcated deep blue fluorescence on the irradiated arterial segment. The laser is Q-switched, with a pulse width of 5 nanoseconds. The laser is spatially filtered to produce a quasi-Gaussian beam profile with maximum energy/pulse of 0.8 mJ (average power of 16 mW at a pulse rate of 20 Hz). Pulsed laser powers are measured with a 380101 detector head (certified by the National Institute of Standards and Technology) connected to a Scientech model 365 power meter (Boulder, Colorado). Dilation of the distal MCA is begun by increasing the average UV beam power to several milliwatts to yield an intensity of about 8 W/cm^2. UV irradiation of segments overlain with dural vessels should be avoided because these vessels can hemorrhage, but often they do not reveal themselves until they *are* irradiated. Irradiation must then be done very carefully or shifted to another segment. The beam is first focused on the proximal border to subtend the blood column and thrombus equally. The thrombus is soon invaded by thin red streaks, which thicken and coalesce, signifying dissolution (ca. 0.5 to 1 minutes). The beam is then advanced distally and the process iterated until the primary thrombus (see the section "MCA Photothrombosis as Mediated by Intravascular Photochemistry *In Situ*" above) is

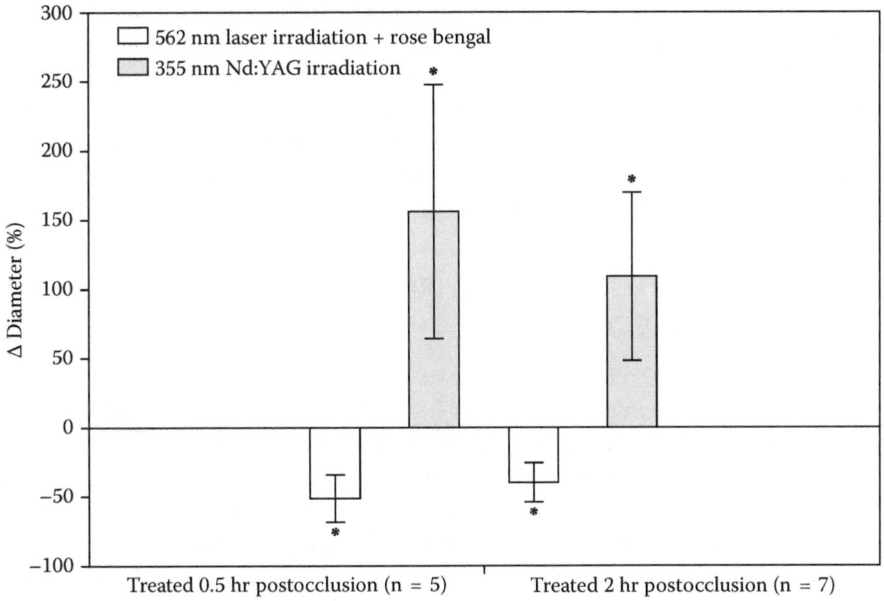

FIGURE 15.11 Percentage change in diameter of the rat distal middle cerebral artery in response to photothrombotic occlusion by platelets (constriction, white bars), followed by 355-nm ultraviolet laser irradiation (dilation, black bars) at intensities of about 8 W/cm^2 at 0.5 or 2 hours after occlusion.[5] Preocclusion baseline arterial diameter averaged 190 ± 40 μm for the 0.5-hour treated group and 170 ± 40 μm for the 2-hour treated group. Coefficients of variation were 22% and 24% for the 0.5- and 2-hour groups, respectively. Values shown are maximum changes in diameter with respect to each rat's preocclusion baseline for the 562-nm irradiation with respect to the postocclusion baseline (0.5-hour group: 90 ± 40 μm; 2-hour group: 110 ± 50 μm) for the 355 nm irradiation. Values were determined from the average of 8 to 16 arterial images per rat and are normalized mean ± standard deviation. *$p < .05$ versus the pre- and postocclusion baseline.

recanalized (Figure 15.9). The extreme variation in diameters induced first by photothrombosis and then by dethrombosis is shown in Figure 15.11.

The secondary thrombus is much less stable and can be disrupted and cleared by the pressure head of the incoming blood. This of course yields emboli that, after their distal deposition, are evidently stable enough to cause reperfusion injury entirely analogous to the clinical experience. The free-radical processes usually invoked to explain reperfusion injury are not involved at all. Although embolus emission is quite often accepted by interventionalists as the price of reperfusion, this consequence can be largely prevented if the segment enveloping the secondary thrombus is UV treated, inasmuch as rethrombosis after recanalization is absent from the UV-treated primary site but does occur in the untreated secondary region.[48]

Recanalization of an arterial segment occluded by a mixed-composition reddish thrombus produced by type I (FMN-mediated) photochemistry can also be achieved by dethrombosis (Figure 15.10). In this case, the segment and thrombus again dilate, but although restoration of flow is observed during UV irradiation, the thrombus appears intact. It is likely that the platelets dissociated to create seepage channels that permitted blood permeation throughout the remaining fibrin-red blood cell mass, which is likely now much more susceptible to rt-PA treatment given the great increase in surface area of interaction inside the thrombus.

Alternatively, recanalization can be obtained more quickly by a method developed by our former colleague H. Yao.[31] This entails focusing a 355-nm Nd:YAG beam at an intensity of 2.3 W/cm^2 onto the entire thrombosed MCA segment by means of a 30-cm f.l. cylindrical lens. Recanalization is effected within a few minutes, thereby enabling more precise assessment of the ischemic time interval. This method cannot be used endovascularly, however.

At the end of treatment, the wounds on the neck and legs are reopened, the catheters removed, and the wounds sutured back. On removal from the ventilator, the rats are extubated and returned to their cages. They are injected with Kefzol® (cefazolin sodium) twice daily for 3 days. After 24 hours, the rats can eat and drink by themselves. Because the animals are studied for survival, all the previously described procedures are performed under aseptic conditions. Instruments are sterilized by leaving them in Cidex (aerated dialdehyde) solution overnight. Operators wear sterile gloves, laboratory coats, masks, and head covers during the operative procedures.

COMMON CAROTID ARTERY PHOTOTHROMBOTIC EMBOLIZATION AND OCCLUSION

Embolization arising from photothrombotic platelet stenosis was originally discovered using argon 514.5 nm irradiation (average intensity 55 W/cm^2) of a CCA in rose bengal-injected rats (40 mg/kg).[19] Here, platelets aggregating in response to a 1-minute period of photochemical injury equilibrate kinetically with previously formed platelet mural thrombi, which are forced by flow pressure to embolize. An improvement in efficiency (50% to 70% stenosis) was soon found by irradiating with 22 W/cm^2 of 562 nm dye laser irradiation at the absorption peak of rose bengal in tissue.[49] Subsequently, argon 514.5 nm irradiation was used at about 50 W/cm^2 in conjunction with erythrosin B at 37 mg/kg.[29] This model does not simulate transient ischemic attack (TIA) because it produces hemodynamic deficits and microfocal ischemia. But, such embolization can precipitate very severe ischemic responses to prior embolic events,[50] strongly suggesting that TIA in humans can potentiate considerable morbidity through enhanced susceptibility to frank stroke.

Male Sprague-Dawley rats weighing 300 to 380 g are anesthetized, ventilated, and placed supine, and then a left paramedian skin incision is made in the ventral aspect of the neck and the underlying soft tissues dissected to expose the left CCA. The artery is carefully dissected from the vagus nerve. The surrounding musculature is retracted to create a watertight cavity. An ultrasonic probe (T206, Transonic Systems USA, 34 Dutch Mill Road, Ithaca, New York 14850, 607-257-5300, listmaster@teamtransonic.com) is placed around the CCA distal to the irradiation site to measure blood flow velocity continuously throughout the experimental protocol. The cavity is filled with warmed saline solution and kept clear of blood in preparation for CCA laser irradiation, essentially with the arrangement in Figure 15.7. The laser beam is first focused at very low power through a 61-cm f.l. spherical lens (needed to form a large-diameter focus) and is reflected downward toward the supine rat via a 90° prism (or mirror) onto the CCA at a distance of 1.5 cm from the sternal-clavicular notch. The saline in the CCA cavity helps dissipate heat during the high-intensity continuous irradiation. In theory, the weak alignment beam is focused optimally by overlapping the artery diameter by 6%,[18] but because the artery will constrict during irradiation, this factor can be ignored. After alignment, the dye of choice is infused via the femoral vein catheter, and irradiation with a laser compatible with the experimental requirements is begun simultaneously.

A less-expensive alternative to the above systems is again a 532 nm DPSS Nd:YAG, for which tunable powers of up to 8 W continuous wave are available (see Laserglow, Table 15.1). For embolus production from a rat CCA, 200 to 300 mW (<25 W/cm^2) should be sufficient in view of the near-resonant absorption of 532 nm light with erythrosin B (>40 mg/kg) (Figure 15.2). We have not tried this combination ourselves but are confident that it will perform as described.

If occlusion per se of a rat CCA (up to 1.2-mm diameter) is desired, this can be done by formation of a pure platelet thrombus in response to peroxidative endothelial damage induced by a type II photochemical reaction.[16,17] We have achieved this by means of the intravascular interaction of rose bengal dye and a rhodamine 560 perchlorate dye laser (Model CR599, Coherent, Santa Clara, California) beam operating at the 562 nm absorption maximum of rose bengal in tissue. An 8-W argon laser (Coherent Innova 90-6) is used to pump the dye laser to a power of up to 2 W in TEM_{00}

(Gaussian, or "spot") mode (Figure 15.7). The two lasers are tuned for maximum output before use to ensure beam-pointing stability and output efficiency.

Initially, male Sprague-Dawley rats weighing 400 to 750 g are anesthetized and ventilated as discussed for embolus production. If x-ray angiography is to be done, the left femoral artery and vein are used for physiological monitoring, while the right femoral artery is reserved for endovascular microcatheter insertion. Next, rose bengal dye (20 mg/mL in 0.9% saline) is infused over a 1-minute interval via the left femoral vein catheter to a body concentration of 20 mg/kg. Dye laser irradiation at a power of 1.5 W (corresponding to an intensity of about 130 W/cm^2) is begun and continued until occlusion. During irradiation, thrombosis (as observed through dye laser goggles, Table 15.1) is characterized initially by vasoconstriction and is accompanied by an intense orange fluorescence from the dye, bound mainly to circulating plasma proteins. This fluorescence decreases during irradiation as blood flow decreases, owing to the formation of a white platelet thrombus in the irradiated arterial segment and to uptake of circulating rose bengal by the liver.[38] Sometimes, however, constriction is so severe that refocusing and repositioning the beam during irradiation is necessary. Induction of CCA occlusion in ventilated animals is nontrivial and can take up to 13 minutes. Because of reactive increases in blood pressure owing to peripheral constriction and central dilation, occlusion is transient and can be followed by as many as six bursts of reperfusion until stability is obtained (Figure 15.12). Evidently, maintenance of respiration is critical to conservation of protective responses, an aspect that is not often considered experimentally in stroke or instituted clinically unless distress is obvious (in about 10% of patients). We were not aware of this protective effect during our initial work and found that the occlusion time in nonventilated rats was much faster (ca. 20 seconds),[17] which we later attributed to the rose bengal-induced hypotension mentioned previously.

Angiographic recordings were also obtained (Figure 15.12) in preparation for endovascular methods of thrombosis and recanalization. An animal is prepared for angiography with the right femoral artery providing the entry point. With the animal in the supine position, an oblique skin incision is made in the right inguinal region and the underlying fascia dissected between the inguinal

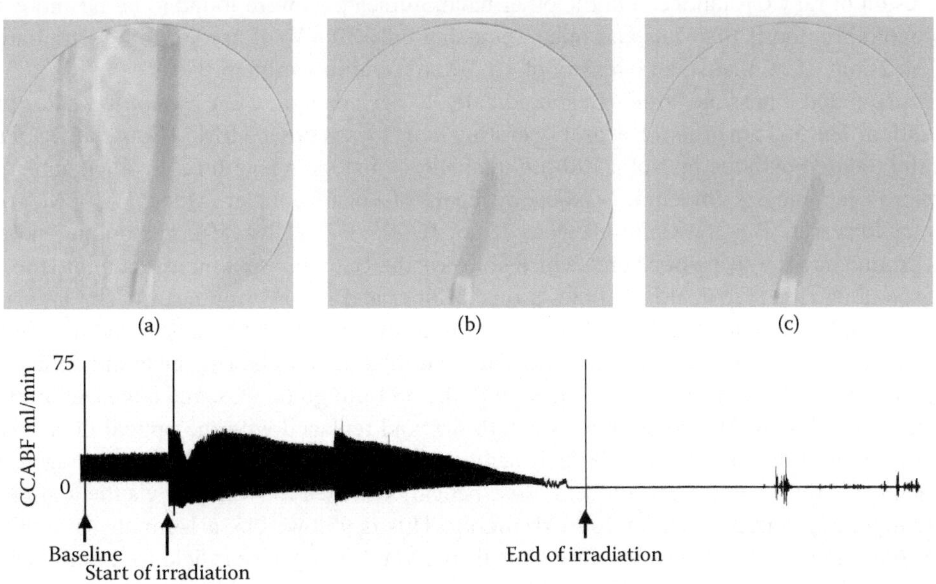

FIGURE 15.12 Angiograms of rat left common carotid artery (LCCA) occluded by *exo*vascular photothrombosis. For 13 minutes, 1 W of 562 nm argon/dye laser irradiation was focused on the LCCA in a rat injected with 40 mg/kg rose bengal. (a) Baseline angiogram; (b) postirradiation angiogram; (c) angiogram taken 30 minutes after saline flush. Bottom: Ultrasonic blood flow trace distal to irradiated LCCA segment showing gradual obstruction of flow by a platelet thrombus.

ligament and the superficial epigastric artery to expose the right femoral vein, artery, and nerve, which are then carefully dissected free from each other. The right femoral artery is temporarily occluded by 3-0 silk sutures, and a small transverse incision is made in the anterior wall by means of microscissors. A custom microcatheter (930-μm outside diameter, 690-μm inside diameter) with a lubricious coating and smooth bullet-shaped head (Opusgen, 1544 NW 89th Court, Doral, Florida 33172, 305-591-7203, www.opusgen.com) is loaded with a 360-μm guide wire (Boston Scientific) and continuously flushed with a heparinized saline solution (100 U/kg) at 15 mL/kg hourly while being introduced transfemorally into the descending aorta, advanced retrogradely under x-ray fluoroscopy, and positioned inside the left CCA. The catheter tip is advanced to the C2–C3 level and radiographic images taken of the guide wire for calibration purposes. The catheter is then withdrawn to the C5 level, the guide wire is removed, and a baseline angiographic image of the CCA segment at the C2–C3 level is obtained by injecting contrast medium (Visipaque 320 mgI/mL) at the CCA blood flow rate (15% of cardiac output assuming a cardiac index [CI] of 250 mL/minute/kg) for 5 seconds. CI was confirmed from the literature and echocardiograms. Angiograms are recorded with a Cohu 1/2-inch CCD (charge-coupled device) camera interfaced to a dimension analyzer (C Squared, Ft. Lauderdale, Florida) via a personal computer and recorded on a monitor. The same system is used to record optical images of the experiment (via a Zeiss operating microscope). Stored images are then analyzed, and the outside and inside CCA diameters are determined with an edge detection program (Image Pro-plus) to establish a baseline for observing diameter changes during subsequent procedures. Images are taken at a rate of 30 frames/second, averaged, and digitally subtracted from the background.

Recently, photothrombotic embolization of mouse CCA in a manner similar to rat[29,50] was attempted by means of the erythrosin B/argon 514-nm interaction, but spontaneous fragmentation of the forming thrombus could not be observed[51]; instead, the formed thrombus was dislodged mechanically. Occlusion of a 400-μm diameter mouse CCA required 165 mW of argon 514 nm laser power, that is, a focused intensity of about 130 W/cm^2, which is comparable to the intensity required for occlusion of rat CCA (above). On the other hand, guinea pigs were found to be far more sensitive to photothrombosis than rats and mice, requiring only 50 mW of argon 514 nm irradiation to occlude a 2-mm CCA (that is, an intensity of 1.7 W/cm^2) within 1 minute.[52]

We also found it possible, although sporadically, to occlude a rat CCA by *endovascular* photosensitization. The 562 nm argon/dye laser operating in its lowest order TEM_{00} (Gaussian) spot mode is initially focused onto the face of a 100-μm diameter, 3 meters in length fused silica optical fiber (Polymicro Technologies, Phoenix, Arizona) by means of a beam coupler (Model PAF-SMA-5-560, Optics for Research, Box 82, Caldwell, New Jersey 07006, 973-228-4480). The output end of the fiber is ground into the shape of a cone, which converts the Gaussian spot input beam into the shape of an expanding ring (Figure 15.13 shows a typical fiber and its UV ring beam). The laser power is then adjusted to produce the desired ring beam intensity, which can be calculated as a function of beam output power, arterial diameter, diameter and thickness of the ring beam, and the conical tip apex angle.[53] The microcatheter is loaded with the 360-μm guide wire and advanced to the site of irradiation (C2–C3). The guide wire is withdrawn and replaced with the optical fiber, inserted through a Touhy-Borst adapter (0.048-inch hemostatic valve), and positioned and imaged at the superior edge of the C3 transverse process. Rose bengal is infused for 1 minute via the femoral vein catheter to a body concentration of 20 to 40 mg/kg. This is followed by a 1-minute saline flush at 0.25 mL/kg/minute to remove excess rose bengal solution, leaving a clear field for laser irradiation of the endothelial-bound rose bengal dye. An atraumatic aneurysm clip is attached just distal to the targeted segment, and the segment is irradiated at 562 nm via the optical fiber for 3 minutes, with the endovascular ring beam intensity at 50 to 100 W/cm^2. The optical fiber is then removed, the catheter is withdrawn into the descending aorta, and a proximal aneurysm clip is attached to trap blood in the irradiated segment for 30 to 45 minutes. This resulted in thrombus formation (Figure 15.14), but thrombus stability was inconsistent. The microcatheter is reinserted into the CCA to

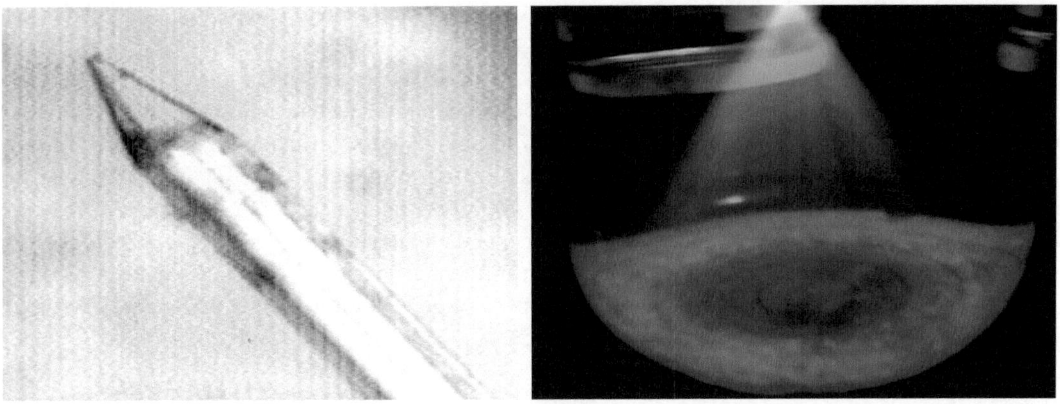

(a) (b)

FIGURE 15.13 Geometric properties of a conical-tip optical fiber. (a) Optical fiber with conical tip ground and polished at an apex angle of 35°. (b) A ring-shaped ultraviolet (UV) laser beam (351 nm argon) suitable for radially symmetric endovascular irradiation produced by this conical-tip fiber and projected into a beaker filled with water. The blue color is due to UV excitation of fluorescent impurities. A conical tip defect (a) permits an axial beam of negligible intensity to appear inside the ring beam in (b). **(See color insert.)**

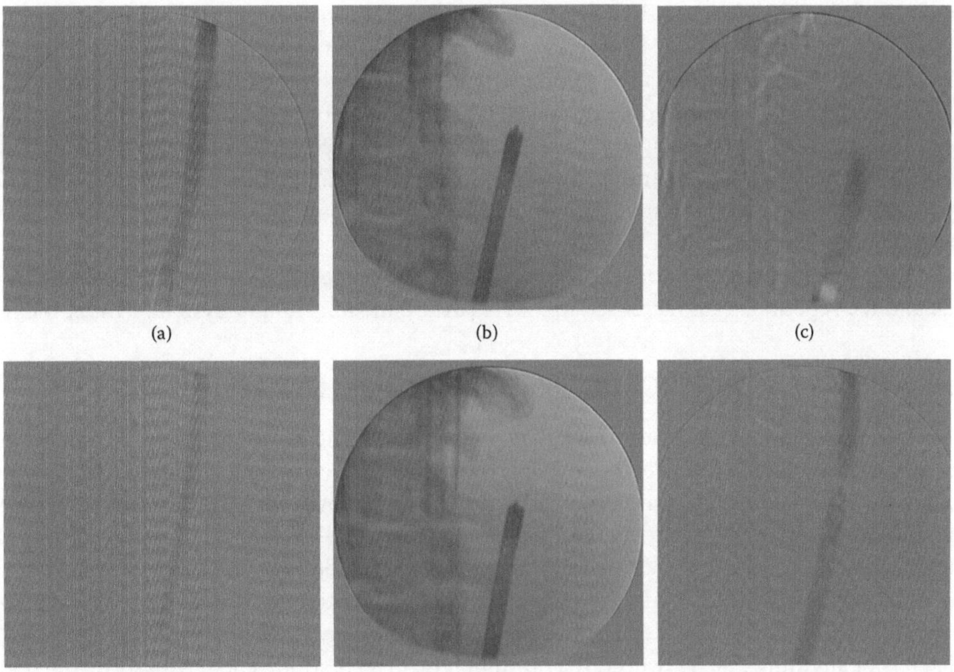

(a) (b) (c)

(d) (e) (f)

FIGURE 15.14 Occlusion of rat common carotid artery by *endo*vascular photothrombosis facilitated by rose bengal-sensitized, endovascular ring beam irradiation (562 nm, 125 W/cm^2) of endothelium for 1 minute via a conical-tip optical fiber delivered by microcatheter. The fiber was interfaced to the incoming laser beam with a beam coupler (Optics for Research, Caldwell, New Jersey). (a), (d) Baseline angiogram (two rats); (b), (e) optical fiber in place; (c), (f) platelet occlusion formed after catheter withdrawal; (c) complete; (f) stenosis (near complete).

the C5 level, and a repeat angiogram is performed to establish the degree of vessel injury (stenosis or occlusion).

It might seem improbable that this technique could be successful given that platelets are prevented from immediate contact with the damaged endothelium. The chief difficulties are (1) maintaining an optically clear irradiation field with a minimum of saline infusion; (2) rapid rate of CCA flow preventing platelet adherence and relative localization of platelet secretions; (3) difficulty of centering the optical fiber inside the artery; and (4) valve leakage, which allowed blood to seep into the irradiation field and coagulate on the fiber tip during irradiation. The last prevented photosensitization of the vessel wall and could cause an off-center fiber tip to attach to the arterial wall. Local heating could also induce proximal coagulation of new blood, which would defeat efforts to recanalize the artery optically (see the next section).

COMMON CAROTID ARTERY RECANALIZATION BY ULTRAVIOLET LASER-FACILITATED DETHROMBOSIS

Animals may then be treated with endovascular UV laser light to dissolve the CCA platelet thrombus by dethrombosis. As explained, an occluded UV-irradiated arterial segment will dilate but so will the enclosed thrombus owing to dissociation of its fibrinogen/platelet GPIIb-IIIa cross-links; the thrombus dissolves concurrently. In previous work, we were able to adapt the conical optical fiber/microcatheter system described to conduct UV laser light endovascularly as a ring beam to irradiate the inner surface of a rat CCA circumferentially (Figure 15.15). Dilations of up to 39% were observed at initial intensities of 20 W/cm^2 (Figure 15.16),[53] and these parameters are commensurate with our previous work in dissolving a platelet thrombus in the rat MCA (Figure 15.11). The animal is again heparinized (100 U/kg), and another conical-tip (apex angle at 31°), 100-μm diameter optical fiber is introduced via the microcatheter, now repositioned two to three arterial diameters from the thrombotic mass to be dissolved. A ring beam formed from the argon 351 nm line (emitted from a Coherent 90-6 laser with special mirrors) is used to irradiate the inner CCA wall. Figure 15.17 shows the effect of argon 351 nm endovascular treatment in a rat with occluded CCA. (Our Q-switched Nd:YAG laser could not be used to conduct powers greater than 3 mW owing to excessive energy per pulse; subsequently, we found that 3 W could be transmitted if the 355 nm pulse length were increased to about 50 nanoseconds, the energy per pulse decreased to less than 0.4 mJ, and the repetition rate increased to 7 kHz [Micro model, Quanta System, Milan, Italy].)

IMPLICATIONS OF DETHROMBOSIS FOR STROKE THERAPY

There is still no treatment for acute stroke that is consistent, efficacious, and safe. The major problem is still arterial obstruction, whether by clot, by spasm, or even by bits of indissoluble plaque. In particular, cross-linked platelets are especially difficult to remove safely by current methods without embolizing them. In our experience, UV laser-induced dilation followed by dethrombosis of the platelet matrix can relieve any such arterial obstruction sufficiently to the point of restoring some degree of blood flow. The mechanism of this method circumvents the cell and artery damage caused by previous laser-based photoacoustic shock methods because only platelet *bonds* are disrupted, not the platelets or anything else. However, if a thrombolytic drug is still needed to dissolve any remaining fibrin, its transport into and throughout the obstruction will be much enhanced by primary administration of dethrombosis and its concomitant creation of clot-permeating microchannels.[5,48] These greatly increase the surface area of rt-PA interaction with fibrin and its efficacy and with a lowered dose decrease the risk of hemorrhage as well.

Dethrombosis is concerned solely with disaggregation of dense platelet masses, which cannot now be achieved consistently with any drug, in part because their entry is physically denied. The dense layers of platelets (20% to 40% of mass) appearing in clinical clots[8] appear to be modeled by

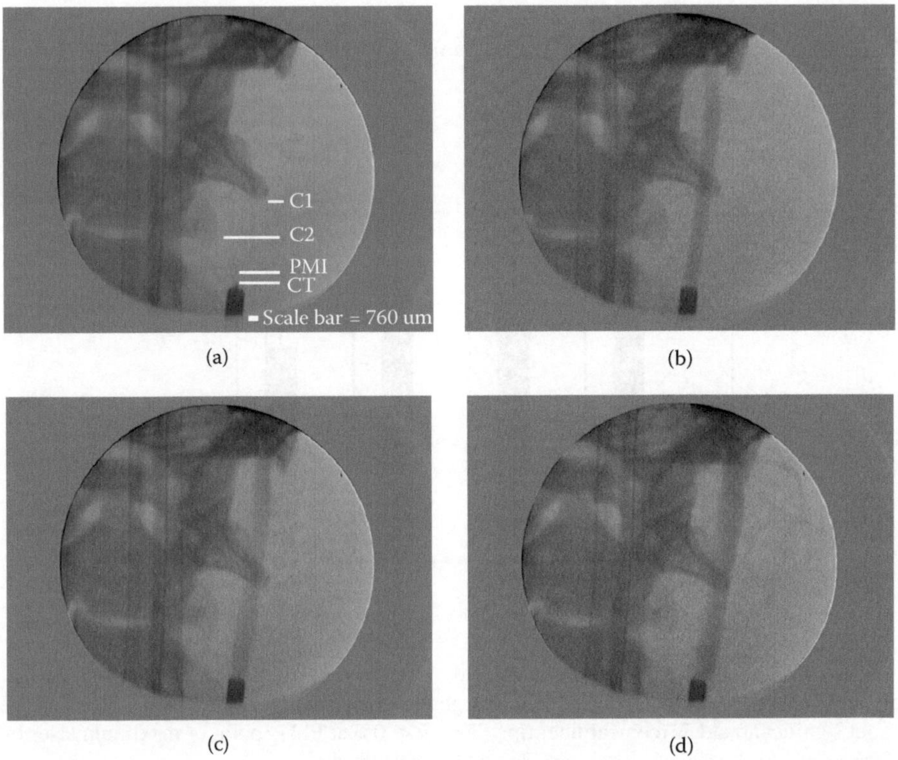

(a) (b) (c) (d)

FIGURE 15.15 Angiograms showing endovascular dilation of rat left common carotid artery (CCA) mediated by ultraviolet (UV) laser irradiation from a conical-tip optical fiber introduced into the CCA via a Roamer™ microcatheter (Opusgen, Doral, Florida). (a) Position of conical fiber tip before irradiation; (b) baseline angiogram before irradiation; (c) dilation observed at 10 minutes after irradiation; (d) dilation observed at 45 minutes after irradiation. The radiographically labeled optical fiber was advanced under fluoroscopic guidance as provided by a specially modified, 50-μm resolution Lixi fluoroscope (Model 85-799). An argon laser beam operating at 351 nm (model 90-6, Coherent) was introduced into the optical fiber with a beam coupler (model PAF-SMA-7-355, Optics for Research, Caldwell, New Jersey), resulting in a ring-shaped output beam. The CCA wall was then irradiated endovascularly at 20 W/cm^2 for 1 minute. Note dilation at 10 minutes, which was increased at 45 minutes.[53]

the arterial photothrombotic models described and likely act as barriers to rt-PA transport. However, if a thrombus is already rent by seepage channels, platelet disaggregation is sporadically possible with thrombin inhibitor drugs, but platelet inhibitor drugs are not effective for acute stroke therapy because they cannot remove platelet-bound fibrinogen. The only kind of substance that can potentially traverse any kind of cell aggregate is a gas. In the case of dethrombosis, it appears that UV laser light induces release of a diffusible agent (likely nitric oxide) from vascular smooth muscle stores of NO adducts,[47] which penetrates the platelet matrix and destabilizes intraplatelet GPIIb-IIIa fibrinogen cross-links.

The UV laser-based method is unique in that the laser beam is *not* aimed directly at the occlusion, so any colored material contained in a thrombus being "dethrombosed" will not be intentionally exposed to laser light. Because platelets are colorless and scatter light far more strongly than they absorb it, they should remain unaffected either way. NO adducts, the putative chemical substrates for mediating dethrombosis, are normally present in the arterial smooth muscle cells and need only be irradiated by the UV laser beam at comparatively low intensities in either continuous-wave or quasi-continuous pulsed mode. To our knowledge, the pathological consequences of middle UV laser irradiation (350 to 365 nm) on arterial smooth muscle cells have not been published. Our

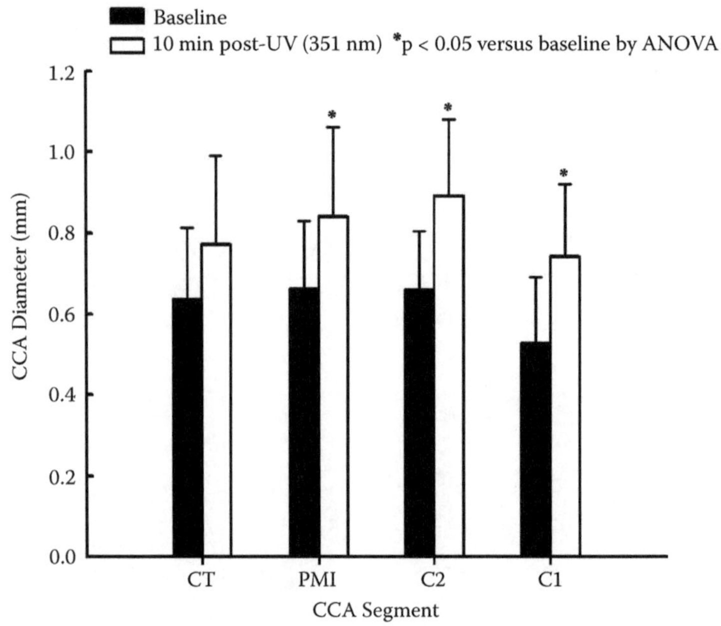

FIGURE 15.16 Dilations of rat left common carotid artery at several anatomical locations by endovascularly administered argon ultraviolet (UV) (351-nm) argon laser ring beam irradiation. Dilations (n = 12) were 20% (n.s., not significant) at CT (conical fiber tip); 27% ($p < .05$) at PM1 (point of maximum laser intensity); 35% and 39% ($p < .05$) at cervical vertebrae C2 and C1, respectively.[53]

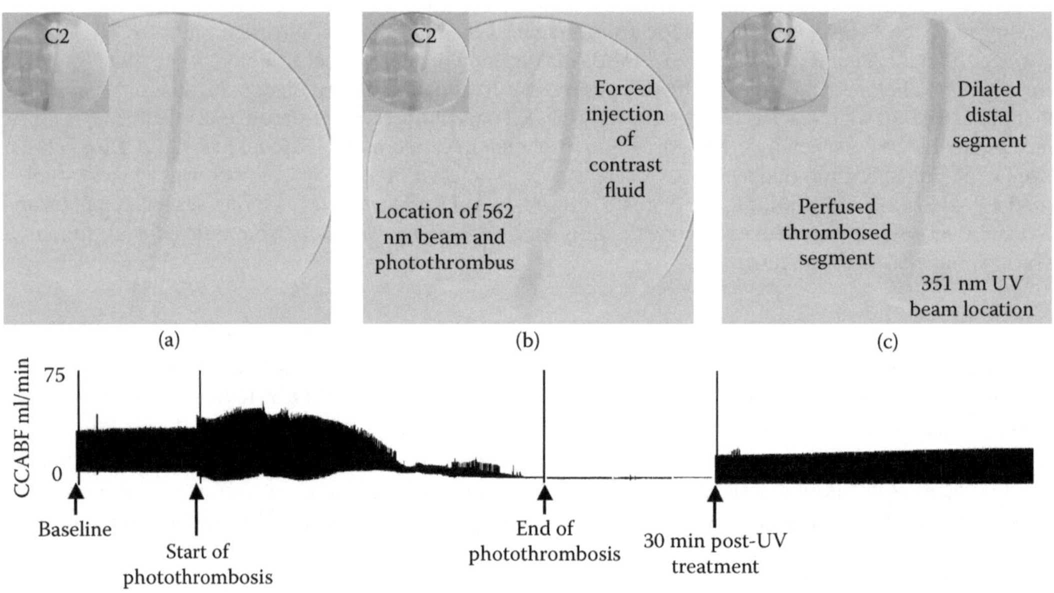

FIGURE 15.17 Photothrombotic occlusion of rat left common carotid artery (CCA) and its subsequent recanalization by ultraviolet (UV) laser-induced *endovascular* dethrombosis. The left CCA was exposed and occluded by photothrombosis induced by 13 minutes of 562 nm externally focused laser irradiation (1 W) in a rat injected with 40 mg/kg rose bengal. (a) Baseline angiogram; (b) 30 minutes postocclusion; (c) the platelet thrombus was aged for 2 hours, then the segment proximal to it was exposed to 3 minutes of 351 nm endovascular ring beam irradiation (12 W/cm^2) and observed 30 minutes later. Dethrombosis (bottom): Ultrasonic blood flow trace showing thrombus formation and dissolution measured just distal to irradiated left CCA segment.

early TEM work on rat MCA revealed no damage to smooth muscle cells or endothelium when irradiated with the 355 nm Nd:YAG laser at 5 W/cm^2, whereas at 20 W/cm^2 vacuolation of smooth muscle cells was observed without loss of function; the concomitant MCA dilation receded over several hours.

The methods described in this chapter, we trust, will prove helpful to readers interested in experimental models of frank thrombotic stroke expressed in several forms and in the mitigation of acute experimental and especially clinical stroke.

ACKNOWLEDGMENTS

The opinions expressed here are our own. This work was supported in part by U.S. National Institutes of Health grants 1 R21 NS48297 and 2 R01 NS23244 to Brant D. Watson, and the latter work resulted in two U.S. patents to Brant D. Watson.[54,55] We thank our numerous colleagues, particularly W. Dalton Dietrich, Hiroshi Yao, Per Wester, and Weigang Gu, for their initiation of or participation in many of the cited studies. Richard A. DeFazio brought the Snake Creek minilasers to our attention. The immunofluorescent thrombus staining in Figure 15.2 was performed by Dr. Andrew J.-W. Huang.

REFERENCES

1. Brott, T. and Bougousslavsky, J., Treatment of acute ischemic stroke, *N Engl J Med*, 343, 710, 2000.
2. Gladstone, D.J., Black, S.E., and Hakim, A.M., Toward wisdom from failure: Lessons from neuroprotective stroke trials and new therapeutic directions, *Stroke*, 33, 2123, 2002.
3. Dietrich, W.D., Prado, R., and Watson, B.D., Photochemically stimulated blood borne factors induce blood-brain barrier alterations, *Stroke,* 19, 857, 1988.
4. Dietrich, W.D. et al., Morphological consequences of early reperfusion following thrombotic or mechanical occlusion of the rat middle cerebral artery, *Acta Neuropathol*, 78, 605, 1989.
5. Watson, B.D. et al., Cerebral blood flow restoration and reperfusion injury following ultraviolet laser-facilitated middle cerebral artery recanalization in rat thrombotic stroke, *Stroke*, 33, 428, 2002.
6. Zhang, K. and Sejnowski, T.J., A universal scaling law between gray matter and white matter of cerebral cortex, *Proc Natl Acad Sci USA*, 97, 5621, 2000.
7. Marder, V. et al., Analysis of thrombi retrieved from cerebral arteries of patients with acute ischemic stroke, *Stroke*, 37, 2086, 2006.
8. Wysokinski, W.E. et al., Atrial fibrillation and thrombosis: immunohistochemical differences between *in situ* and embolized thrombi, *J Thromb Haemost*, 2, 1637, 2004.
9. Daffertshofer, M. et al., Transcranial low-frequency ultrasound-mediated thrombolysis in brain ischemia, *Stroke,* 36, 1441, 2005.
10. Zausingera, S., Baethmann, A., and Schmid-Elsaessera, R., Anesthetic methods in rats determine outcome after experimental focal cerebral ischemia: Mechanical ventilation is required to obtain controlled experimental conditions, *Brain Res Prot*, 9, 112, 2002.
11. Jovin, T.J. et al., The cortical ischemic core and not the consistently present penumbra is a determinant of clinical outcome in acute middle cerebral artery occlusion, *Stroke*, 34, 2426, 2003.
12. Palesch, Y. et al., The ALIAS pilot trial. II, *Stroke*, 37, 2107, 2006.
13. Watson, B.D. et al., Induction of reproducible brain infarction by photochemically initiated thrombosis, *Ann Neurol*, 17, 497, 1985.
14. Rosenblum, W.I. and el-Sabban, F., Platelet aggregation in the cerebral microcirculation: Effect of aspirin and other agents, *Circ Res*, 40, 320, 1977.
15. Gandin, E., Lion, Y., and Van de Vorst, A., Quantum yield of singlet oxygen production by xanthene derivatives, *Photochem Photobiol*, 37, 271, 1973.
16. Watson, B.D., Animal models of photochemically induced brain ischemia and stroke. In: Ginsberg, M.D. and Bogousslavsky, J., eds., *Cerebrovascular Disease—Pathophysiology, Diagnosis and Treatment*, Blackwell Science, Cambridge, U.K., 1998, chap. 4.
17. Watson, B.D. et al., Concepts and techniques of experimental stroke induced by cerebrovascular photothrombosis. In: *Central Nervous System Trauma: Research Techniques*, Ohnishi, S.T. and Ohnishi, T., eds., CRC Press, Boca Raton, Florida, 1995, chap. 12.

18. Watson, B.D. et al., Argon laser-induced arterial photothrombosis: characterization and possible application to therapy of arteriovenous malformations, *J Neurosurg*, 66, 748, 1987.
19. Futrell, N. et al., Embolic stroke from a carotid arterial source in the rat: Pathology and clinical implications, *Neurology,* 39, 1050, 1989.
20. Wester, P. et al., A photothrombotic "ring" model of rat stroke-in-evolution displaying putative penumbral inversion, *Stroke*, 26, 444, 1995.
21. Wysokinski, W. et al., Reversibility of platelet thrombosis *in vivo*: Quantitative analysis in porcine carotid arteries, *Thromb Haemost*, 76, 1108, 1996.
22. Gold, H.K. et al., Restoration of coronary flow in myocardial infarction by intravenous chimeric 7E3 antibody without exogenous plasminogen activators, *Circulation*, 95, 1755, 1997.
23. Kontos, H.A., Oxygen radicals in cerebral vascular injury, *Circ Res*, 57, 508, 1985.
24. Fungaloi, P. et al., Photochemically modulated endothelial cell thrombogenicity via the thrombomodulin–tissue factor pathway, *Photochem Photobiol*, 78, 475, 2003.
25. Maeda, M. et al., FK419, a nonpeptide platelet glycoprotein IIb/IIIa antagonist, ameliorates brain infarction associated with thrombotic focal cerebral ischemia in monkeys: Comparison with tissue plasminogen activator, *J Cereb Blood Flow Metab*, 25, 108, 2005.
26. Lambert, C.R. and Kochevar, I.E., Does rose bengal triplet generate superoxide anion? *J Am Chem Soc*, 118, 3297, 1996.
27. Boquillon, M., Boquillon, J.P., and Bralet, J., Photochemically induced, graded cerebral infarction in the mouse by laser irradiation evolution of brain edema, *J Pharmacol Toxicol Meth*, 27, 1, 1992.
28. Sieve, P.D., Solomon, H.M., and Krevans, J.R., The effect of hematoporphyrin and light on human platelets. I. Morphologic, functional, and biochemical changes, *J Cell Physiol*, 67, 271, 1966.
29. Danton, G. et al., Endothelial nitric oxide synthase pathophysiology following non-occlusive thromboembolic stroke in rats, *J Cereb Blood Flow Metab*, 22, 612, 2002.
30. Yao, H. et al., Simplified model of krypton laser-induced thrombotic distal middle cerebral artery occlusion in spontaneously hypertensive rats, *Stroke,* 27, 333, 1996.
31. Yao, H. et al., Photothrombotic middle cerebral artery occlusion and reperfusion laser system in spontaneously hypertensive rats, *Stroke*, 34, 2716, 2003.
32. Nowicki, J.P. et al., Nitric oxide mediates neuronal death after focal cerebral ischemia in the mouse, *Eur J Pharmacol*, 204, 339, 1991.
33. Kuluz, J. et al., The effect of nitric oxide synthase inhibition on infarct volume following reversible focal cerebral ischemia in conscious rats, *Stroke*, 24, 2023, 1993.
34. Hara, H. et al., Reduced brain edema and infarction volume in mice lacking the neuronal isoform of nitric oxide synthase after transient MCA occlusion, *J Cereb Blood Flow Metab*, 16, 605, 1996.
35. Kontos, H.A., private communication, 1992.
36. Levitan, R.M., Design rationale and intended use of a short optical stylet for routine fiberoptic augmentation of emergency laryngoscopy, *Am J Emerg Med*, 24, 490, 2006.
37. Busto, R. et al., Small differences in intraischemic brain temperature critically determine the extent of ischemic neuronal injury, *J Cereb Blood Flow Metab*, 7, 729, 1987.
38. Klaassen, C.D., Pharmacokinetics of rose bengal in the rat, rabbit, dog and guinea pig, *Toxicol Appl Pharmacol*, 38, 85, 1976.
39. Lee, V.M. et al., Evolution of photochemically induced focal cerebral ischemia in the rat: magnetic resonance imaging and histology, *Stroke,* 27, 2110, 1996.
40. Belayev, L. et al., Postischemic administration of HU-211, a novel non-competitive NMDA antagonist, protects against blood–brain barrier disruption in photochemical cortical infarction in rats: a quantitative study, *Brain Res*, 702, 266, 1995.
41. Dietrich, W.D. et al., Photochemically induced cerebral infarction. I. Early microvascular alterations, *Acta Neuropathol (Berl)*, 72, 315, 1987.
42. Verlooy, J. and Van Reempts, J., The blood–brain barrier in trauma, stroke and edema, *Int Congress Ser,* 1277, 227, 2005.
43. Watson, B.D., Prado, R., and Dietrich, W.D., unpublished observations, 1988.
44. Hu, X.-L. et al., Progressive and reproducible focal cortical ischemia with or without late spontaneous reperfusion generated by a ring-shaped, laser-driven photothrombotic lesion in rats, *Brain Res Prot,* 7, 76, 2001.
45. Gu, W., Brännström, T., and Wester, P., Cortical neurogenesis in adult rats after reversible photothrombotic stroke, *J Cereb Blood Flow Metab*, 20, 1166, 2000.

46. Watson, B.D., Prado, R., and Dietrich, W.D., Q-switched Nd:YAG laser irradiation at 355 nm mediates rapid focal dilation of rat middle cerebral artery, *Soc Neurosci Abstr*, 24(Pt 1), 1171, 1998.
47. Ng, E.S. et al., Nitrosothiol stores in vascular tissue: modulation by ultraviolet light, acetylcholine and ionomycin, *Eur J Pharmacol,* 560, 183, 2007.
48. Watson, B.D. et al., A tissue plasminogen activator (reteplase) augments the efficacies of UV laser-facilitated dethrombosis in recanalizing aged platelet and platelet-rich occlusive thrombi in rat middle cerebral artery, *J Cereb Blood Flow Metab, 23*(Suppl 1), 279, 2003.
49. Wester, P. et al., Measurement of serotonin in plasma by in vivo microdialysis during photochemically induced thrombosis—Methodological aspects. In: *Role of Neurotransmitters in Brain Injury*, Globus, M.Y.-T. and Dietrich, W.D., eds., Plenum Press, New York, 1992, p. 153.
50. Danton, G. et al., Temporal profile of enhanced vulnerability of the post-thrombotic brain to secondary embolic events, *Stroke,* 33, 1113, 2002.
51. Lozano, J.D. et al., Characterization of a thromboembolic photochemical model of repeated stroke in mice, *J. Neurosci. Meth.,* 162, 244, 2007.
52. Watson, B.D., Leamy, A., and Feuerstein, G.Z., unpublished data, 1993.
53. Watson, B.D. et al., Endovascular carotid artery dilation by ultraviolet laser light transmitted by micro-catheterized optical fiber, Abstract Viewer, *CD-ROM, Brain 05*, abstract 593, 2005.
54. Watson, B.D., Method for the Permanent Occlusion of Arteries, U.S. Patent No. 5,056,006, 1991.
55. Watson, B.D., Dethrombosis Facilitated by Vasodilation, U.S. Patent No. 6,539,944, 2003.

16 Induction of Asphyxia Cardiac Arrest in a Rat as a Model of Global Cerebral Ischemia

Kunjan R. Dave, Ricardo Prado, and Miguel A. Perez-Pinzon

CONTENTS

INTRODUCTION

Cardiopulmonary arrest remains one of the leading causes of death and disability in the United States. Despite quick emergency responses and better techniques of defibrillation, the chances of survival following cardiac arrest (CA) are still poor. The prevailing quandary in this field is that CA is multifactorial in that it results in whole-body ischemia, which compromises systemic blood parameters and cerebral, renal, and cardiac functions. In fact, of the 70,000/year patients who are resuscitated after CA, 60% die from extensive brain injury, and only 3% to 10% are able to resume their former lifestyles.[1] Thus, development of novel therapies that may be common to the organs most sensitive to CA, such as heart and brain, is a key to improvement in survival and better outcome following such a devastating event. By isolating the effects of global ischemia on the brain, two-vessel occlusion (2-VO) with hypotension and four-vessel occlusion (4-VO) are the most widely used animal models that simulate the effect of CA on the brain.[2] The CA model adds cardiovascular variables that may play additional roles in the development of pathology that ensues in the brain after CA. The model described here produces systemic hypoxia/ischemia, leading to arterial hypotension and eventual CA. The type of CA described here represents a frequent cause (hypoxia-induced CA) of CA in adults.[3] This CA model, besides sudden disruption of cerebral blood flow, adds additional cardiovascular variables such as lower tissue pH and a greater accumulation of carbon dioxide in tissues.

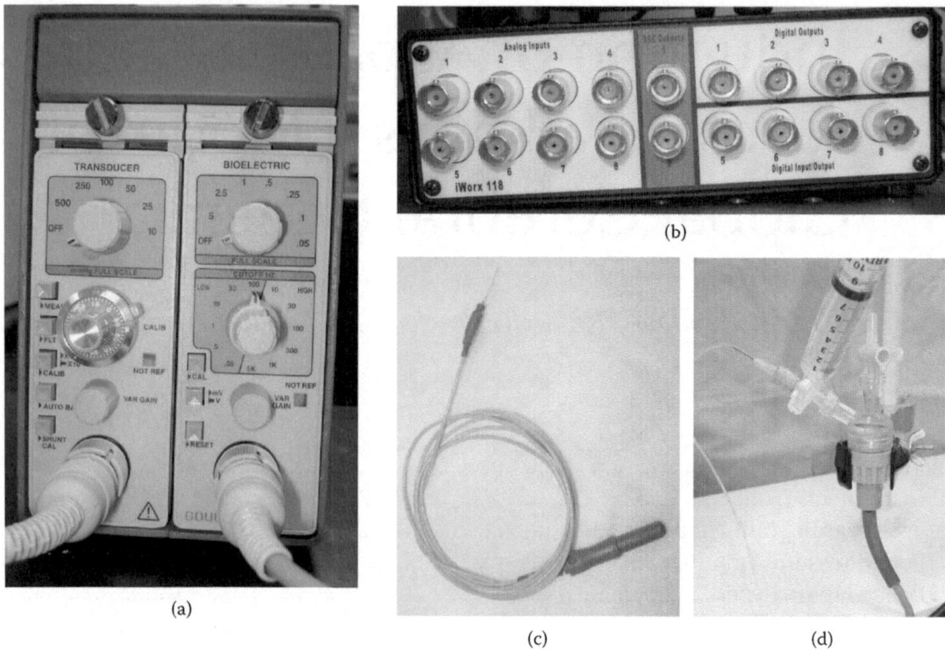

FIGURE 16.1 Devices for monitoring the electrocardiogram (ECG) and arterial pressure in the rat. (a) Blood pressure amplifier (Cat. No. 13-6615-50, Gould Instrument Systems, Valley View, Ohio) (left), bioelectric amplifier (Cat. No. 13-6615-58, Gould Instrument Systems) (right), and housing box (Cat. No. 11-4123-09, Gould Instrument Systems); (b) analog-to-digital converter (AD converter) (Cat. No. iWorx 118, CB Sciences, Dover, New Hampshire); (c) subdermal electrodes (Cat. No. F-E2, Astro-Med, West Warwick, Rhode Island); and (d) blood pressure transducer (Cat. No. P23XL, Gould Instrument Systems).

SPECIAL INSTRUMENTS

Electrocardiogram Amplifier

It is essential to monitor the electrocardiogram (ECG) before, during, and after asphyxia CA. For this purpose, ECG monitors have been used by several investigators. ECG amplifiers that are commonly used can monitor ECG using either three or five leads that can be placed into the subject. In our lab, we are using a bioelectric amplifier (Cat. No. 13-6615-58, Gould Instrument Systems, Valley View, Ohio) (Figure 16.1a). The amplifier is fixed in a housing (Cat. No. 11-4123-09, Gould Instrument Systems) (Figure 16.1a) to which a power supply and connectors for signal output can be connected. The amplifier can be connected to the rat via a cable (Cat. No. FSR1387, Gould Instrument Systems), which can be attached with subdermal electrodes (Cat. No. F-E2, Astro-Med, West Warwick, Rhode Island) (Figure 16.1c). The analog output from the amplifier housing box can be connected to an analog-to-digital converter (AD converter) (Cat. No. iWorx 118, CB Sciences, Dover, New Hampshire) (Figure 16.1b). This AD converter can be connected to a personal computer (PC) via a USB (universal serial bus) port.

Blood Pressure Amplifier

Monitoring of blood pressure during induction of asphyxia CA is very important because (1) after induction of asphyxia, blood pressure reaches zero between 3 and 4 minutes, and to get consistent brain pathology it is necessary to monitor the time when the blood pressure reaches zero; (2) at the time of animal resuscitation, the intensity of chest massage should be sufficient enough to increase diastolic pressure to about 50 mm Hg to enable monitoring of blood pressure during resuscitation;

and (3) restoration of spontaneous circulation (ROSC) rapidly increases the blood pressure above normal levels and thus can be a useful indicator of when to stop chest massage. For this purpose, we use a blood pressure amplifier (Cat. No. 13-6615-50, Gould Instrument Systems) (Figure 16.1a). A femoral arterial catheter can be connected to the blood pressure amplifier via a blood pressure transducer (Cat. No. P23XL, Gould Instrument Systems) (Figure 16.1d). These transducers are available with different connectors, which must be compatible with the 14-pin connector in the amplifier. Similar to the ECG amplifier, the blood pressure amplifier can be connected to a PC via its housing and an AD converter.

To record ECG and blood pressure, we use Labscribe software (CB Sciences). This software is particularly advantageous over others because it can record data at 20 kHz. To observe all the waveforms in a rat ECG signal, it is recommended to record data at 1 kHz. Labscribe has the capacity to record both blood pressure and ECG channels in a single data file.

Blood Gas Analyzer

It is well known that to minimize variation in cerebral pathology, it is very important to keep parameters such as pO_2, pCO_2, and pH in their normal physiological ranges.[4] To monitor blood gases, we are using the ABL50 (Radiometer, Copenhagen, Denmark) (Figure 16.2a). This instrument is ideal for repeated blood gas measurements in the rat because the volume required is about 70 μL. Once a sample is inserted into the instrument, results can be obtained in 1 to 3 minutes.

Rodent Ventilator

Initially, rats are maintained under isoflurane anesthesia on a face mask for vascular access. However, later they are connected to a rodent ventilator because (1) to induce asphyxia we paralyze animals so they do not breathe on their own, and thus a ventilator is essential for respiration of the paralyzed animal; (2) blood gases can be controlled throughout the experimental procedure by changing ventilator settings; and (3) 100% oxygen can be delivered directly into the lungs of the animal at the time of resuscitation to improve the survival rate. We use an Ugo rodent respirator (Cat. No. 7025, Ugo Basile, Comerio, Varese, Italy) (Figure 16.3a) with a cylinder volume of 1 to 10 mL. From our experience, we keep the ventilator volume to 1 mL/100 g of rat body weight. If the rat weighs more than 300 g, we do not increase this volume proportionate to the body weight because our experience suggests that a ventilator volume of more than 3.2 mL damages the lungs.

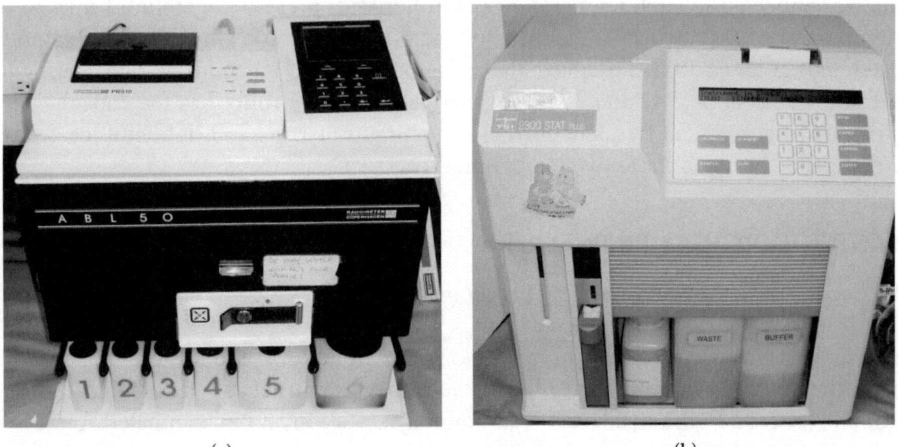

(a) (b)

FIGURE 16.2 Blood gas analyzer (ABL50 instrument, Radiometer, Copenhagen, Denmark) and blood glucose analyzer (2300 STAT Plus, YSI, Yellow Springs, Ohio).

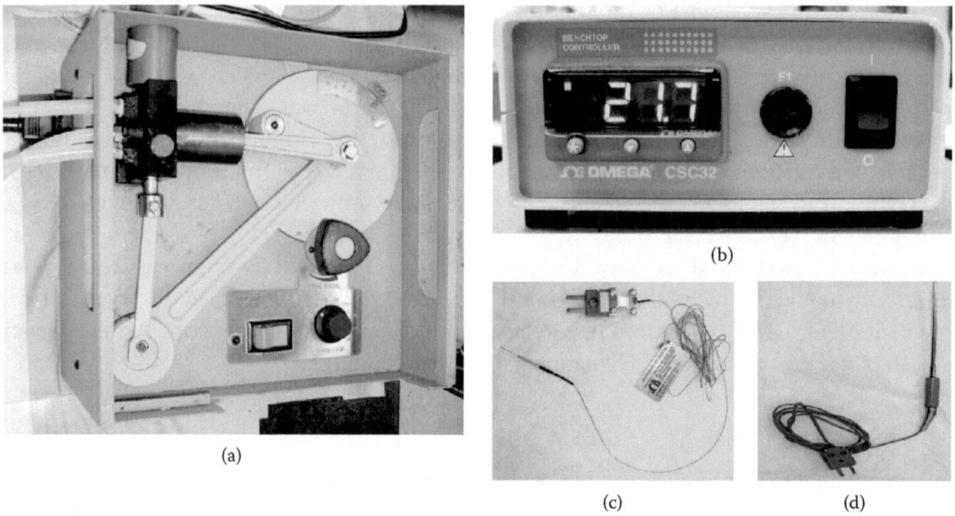

FIGURE 16.3 (a) Ugo rodent respirator (Cat. No. 7025, Ugo Basile, Comerio, VA, Italy); (b) Omega temperature controllers (Cat. No. CSC32, Omega Engineering, Stamford, Connecticut); (c) thermal probe (Cat. No. Hypo-33-1-T-G-60-SMP-M, Omega Engineering) for head; and (d) thermal probe (Cat. No. VIP-T, Omega Engineering) for rectum.

Head and Rectal Temperature Controller with Probes

It is well documented that small changes in body and head temperature can result in large variations in brain pathology following ischemia; that is, hypothermia causes less damage, while hyperthermia causes more damage compared to normothermia (37°C) (see 1999 review by Lipton[5]). We use Omega temperature controllers (Cat. No. CSC32, Omega Engineering, Stamford, Connecticut) (Figure 16.3b) connected to the animal via thermal probes (Cat. No. Hypo-33-1-T-G-60-SMP-M for head and Cat. No. VIP-T for rectum) (Figure 16.3c and 16.3d). The head probe is placed in the temporalis muscle (Figure 16.4a), the temperature of which changes in parallel with the brain temperature. The brain temperature is about 0.3°C to 0.5°C lower than that of the temporalis muscle. The head probe can be sterilized by cleaning it with alcohol swabs. For easy penetration, it is recommended that the rectal temperature probe be lubricated with water-soluble KY jelly. Heating lamps can be connected to each temperature controller with preset temperature limits to turn the heating lamp on or off as needed. Generally, we use 60-W light bulbs in the heating lamp, and we keep these lamps at about 18 inches above the rat's body to avoid burn injury. The heating lamp on and off points are set at 36.5°C and 37°C, respectively. The controller has the capacity to set this interval at 0.1°C, but to avoid frequent current switching transients and light bulb destruction, we keep the difference between these two temperatures at 0.5°C.

STEP-BY-STEP SURGICAL PROCEDURE

The following is the step-by-step surgical procedure[6,7]:

- Male Sprague-Dawley rats weighing 250 to 350 g are used. If the rat weight is more than 350 g, the mortality of animals is high because they are difficult to resuscitate.
- Animals ordered from outside the local animal facility must be acclimated for at least a week.
- Rats are fasted overnight before the surgery. Fasting is important for minimizing variations in blood glucose levels. It has been shown that a high blood glucose concentration results in more ischemic damage.[8,9]

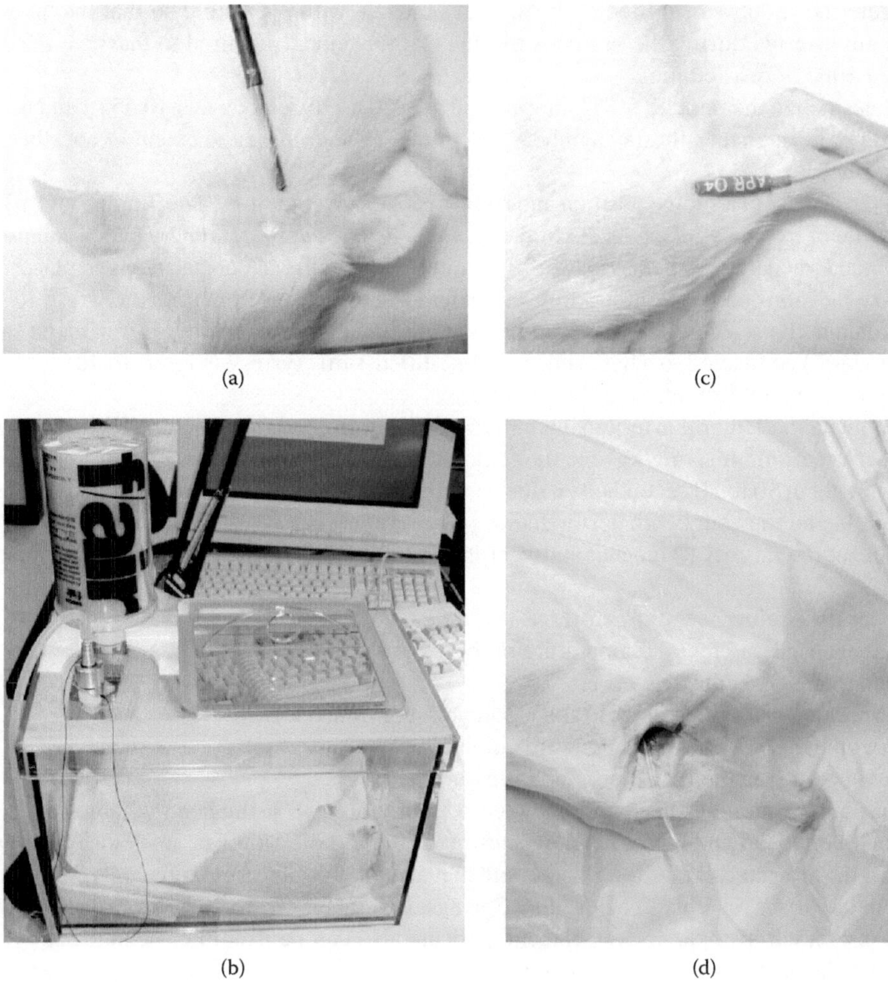

FIGURE 16.4 (a) Position of head temperature probe in temporalis muscle; (b) rat in an anesthesia chamber; (c) electrocardiogram (ECG) electrode inserted in the left leg of the rat; and (d) rat with femoral artery and vein cannulas in right leg.

- Anesthetize with 5% isoflurane and a 30:70 mixture of oxygen (0.45 L) and nitrous oxide (1.0 L) in an anesthesia chamber (Figure 16.4b) for 10 minutes. This chamber can be attached to an anesthesia scavenger. Keep observing the respiration rate.
- Shave hair from the surgery site (skin connecting right hindlimb and abdomen) and from limbs (right and left forelimbs and left hindlimb) before inserting ECG needle electrodes (Cat. No. F-E2, Astro-Med) (Figure 16.4c).
- Intubate the animal with an endotracheal tube (Cat. No. 381467, BD Infusion Therapy Systems, Sandy, Utah). A custom-made blunt spinal needle is used as a guide (Cat. No. 7427, Popper and Sons, New York, New York). With practice, we are able to do it blind. During early attempts, it may be necessary to incise the neck area to make sure that the endotracheal tube is being inserted correctly (see Chapter 15). A quick assessment is made by holding a stainless steel instrument in front of the tube so condensation of respiratory moisture can be observed. If we do not observe condensation of moisture, it means that the tube is not in the trachea.

- Secure the endotracheal tube to the animal's muzzle with 3-0 suture so that the tube will remain inserted during the vigorous mechanical movement required to massage the chest at the time of resuscitation.
- Reanesthetize the rat with 5% isoflurane and a 30:70 mixture of oxygen (0.45 L) and nitrous oxide (1 L) in an anesthesia chamber for 5 minutes. Determine the extent of anesthesia as described.
- Put the animal in a supine position and reanesthetize by delivering 2% isoflurane and a 30:70 mixture of oxygen via face mask. Make sure that an anesthesia scavenging device (e.g., Lab Animal Evacuation System, Harvard Apparatus, Holliston, Massachusetts) is in place.
- Make incisions on the right hindlimb for a femoral artery and vein access (Figure 16.4d).
- Cannulate the selected femoral vein using a PE-50 catheter filled with heparinized saline (3.3 U/mL) connected to a heparinized saline-filled 1-mL syringe (Figure 16.4d). This will help prevent clot formation in the catheter.
- Cannulate the femoral artery using a PE-50 catheter in the same fashion (Figure 16.4d).
- Inject vecuronium (2 mg/kg) via the femoral vein. Make sure the animal stops breathing within about 10 to 20 seconds. If it does not, than make sure that the vein catheter is inside the vein and the vein is not perforated.
- Connect the animal to mechanical ventilation (60 breaths/minute and stroke volume 3.0 mL).
- Lower the isoflurane to 0.5% to 1.0%.
- Connect head and rectal temperature probes.
- Connect ECG needle electrodes.
- Connect the arterial catheter to the blood pressure transducer.
- Turn on the ECG and blood pressure amplifiers.
- Start monitoring the ECG and blood pressure on a PC.
- Maintain the animal's head and body temperatures at 37°C using heating lamps.
- When both head and rectal temperatures reach 37°C, take blood samples for blood gases from the arterial catheter for glucose and hematocrit. For blood glucose analysis, we use a model 2300 STAT Plus (YSI, Yellow Springs, Ohio) (Figure 16.2b). However, a personal glucose monitoring device for diabetic patients also can be used. For hematocrit, we use regular hematocrit centrifuge and tubes.
- If blood gas values are not in the normal range, adjust ventilator settings to normalize them.
- If blood gas values are normal, inject vecuronium (2 mg/kg). Make sure to flush the catheter with heparinized saline.
- Lower the isoflurane to zero.
- Wait for 2 minutes.
- Induce asphyxial CA by disconnecting the ventilator from the endotracheal tube.
- In the meantime, decrease the nitrous oxide to zero and increase the oxygen to 2 L.
- Increase the rate of mechanical ventilation to 80 breaths/minute.
- After 6 or 8 minutes (depending on the desired duration of asphyxia CA), connect the ventilator to the endotracheal tube.
- Administer a bolus injection of epinephrine (0.005 mg/kg i.v.) and sodium bicarbonate (1 mEq/kg i.v.) and begin manual chest compressions (see diagrammatic representation of how to apply chest compressions in Figure 16.5) at the rate of 200/minute until mean arterial pressure (MAP) reaches 50 mm Hg and is maintained by the spontaneously beating heart for more than 10 seconds. If ROSC is not achieved within 2 minutes, we discard the animal.
- Measure arterial blood gases at 10 minutes of reperfusion.
- If any corrections in acid-base status are necessary, give sodium bicarbonate or adjust the ventilator settings.

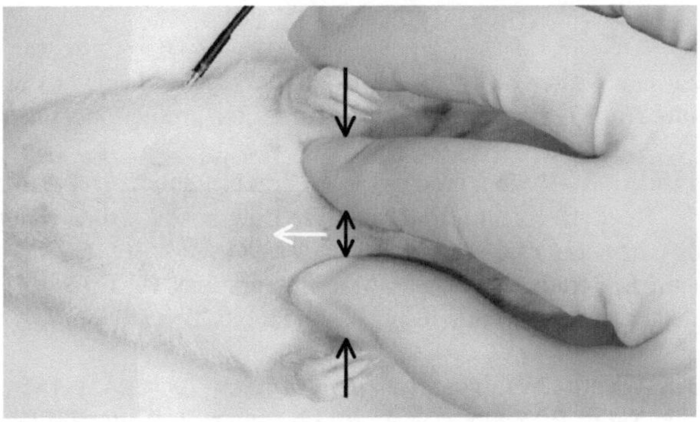

FIGURE 16.5 Direction of figure movement to resuscitate an animal. Black arrows indicate movement of thumb and index fingers from outside to inside. When the distance between both thumb and finger is minimized on the rat, move hand in forward direction (white arrow).

- After 10 minutes of ROSC, decrease the ventilator rate to 60 breaths/minute.
- Lower oxygen to 30% in a mixture with N_2O.
- Remove the catheters.
- Suture the incision with wound clips or 3-0 suture.
- Try disconnecting the animal from the ventilator. If the animal breathes by itself, leave it disconnected; if not, reconnect the animal to the ventilator and try the same after 5 minutes.
- Once the animal can breathe by itself, remove the endotracheal tube.
- Maintain the animal at 37°C for 45 to 60 minutes.
- Put the animal in a humidified fetal incubator.
- Provide liquefied rat chow in a Petri dish and provide a water bottle.
- Keep monitoring the animal for the first 6 hours.
- Leave the animal in the fetal incubator overnight.
- The rat can be returned to its normal cage the next morning.

PROBLEMS

Sometimes an animal will start to breathe by itself during the induction of asphyxia. This happens if the duration of vecuronium injection and induction of asphyxia is longer than 2 minutes because vecuronium is a short-lasting drug compared to pancuronium. We do not use pancuronium because in our experience the rate of resuscitation is poor. First, when vecuronium is injected at the time of connecting the animal to the ventilator following vascular access, we repeat the same dose of vecuronium every 10 minutes. Failure to do so results in spontaneous breathing of the animal during induction of asphyxia.

Another problem is the rate of successful resuscitation. Generally, the average rate of resuscitation is about 80%. With skilled persons or persons having more experience at resuscitation, this rate increases; persons with no experience will evince lower survival rates. With practice, this rate improves. Another compromising factor is the weight of the animal. In our experience, if the weight is more than 350 g, the rate of successful resuscitation drops drastically.

Generally, rats remain in a coma for about 6 to 12 hours following resuscitation. The next morning, the animal should be out of the coma. However, sometimes (about 20%) in the case of 6-minute CA and most of the time (about 60% to 80%) in the case of 8-minute CA, the animal cannot feed itself once out of the coma. In such cases, we gavage feed the animal. We soak a pellet of rat chow in the drinking water (two pellets in about 50 mL of water). We feed the animal by gavage with this semiliquid. We also inject 3.0 mL saline i.p. twice a day and provide a Petri dish with water-soaked rat chow pellets inside the cage. Generally, at 48 hours of resuscitation, the animal can feed itself.

REFERENCES

1. Krause, G.S., Kumar, K., White, B.C., Aust, S.D., and Wiegenstein, J.G., Ischemia, resuscitation, and reperfusion: Mechanisms of tissue injury and prospects for protection, *Am Heart J,* 111(4), 768–780, 1986.
2. Ginsberg, M.D. and Busto, R., Rodent models of cerebral ischemia, *Stroke,* 20(12), 1627–1642, 1989.
3. Katz, L.M., Wang, Y., Rockoff, S., and Bouldin, T.W., Low-dose Carbicarb improves cerebral outcome after asphyxial cardiac arrest in rats, *Ann Emerg Med,* 39(4), 359–365, 2002.
4. Takeuchi, T., Horiuchi, J., Terada, N., Nagao, M., and Terajima, H., Effects of hypoxia, hyperoxia and hypercapnia on graded cerebral ischemic responses in rabbits, *Am J Physiol,* 263(6 Pt 2), H1839–H1846, 1992.
5. Lipton, P., Ischemic cell death in brain neurons, *Physiol Rev,* 79(4), 1431–1568, 1999.
6. Katz, L., Ebmeyer, U., Safar, P., Radovsky, A., and Neumar, R., Outcome model of asphyxial cardiac arrest in rats, *J Cereb Blood Flow Metab,* 15(6), 1032–1039, 1995.
7. Dave, K.R., Raval, A.P., Prado, R., Katz, L.M., Sick, T.J., Ginsberg, M.D., Busto, R., and Perez-Pinzon, M.A., Mild cardiopulmonary arrest promotes synaptic dysfunction in rat hippocampus, *Brain Res,* 1024(1–2), 89–96, 2004.
8. Siemkowicz, E., Hyperglycemia in the reperfusion period hampers recovery from cerebral ischemia, *Acta Neurol Scand,* 64(3), 207–216, 1981.
9. Siemkowicz, E. and Gjedde, A., Post-ischemic coma in rat: effect of different pre-ischemic blood glucose levels on cerebral metabolic recovery after ischemia, *Acta Physiol Scand,* 110(3), 225–232, 1980.

17 Four-Vessel Occlusion Stroke Model in Rats

Yanlin Wang-Fischer and Lee Koetzner

CONTENTS

RATIONALE

The animal model of four-vessel occlusion (4-VO) is to induce global ischemia by occluding four major vessels (two vertebral and two common carotid arteries [CCAs]) that supply blood to the brain. This model mimics human cardiac arrest. The technique does not cause systemic hypoxia or hypotension and produces a sufficiently low incidence of seizures.[1–4] The surgical procedure is relatively simple compared to other stroke models. Large numbers of animals can be assessed statistically. Some brain areas predictably show ischemic neuronal damage. Areas that are most vulnerable to ischemic damage in this model included multiple regions of the hippocampus, the posterior neocortex, and the striatum. This selective vulnerability is similar to the vulnerability seen in other animal models of hypoxia-ischemia.[1,5] The technique of permanent occlusion of the vertebral arteries (VAs) allows control of cerebral circulation by the occlusion of the CCAs alone or with the occlusion of cervical branches of the subclavian arteries. The surgical approach and electrocauterization of the VAs through the alar foramen or groove of the first cervical vertebra are relatively simple and atraumatic. Vertebral artery occlusion can be completed before carotid artery occlusion. This procedure can be used for other animals which have a complete circle of Willis to provide collateral circulation to maintain vital brainstem centers. In studies on dogs[6,7] or monkeys,[8] occlusion of the VAs did not damage cardiorespiratory centers, which means animals will tolerate permanent VA occlusion.

STUDIES BASED ON THIS MODEL

Research on ischemic brain damage in small animals has been complicated by factors such as anesthetics, systemic hypoxia, hypotension, and convulsions. Gerbils were first used for global ischemia models.[9,10] The advantage of using Mongolian gerbils is that ischemic brain damage can be induced by occlusion of the carotid arteries due to a functionally incomplete circle of Willis.[9,10] The disadvantage of using gerbils is that only 30% to 40% of gerbils will show clinical signs of hemispheric ischemia after carotid occlusion (due to variations in the circle of Willis). Genetically, the gerbil is predisposed to seizures induced by mild stimuli,[11,12] and 75% of gerbils developed generalized seizures after unilateral carotid artery ligation.[11–13] It is well known that seizures increase cerebral metabolism,[14] which may damage neurons during the ischemic study. In human ischemic stroke, seizures are a rare (<10%) complication.[15,16] Avoiding animals with seizures is justified in studies of ischemic brain damage.

There are several methods available for inducing global cerebral hypoxia-ischemia in small animals. The methods fall into two categories: systemic-global ischemia and brain-focused global ischemic damage.

1. The systemic-global models for bilateral hemispheric ischemia in the rodent include:
   a. Ligating both carotid arteries and withdrawing blood to reduce blood pressure.[17]
   b. Injecting KCl intracardially to induce cardiac arrest.[18]

Since most animals have a well-developed circle of Willis, unilateral or bilateral carotid artery ligations have to be combined with hypoxia or hypotension to consistently interrupt cerebral metabolism and induce ischemic damage. The systemic-global models have one disadvantage: Systemic hypoxia and hypotension may damage other vital organs (heart, kidney, and liver) and thereby may change physiological mechanisms of postischemic survival. These models are less invasive surgically but lack clinical relevance.

2. The local global brain ischemic models include:
   a. Injecting artificial cerebrospinal fluid to increase the intracranial pressure.[19]
   b. Occluding four major vessels to block the blood supply to the brain.

The 4-VO model described in 1979 by Pulsinelli and Brierley[1] was the most-used technique and contributed greatly to research on global ischemia. This technique included two steps. Under anesthesia, the rat's VAs were electrocauterized through the alar foramina of the first cervical vertebra, and reversible clasps were placed loosely around the CCAs. The second step was carried out 24 hours later; the rats were restrained without anesthesia, and the carotid clasps tightened to induce 4-VO. The carotid clasps were removed after different durations of 4-VO (10, 20, or 30 minutes). The animals were sacrificed 72 hours later. Their results showed (1) All rats subjected to 20 or 30 minutes of 4-VO demonstrated ischemic neuronal damage. (2) The paramedian hippocampus, striatum, and layers 3, 5, and 6 of the posterior neocortex were the regions most frequently damaged.

The advantages of this model are (1) relative ease of preparation of large numbers of animals; (2) a high positive rate of ischemic neuronal injury; (3) a low incidence of seizures; and (4) less use of anesthesia.

The disadvantages of the two-step method are incomplete occlusion and the required two surgical procedures at a 24-hour interval, which may cause preconditioning and development of collateral circulation. This model was modified to produce a more consistent reduction in cerebral blood flow (CBF),[2,3] but the surgery itself is invasive and technically demanding. The surgical complications include leakage of cerebrospinal fluid, resulting in a decompression effect. Another complication is excessive tissue damage, which may cause neurological damage not associated with global ischemia.

Yamaguchi et al.[4] modified this model in 2005. They used a one-stage anterior approach to occlude the CCAs and VAs. This approach is a safe, easy, and less-invasive technique for producing bilateral hemispheric ischemia than the original 4-VO model. The VAs and CCAs can be completely occluded at the same time and reperfused for a desired duration. Therefore, consistent neurological deficits and morphological changes in bilateral hemisphere are obtained. The significantly reduced CBF produces the ischemic lesions in this modified 4-VO, and the lesions are consistent with that in the original 4-VO.[1]

STEP-BY-STEP SURGICAL PROCEDURES ON RATS

Here, both Pulsinelli and Brierley's two-stage method[1] and Yamaguchi et al.'s one-stage method[4] for global ischemia in rats are described.

Pulsinelli and Brierley's Two-Stage Method

Stage I

1. Rats weighing between 250 and 300 g are anesthetized.[1]
2. Both CCAs are isolated via a ventral midline cervical incision.
3. An arterial clasp or a size 2-0 silk suture is placed loosely around each CCA without interrupting carotid blood flow, and the incision is closed with a clip.
4. A second incision, about 1 cm long, is made behind the occipital bone directly overlying the first two cervical vertebrae.
5. The paraspinal muscles are separated from the midline, and the right and left alar foramina of the first cervical vertebrae are exposed (Figure 17.1).
6. A 0.5- to 1-mm-diameter electrocautery probe (Bovie 1250) (see Chapter 12, Figure 12.2) is inserted through each alar foramen, and both VAs are electrocauterized and permanently occluded. The electrocautery is grounded through the animal's foreleg with operational settings that produce minimal local muscle contraction.
7. The rats are allowed to recover from anesthesia and surgery for 24 hours.

Stage II

1. The rats are restrained by handhold (without anesthesia); the ventral neck clip or suture is removed, and both carotid clasps are tightened to produce 4-VO.
2. Carotid clasps are removed after 10, 20, or 30 minutes of 4-VO (the time is dependent on your study purpose), and restoration of carotid blood flow is verified by direct observation.

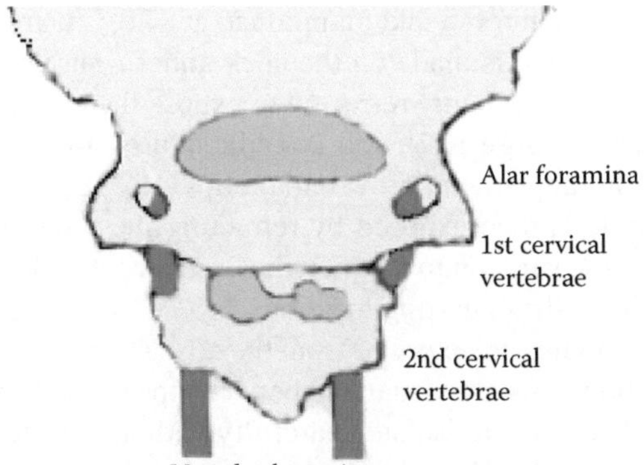

FIGURE 17.1 Schema showing the dorsal view of a rat skull with first and second cervical vertebrae and alar foramina of the first cervical vertebra. Vertebral arteries pass through the vertebral canal and beneath the alar foramina.

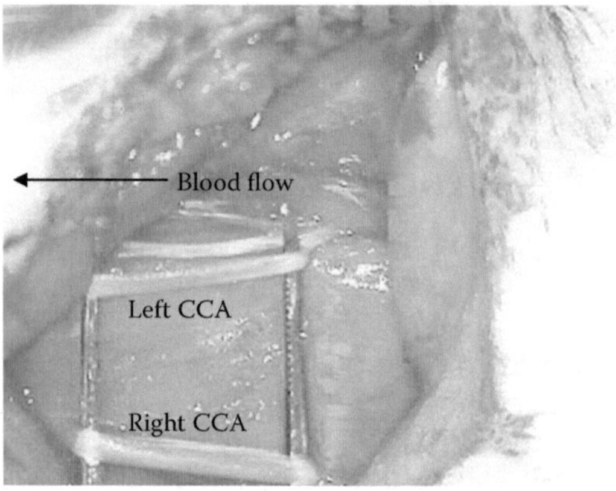

FIGURE 17.2 A 2-cm midline skin incision was made on the neck. The connective tissue and muscles were gently retracted. Both common carotid arteries (CCAs) were isolated.

3. The animals are observed at cage side for the level of consciousness, the presence or absence of a corneal reflex, and the ability to walk and climb. The body temperature is monitored and maintained at 37.0°C using a heating pad during the procedure.
4. Animals with sham operation undergo the same procedures with the exception of occlusion of vessels.

At 72 hours after ischemia, the rats are anesthetized and perfused transcardially with heparinized physiological saline (0.9% normal saline) to wash out the cephalic circulation (abdominal aorta clamped). The rats are perfusion fixed with PFA (10% paraformaldehyde). The brains are left *in situ* for 1 to 4 hours at 4°C before removal and then stored in PFA for histological analysis or for other immunohistochemical stain (refer to Chapter 19).

Yamaguchi et al.'s One-Stage Method

1. Rats weighing between 250 and 390 g are selected (body weight is based on the study project, but it is suggested the weight range be no more than 30 g in the same project). The rats are anesthetized and placed in the supine position. A rectal temperature probe is inserted, and body temperature is monitored and maintained at 37.0°C using a heating pad.[4]
2. A 2-cm midline skin incision is made on the neck under a surgical microscope; the connective tissue and muscles are gently retracted to expose the trachea and thyroid.
3. Both CCAs are isolated (Figure 17.2), and 2-0 silk sutures are placed loosely around the arteries as landmarks for applying microvascular clips (No. 18055-04, Fine Science Tools). The longus colli muscle is then exposed by retracting the trachea and esophagus to the right side. This allows the surgeon to identify the anterior tubercle of the atlas through the muscle, thereby verifying the operating location.
4. The cervical vertebral bodies are exposed from the atlas to the upper half of the fourth cervical vertebral body; both VAs are visualized between the second and third transverse processes (Figure 17.3). The VAs are isolated carefully without interfering with blood flow.
5. The right CCA, left CCA, right VA, and left VA are occluded in turn with microvascular clips.
6. The arteries are allowed to reperfuse for 10, 20, or 30 minutes after occlusion. The time depends on your study purpose.
7. Animals with sham operation undergo the same procedures with the exception of the vessel occlusion.

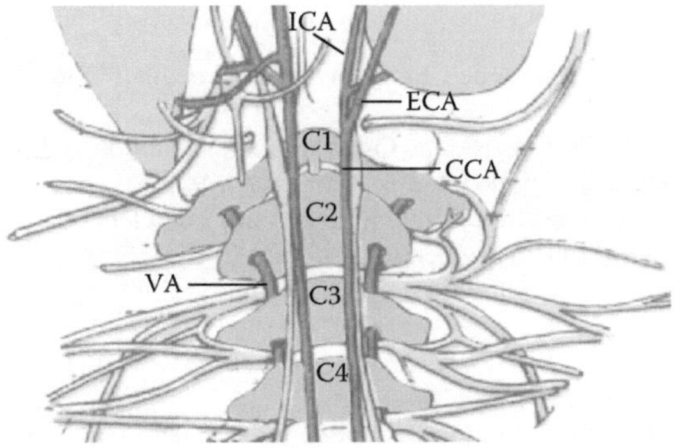

FIGURE 17.3 Schema showing an anterior view of the vertebral arteries (VAs), common carotid arteries (CCAs), and C1 to C4 vertebral bodies. The vertebral arteries pass through the vertebral canal between the transverse processes of C1 to C4. ECA, external carotid artery; ICA, internal carotid artery.

Combined Method

The combined method technique was also modified in 2004 by Poder's group[20]; they combined both Pulsinelli and Brierley's[1] and Yamaguchi et al.'s[4] methods. Next is a brief description of their method:

1. Rats are anaesthetized and intubated. Body temperature is monitored and maintained at 37.0°C using a heating pad.
2. The rats are placed on a head holder and tilted downward by approximately 30°.
3. The animals are ventilated during the surgical procedure with a respirator (TSE Animal Respirator "Advanced").[21]
4. The VAs are coagulated through the alar foramina of the first cervical vertebra (like Pulsinelli and Brierley's method).
5. Both CCAs are exposed via a ventral midline incision, and transient global forebrain ischemia is induced for 10 minutes by clamping both CCAs with microvascular clamps.
6. The animals are monitored for body temperature, respiration pattern, loss of righting reflex, unresponsiveness, corneal reflexes, as well as fixed and dilated pupils during ischemia.
7. Restoration of blood flow in the carotid arteries is confirmed visually. The incision is closed.
8. In the sham surgery group, rats undergo the same procedure except without vessel occlusion. In Poder et al.'s method, they cauterized both VAs, and both CCAs were exposed but not clamped.

PROBLEMS AND SOLUTIONS

Incomplete Occlusion

Incomplete occlusion occurs when VAs are cauterized incorrectly or CCAs are clamped incorrectly. Carefully checking the occluded vessels can reduce this problem.

Surgical Complications

Surgical complications include leakage of cerebrospinal fluid, resulting in a decompression effect and excessive tissue damage. A well-trained surgeon is required to perform this surgical procedure.

REFERENCES

1. Pulsinelli, W.A. and Brierley, J.B., A new model of bilateral hemispheric ischemia in the unanesthetized rat, *Stroke,* 10, 267, 1979.
2. Kameyama, M., Suzuki, J., Shirane, R., and Ogawa, A., A new model of bilateral hemispheric ischemia in the rat—Three vessel occlusion model, *Stroke,* 16, 489, 1985.
3. Shirane, R. et al., A new method for producing temporary complete cerebral ischemia in rats, *J Cereb Blood Flow Metab,* 11, 949, 1991.
4. Yamaguchi, M. et al., One-stage anterior approach for four-vessel occlusion in rat, *Stroke,* 36(10), 2212, 2005.
5. Brierley, J.B., Cerebral hypoxia. In: *Greenfield's Neuropathology*, Blackwood, W. and Corsellis, A.N., eds., Edward Arnold, London, 1976, pp. 43–85.
6. Kabat, H. and Dennis, C., Decerebration in the dog by complete temporary anemia of the brain, *Proc Soc Exp Biol Med,* 38, 864, 1938.
7. Boyd, R. and Connally, J., Total cerebral ischemia in the dog, *Arch Surg*, 84, 434, 1962.
8. Donald, D. and White, T., Temporary bilateral occlusion of the common carotid and vertebral arteries in the monkey at normal body temperature, *Neurology (Minneap),* 11, 836, 1961.
9. Levine, S. and Sohn, D., Cerebral ischemia in infant and adult gerbils. Relation to incomplete circle of Willis, *Arch Pathol,* 87, 315, 1969.
10. Cappuccio, I. et al., Induction of Dickkopf-1, a negative modulator of the Wnt pathway, is required for the development of ischemic neuronal death, *J Neurosci,* 25(10), 2647, 2005.
11. Loskota, W. and Lomax, P., The Mongolian gerbil as a model for the study of epilepsies, *Electroenceph Clin Neurophysiol,* 381, 597, 1975.
12. Seaman, R., Seaman, S., and Sun, A., Neurochemical correlates in seizure prone gerbils, Society for Neuroscience, *Abstr 7th Ann Meeting,* 3, 145, 1977.
13. Yanigihara, T., Experimental stroke in gerbils: correlation of clinical, pathological and electroencephalographic findings and protein synthesis, *Stroke,* 9, 155, 1978.
14. Howse, D., Caronna, J., Duffy, T., and Plum, F., Cerebral energy metabolism, pH and blood flow during seizures in the cat, *Am J Physiol,* 227, 1444, 1974.
15. Richardson, E. and Dodge, P., Epilepsy in cerebrovascular disease, *Epilepsia 3* (3rd series), 49, 1954.
16. Louis, S. and McDowell, F., Epileptic seizures in nonembolic cerebral infarction, *Arch Neurol*, 17, 414, 1967.
17. Eklof, B. and Siesjo, B.K., The effect of bilateral carotid artery ligation upon the blood flow and the energy state of the rat brain, *Acta Physiol Scand*, 86, 155, 1972.
18. Blomqvist, P. and Wieloch, T., Ischemic brain damage in rats following cardiac arrest using a long-term recovery model, *J Cereb Blood Flow Metab*, 5, 420, 1985.
19. Ljunggren, B., Schutz, H., and Siesjo, B.K., Changes in energy state and acid-base parameters of the rat brain during complete compression ischemia, *Brain Res,* 73, 277, 1974.
20. Poder, P. et al., An antioxidant tetrapeptide UPF1 in rats has a neuroprotective effect in transient global brain ischemia, *Neurosci Lett,* 370, 45, 2004.
21. Jou, I.M. et al., Simplified rat intubation using a new oropharyngeal intubation wedge, *J Appl Physiol,* 89, 1766, 2000.

18 Brain Hemorrhage Models in Rodents

Yanlin Wang-Fischer and Lee Koetzner

CONTENTS

RATIONALE

Brain hemorrhage includes subarachnoid and intracerebral hemorrhage. The aim of surgically produced brain hemorrhage in animals is to model the neurobiological effects of brain hemorrhage in humans. The most common hemorrhage model is subarachnoid hemorrhage (SAH). There are different approaches for inducing this condition, including prechiasmatic or intracisternal administration of blood and endovascular puncture of a brain artery by a nylon filament suture (refer to the complication in Chapter 13 filament suture-induced stroke model). The prechiasmatic method has the advantage that the injected blood is mostly distributed around the basal arteries, which is clinically relevant, and the amount of blood is controllable, while the intracisternal approach leads to lower mortality. Endovascular perforation of a brain artery mimics the clinical situation of aneurysm rupture, which is associated with high mortality, but this model has the disadvantage of uncontrollable bleeding. SAH is accompanied by acute ischemic injury at the time of the initial hemorrhage (48 hours),[1–3] followed by delayed vasospasm, which leads to the development of ischemia at 3 to 7 days after the initial bleed.[4–6] Acute ischemic brain injury and reduced cerebral blood flow (CBF) after SAH are the most important factors leading to a poor outcome after hemorrhage.[7–9] Both in humans and in experimental models, there is a marked increase in the permeability of cerebral capillary microvessels during the acute phase of SAH.[10–14] This permeability change has been correlated with the development of delayed cerebral ischemia and poor clinical outcome.[11,12,15–18] Animal models are important for (1) understanding the mechanisms underlying SAH in the nervous system and (2) the discovery of drugs to reduce or prevent neuronal damage and SAH-induced deficits in motor activity and cognitive functions.

STUDIES USING THIS MODEL

Approximately 40% of patients who suffer from SAH do not survive; the surviving patients have a high incidence of important neurological and cognitive disturbances.[19,20] Most of the deaths occur within hours after the bleeding,[7] and it is well established that the events taking place during the first minutes after SAH are critical for the subsequent outcome.[19,21–23] Pathological examinations of the brains from the patients who died soon after SAH showed extensive ischemic injury.[24] The damage from SAH is similar to other brain injuries leading to secondary injuries that disturb intracranial dynamics,[25] including increased intracranial pressure (ICP), decreased CBF, and disordered cerebral metabolism. Because scientific investigations of the condition occurring immediately after SAH are not possible with patients and there are no computerized models for this condition, animal models are critical. Rats have become increasingly popular due to cost issues.

Varsos et al. (1983)[26] induced an SAH model in dogs by cisternal double injection of blood 48 hours apart; this idea was adapted for modeling SAH in rats.[27] The ideal model should meet several criteria, including (1) the mechanisms of hemorrhage and blood distribution should be consistent with those of aneurysm rupture in humans; (2) the model should be highly reproducible with a reasonable rate of pathological and pathophysiological findings; (3) the model should be suitable for study of both the acute and late phases of SAH; (4) the surgical procedure should be easy to perform; and (5) the mortality rate should be acceptable.

The three most common techniques used to model brain hemorrhage are:

1. *The vessel disruption model or intraluminal perforation model*, both forms of SAH, is produced via intracranial puncture of the basilar artery[28] or internal carotid artery.[29,30]
2. *The blood injection model* is produced by injecting blood into the brain either at a selected pressure, preferably the mean arterial blood pressure (MABP),[31–33] or at an arbitrary pressure (in a preselected amount of blood) into the subarachnoid space—preferably the cisterna magna.[34–38] This technique is the most commonly used in the SAH model.
3. *The collagenase injection model* is produced by injecting collagenase into the brain to induce intracerebral hemorrhage.[39]

After experimental hemorrhage in rats, CBF has been observed to decrease for at least 1 hour.[40,41]

Prunell et al. (2003)[32] compared three different procedures in rats: (1) intraluminal perforation model; (2) blood injection to the prechiasmatic cistern; and (3) blood injection to the cisterna magna. They investigated which of the three methods is the most suitable for studying pathological and pathophysiological processes after SAH and the relationships between subarachnoid blood volume, CBF, MABP, ICP, neuronal death, and animal death after SAH. They measured the amount of blood in the subarachnoid space and CBF changes during the first 90 minutes of SAH. They showed that severity of damage is as follows: intraluminal perforation model > blood injection (200 μL) in the prechiasmatic cistern > blood injection into the cisterna magna (300 μL). They concluded that endovascular perforation SAH mimicked clinical SAH but had low reproducibility and lacked adequate controls; the cisterna magna SAH mimicked posterior fossa SAH but only mildly affected CBF and did not cause histological damage; the prechiasmatic SAH was highly reproducible and led to significant CBF and histological changes reminiscent of clinical supratentorial SAH.[32]

In 2004, Prunell et al.[42] further compared the metabolic changes in the brain and their connections to CBF changes during the decisive first minutes after SAH with these three different methods. They hypothesized (1) that experimental SAH causes acute metabolic depression and global ischemia and (2) that the metabolic changes are, at least in part, independent of the CBF changes. They measured cerebral perfusion pressure (CPP), CBF, oxygen tension, cerebral arteriovenous difference of oxygen ($AVDO_2$), and brain energy metabolite levels in the acute phase of SAH and the relationships between them. The comparison of three different methods can assist future researchers in the area to choose the optimal method and to familiarize with the bias incurred by using any single method. Their results showed

1. Following a transient reduction, CPP normalized within 5 minutes after SAH in all methods.
2. There was a transient global decrease in CBF after SAH; its duration depended on the severity of the hemorrhage (CBF of less than 20% of baseline was observed for at least 15 minutes in 25% of the animals after endovascular perforation and 14% of the animals after prechiasmatic SAH).
3. In all SAH models, O_2 tension was suddenly reduced to approximately 40% of baseline and gradually increased, reaching 70% to 90% of baseline at 90 minutes after SAH.
4. The cerebral metabolic rate of O_2 was decreased only at 15 minutes after endovascular perforation and prechiasmatic SAH, but arteriovenous O_2 difference was normal in all models.
5. In the endovascular perforation SAH, glucose decreased 50%, and lactate and pyruvate levels increased threefold during the 30 minutes after perforation. They concluded that SAH induced an acute global decrease in CBF together with a depression in the cerebral metabolism. The degree of the changes was related to the severity of the hemorrhage. The metabolic derangements were not always explained by ischemic episodes.[42]

Structural changes in cerebral microvessels (≤100 μm) after SAH have been evaluated by Sehba et al. (2004).[14] Cerebral capillary microvessels consist of endothelial cells, basal lamina, a constituent of the vascular extracellular matrix, and astrocyte end feet. Collagen IV is up to 90% of the total protein of the basal lamina and contributes structural integrity to the vessel wall.[43,44] The microglia, astrocytes, and endothelial cells in the brain can secrete matrix metalloproteinases (MMPs) and serine proteases.[45] Among the MMP family, gelatinase A (MMP-2) and gelatinase B (MMP-9) can digest the vascular basal lamina.[46] The substrates of MMP-2 and MMP-9 include gelatin, type IV collagen, fibronectin, and elastin.[47,48] Sehba and colleagues found a marked loss of collagen IV from basal lamina together with increases in both microvascular MMP-9 levels and collagenase activity in the first 6 hours after brain hemorrhage. Their study showed that microvascular collagenases were released and activated immediately after brain hemorrhage; this pathological change may affect acute destruction of the basal lamina and play an important role in the degradation of parenchymal vascular function. The technique of injecting collagenase into the brain to produce intracerebral hemorrhage model is based on the function of the MMP family.

STEP-BY-STEP SURGICAL PROCEDURES ON RATS

ANIMALS

Male Sprague-Dawley rats or Wistar rats weighing 300 to 400 g are used in the experiments.

Anesthesia and Catheterization

1. Animals are anesthetized with isoflurane (3% to 5% via an induction chamber and maintained with 1.5% to 2% via an anesthetic gas tubing attached to the mouthpiece on the stereotaxic frame) in a mixed gas of 70% nitrous oxide and 30% oxygen. Alternate anesthetics include pentobarbital (50 mg/kg i.p.) or ketamine (60 mg/kg) plus xylazine (10 mg/kg i.m.); the choice of anesthetics is based on your study need.
2. Body temperature is maintained at 37°C ± 0.5°C with an automatic heating pad.
3. The animals are placed in a stereotaxic frame (David Kopf Instruments). The frame was modified to allow longitudinal rotation to permit surgery in the supine or prone position.
4. The femoral arteries are cannulated with PE-50 catheters in acute experiments (if animals will be killed in the 90 minutes after SAH) for measurements of blood gas values, blood pH, hematocrit values, and MABP.
5. The tail artery can be cannulated with a PE-10 catheter in long-term experiments (if animals will be sacrificed in the 7 days after SAH). One femoral vein (in acute experiments) or a tail vein (in long-term experiments) is cannulated for infusion of drugs.
6. All animals receive 5 mL of physiological saline solution intraperitoneally after the surgical procedure.

Intracranial Pressure

Intracranial pressure (ICP)[32] is measured with a polyethylene catheter (1.22-mm outer diameter, 0.75-mm inner diameter) implanted in the cisterna magna.

1. The atlantooccipital membrane is exposed, and a small longitudinal incision is made, including the arachnoid membrane.
2. The tip of the catheter is heated to yield a collar-like form. The collar is inserted into the incision, like a button into a buttonhole.
3. Super Glue (Office Max) or histoacryl (B. Braun Surgical, Melsungen, Germany) is used to prevent leakage of cerebrospinal fluid (CSF).
4. In the cisterna magna SAH group, an additional catheter is implanted in the cisterna magna for administration of blood or saline solution.

Cerebral Blood Flow (CBF)

CBF is measured by laser Doppler flowmetry (LDF). L*aser Doppler blood flow monitors and imagers* were purchased from Moor Instruments (www.moor.co.uk). The data collector (Model No. ML880, PowerLab 16/30) and software (Chart 5) for the system were purchased from AD Instruments (www.adinstruments.com).

1. Small holes (2-mm diameter) are drilled bilaterally over the frontal cortices (3 mm anterior to the coronal suture and 2 mm lateral to the midline), and the optical fibers (probes) are placed at the epidural surface, away from large vessels.
2. The probes are glued to the dura with histoacryl or Super Glue. The signals are monitored with a laser Doppler monitor (for detailed information, refer to Chapter 19).

Surgical Procedures for Prechiasmatic Subarachnoid Hemorrhage

The following are the surgical procedures for prechiasmatic SAH[49]:

1. The anesthetized animal is mounted in a stereotaxic frame.

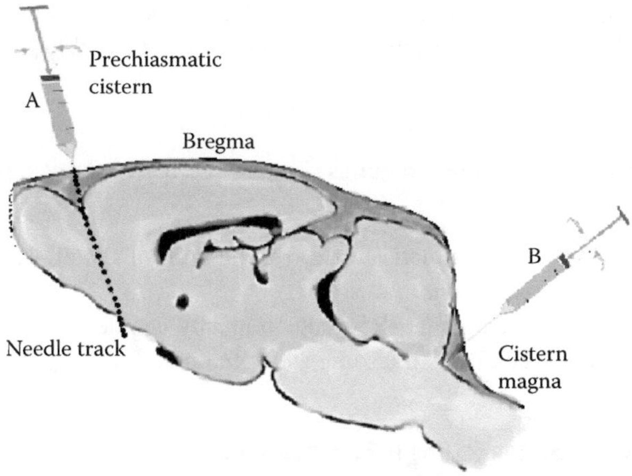

FIGURE 18.1 Diagram of needle tracks for inducing prechiasmatic subarachnoid hemorrhage (SAH) (needle A) and of cisternal puncture and injection of blood for cisterna magna SAH (needle B).

2. An incision is made in the top of the skull; a hole (2-mm diameter) is made using a dental drill at the site 7.5 mm anterior to the bregma, 1 mm lateral to the midline.
3. A 27-gauge needle is inserted into the prechiasmatic cisterna. The needle is tilted 30° in the sagittal plane with the bevel of the needle angle caudal to the hub and lowered until the tip reaches the base of the skull, 2 to 3 mm anterior to the chiasm (Figure 18.1A).
4. Then, 300 to 400 µL of autologous blood (taken from the orbital sinus of the eye or an arterial cannula) or saline (sham surgery) are injected manually over 2 minutes.
5. The needle is removed and the hole filled with bone wax or Super Glue to prevent CSF leakage.
6. The wound is sealed, and animals are returned to their cages.

All procedures will take about 20 to 30 minutes. Based on our previous experience, most animals can survive these procedures. The immediate mortality is less than 10%, and the 7-day mortality is 25%.[31]

Procedures for Cisterna Magna Subarachnoid Hemorrhage

There are two ways to puncture the cisterna magna: one surgical and the other not.

Nonsurgical Procedure

1. After the rat is anesthetized, it is placed in a side position with the head bowed toward the chest to reveal a prominent external occipital protuberance in the neck region.
2. Directly caudal to this, a depression can be felt between the protuberance and the spinous process of the atlas; this locates the atlantooccipital membrane.
3. A 25-gauge butterfly needle attached to a polyethylene tube and connected to a 1-mL syringe is slowly pushed into the center of this depression. On entrance to the cisterna magna, a sudden decrease in resistance should be felt (Figure 18.1B).
4. CSF flows immediately; 0.1 mL of CSF is then gently aspirated.
5. Freshly drawn blood (0.2 to 0.3 mL) from the orbital sinus or an arterial catheter or the same amount of physiological saline (for sham control) is then injected into the cisterna magna over a period of 2 to 3 minutes.
6. Immediately after the injection of blood, the hole is sealed with glue to prevent fistula.

7. To permit blood distribution around the basal arteries, the animal is tilted at 20° for 30 minutes in a head-down position, and then the animal is returned to its cage.

Surgical Procedure

Because the nonsurgical cisterna puncture requires skill, an alternative surgical procedure can be used.

1. A small suboccipital incision is made to expose the arch of the atlas, the occipital bone, and the atlantooccipital membrane.
2. The cisterna magna is tapped using a 25-gauge butterfly needle.
3. The remainder of the procedure is the same as for the nonsurgical procedure.

Surgical Procedures for Subarachnoid Hemorrhage by Internal Carotid Artery Perforation

The perforation SAH model was established according to the techniques described by Bederson et al.[29] and Veelken et al.[30] with a slight modification. Basically, the surgical steps are the same as the filament ischemic stroke model (see Chapter 13) except that the filament has a sharpened end and is not coated.

1. The bifurcation of the right common carotid artery and the external and internal carotid arteries is exposed, and the external carotid artery is ligated and dissected distally.
2. A small cut is made in the external carotid artery stump; a size 3-0 monofilament suture (Ethicon, Somerville, New Jersey) is advanced into the internal carotid through the external carotid artery approximately 21 to 24 mm from the common carotid bifurcation and then withdrawn.
3. Sham-operated rats undergo the same procedure, but the suture is advanced only 18 mm (Figure 18.2).

Collagenase-Induced Intracerebral Hemorrhage

Type IV collagen is the predominant component of brain basal lamina. Collagenase occurs in an inactive form in cells; brain injury induces collagenase release and activation, which leads to digestion of the extracellular matrix. Collagenase-induced intracerebral hemorrhage is a reproducible animal model for the study of the effects of the hematoma and brain edema. We introduce Rosenberg et al.'s (1990) technique[39] for collagenase-induced SAH.

Surgical Procedures

1. Under anesthesia, rats are placed into a stereotaxic apparatus (David Kopf Instruments, Tujunga, California).
2. A 23-gauge needle is implanted into the caudate nucleus at coordinates (A5.8, L3.0, H1.0).
3. Infuse 2 μL of saline containing 0.01 to 1 unit bacterial collagenase (type XI or type VII, Sigma Chemical, St. Louis, Missouri) by a microinfusion pump (Harvard Apparatus, South Natick, Massachusetts) over 9 minutes.
4. Control rats are infused with 2 μL normal saline.
5. After infusion, the needle is removed and the wound sutured.
6. The rats are allowed to recover from surgery in a warm place with access to food and drink.
7. Sacrifice is by intracardiac injection of KCl.

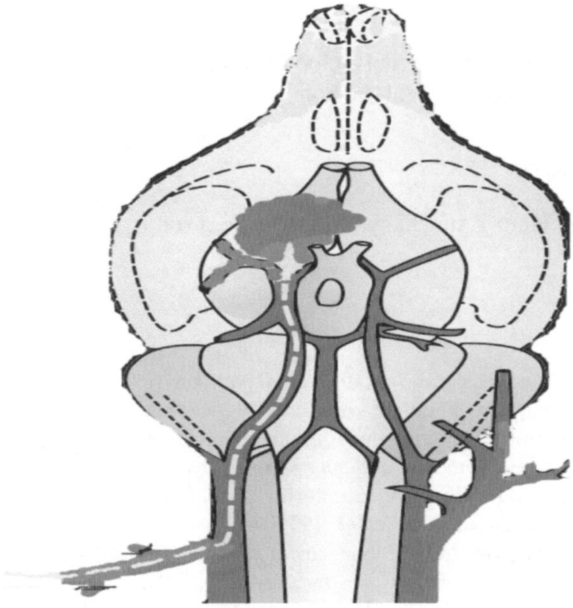

FIGURE 18.2 Diagram of the endovascular suture technique for inducing subarachnoid hemorrhage in one hemisphere. The technique is similar to the suture stroke model. A 3-0 monofilament suture sharpened at one end is introduced into the internal carotid artery and is pushed into the internal carotid artery (ICA) wall near its bifurcation with the middle cerebral artery. Withdrawal of the suture results in subarachnoid hemorrhage.

Changes after Collagenase Injection

Regarding changes after collagenase injection,[39] initially type XI collagenase with some protease contamination was used. Subsequent studies were done with type VII collagenase that was essentially free of proteases.

1. Within the first hour; erythrocytes were seen around blood vessels at the needle puncture site.
2. At 4 hours after injection, hematomas had occurred; the size depended on the amount of collagenase injected.
3. At 24 hours, necrotic masses containing fluid, blood cells, and fibrin were seen.
4. At 7 days, lipid-filled macrophages were observed.
5. At 3 weeks, cysts were observed.
6. Water content was significantly increased at the needle puncture site at 4, 24, and 48 hours after infusion and at 24 hours after injection in posterior brain sections.
7. Behavioral abnormalities were present for 48 hours, with recovery of function occurring during the first week.

PROBLEMS AND SOLUTIONS

Mortality

There is a high mortality rate in perforation SAH, up to 44% in 24 hours. A careful choice of model is necessary before beginning a study.

Blood Leakage

In cisterna and prechiasmatic SAH models, blood leakage occurs after the injection. Super Glue or bone wax is immediately needed to seal the hole.

ACKNOWLEDGMENTS

We thank Drs. Zhihong Haung and Jianya Ma from Johnson & Johnson, PRD, for their technical assistance and advice.

REFERENCES

1. Jackowski, A. et al., The time course of intracranial pathophysiological changes following experimental subarachnoid hemorrhage in the rat, *J Cereb Blood Flow Metab*, 10, 835, 1990.
2. Jakobsen, M., Role of initial brain ischemia in subarachnoid hemorrhage following aneurysm rupture, a pathophysiological survey, *Acta Neurol Scand Suppl*, 141, 1, 1992.
3. Persson, L. and Hillered, L., Chemical monitoring of neurosurgical intensive care patients using intracerebral microdialysis, *J Neurosurg*, 76, 72, 1992.
4. Al-Yamany, M. and Wallace, M.C., Management of cerebral vasospasm in patients with aneurysmal subarachnoid hemorrhage, *Intensive Care Med,* 25, 1463, 1999.
5. Grosset, D.G. et al., Angiographic and Doppler diagnosis of cerebral artery vasospasm following subarachnoid hemorrhage, *Br J Neurosurg,* 7, 291, 1993.
6. Weir, B. et al., Time course of vasospasm in man, *J Neurosurg,* 48, 173, 1978.
7. Broderick, J.P. et al., Initial and recurrent bleeding are the major causes of death following subarachnoid hemorrhage, *Stroke,* 25, 1342, 1994.
8. Findlay, J.M. and Deagle, G.M., Causes of morbidity and mortality following intracranial aneurysm rupture, *Can J Neurol Sci,* 25, 209, 1998.
9. Roos, Y.B. et al., Complications and outcome in patients with aneurysmal subarachnoid haemorrhage: A prospective hospital based cohort study in the Netherlands, *J Neurol Neurosurg Psychiatry,* 68, 337, 2000.
10. Doczi, T. et al., Blood–brain barrier damage during the acute stage of subarachnoid hemorrhage, as exemplified by a new animal model, *Neurosurgery,* 18, 733, 1986.
11. Doczi, T. et al., Increased vulnerability of the blood–brain barrier to experimental subarachnoid hemorrhage in spontaneously hypertensive rats, *Stroke,* 17, 498, 1986.
12. Germano, A. et al., Time-course of blood-brain barrier permeability changes after experimental subarachnoid haemorrhage, *Acta Neurochir,* 142, 575, 2000.
13. Peterson, J.W. et al., Evidence of the role of hemolysis in experimental cerebral vasospasm, *J Neurosurg,* 72, 775, 1990.
14. Sehba, F.A. et al., Acute alterations in microvascular basal lamina after subarachnoid hemorrhage, *J Neurosurg,* 101, 633, 2004.
15. Doczi, T., The pathogenetic and prognostic significance of blood–brain barrier damage at the acute stage of aneurysmal subarachnoid haemorrhage. Clinical and experimental studies, *Acta Neurochir,* 77, 110, 1985.
16. Germano, A. et al., Blood–brain barrier permeability changes after experimental subarachnoid hemorrhage, *Neurosurgery,* 30, 882, 1992.
17. Imperatore, C. et al., Effects of the radical scavenger AVS on behavioral and BBB changes after experimental subarachnoid hemorrhage, *Life Sci,* 66, 779, 2000.
18. Symon, L., Disordered cerebro-vascular physiology in aneurismal subarachnoid haemorrhage, *Acta Neurochir,* 41, 7, 1978.
19. Ljunggren, B. et al., Aneurysmal subarachnoid hemorrhage: total annual outcome in a 1.46 million population, *Surg Neurol,* 22, 435, 1984.
20. van Gijn, J., and Rinkel, G.J., Subarachnoid haemorrhage: Diagnosis, causes and management, *Brain,* 124, 249, 2001.
21. Edner, G., Kagstrom, E., and Wallstedt, L., Total overall management and surgical outcome after aneurysmal subarachnoid haemorrhage in a defined population, *Br J Neurosurg,* 6, 409, 1992.
22. Jakobsen, M., Skjodt, T., and Enevoldsen, E., Cerebral blood flow and metabolism following subarachnoid haemorrhage: Effect of subarachnoid blood, *Acta Neurol Scand,* 83, 226, 1991.

23. Säveland, H. and Brandt, L., Which are the major determinants for outcome in aneurysmal subarachnoid hemorrhage? A prospective total management study from a strictly unselected series, *Acta Neurol Scand,* 90, 245, 1994.
24. Crompton, M.R., The pathogenesis of cerebral infarction following the rupture of cerebral berry aneurysms, *Brain,* 87, 491, 1964.
25. Knuckey, N.W., Fox, R.A., Surveyor, I., and Stokes, B.A., Early cerebral blood flow and computerized tomography in predicting ischemia after cerebral aneurysm rupture, *J Neurosurg,* 62, 850, 1985.
26. Varsos, V.G. et al., Delayed cerebral vasospasm is not reversible by aminophylline, nifedipine, or papaverine in a "two-hemorrhage" canine model, *J Neurosurg,* 58, 11, 1983.
27. Takeuchi, K. et al., Reversal of delayed vasospasm by an inhibitor of the synthesis of 20-HETE, *Am J Phys Circ Physiol,* 289, H2203, 2005.
28. Barry, K.J., Gogjian, M.A., and Stein, B.M., Small animal model for investigation of subarachnoid hemorrhage and cerebral vasospasm, *Stroke,* 10, 538, 1979.
29. Bederson, J.B., Germano, I.M., and Guarino, L., Cortical blood flow and cerebral perfusion pressure in a new noncraniotomy model of subarachnoid hemorrhage in the rat, *Stroke,* 26, 1086, 1995.
30. Veelken, J.A., Laing, R.J., and Jakubowski, J., The Sheffield model of subarachnoid hemorrhage in rats, *Stroke,* 26, 1279, 1995.
31. Klinge, P.M. et al., Induction of heat shock protein 70 in the rat brain following intracisternal infusion of autologous blood: Evaluation of acute neuronal damage, *J Neurosurg,* 91, 843, 1999.
32. Prunell, G.F. et al., Experimental subarachnoid hemorrhage: Subarachnoid blood volume, mortality rate, neuronal death, cerebral blood flow, and perfusion pressure in three different rat models, *Neurosurgery,* 52, 165, 2003.
33. Zhao, W. et al., Sudden death in a rat subarachnoid hemorrhage model, *Neurol Med Chir (Tokyo),* 39, 735, 1999.
34. Delgado, T.J., Brismar, J., and Svendgaard, N.A., Subarachnoid haemorrhage in the rat: Angiography and fluorescence microscopy of the major cerebral arteries, *Stroke,* 16, 595, 1985.
35. Lacy, P.S. and Earle, A.M., A small animal model for electrocardiographic abnormalities observed after an experimental subarachnoid hemorrhage, *Stroke,* 14, 371, 1983.
36. Marzatico, F. et al., Experimental isobaric subarachnoid hemorrhage: Regional mitochondrial function during the acute and late phase, *Surg Neurol,* 34, 294, 1990.
37. Piepgras, A., Thome, C., and Schmiedek, P., Characterization of an anterior circulation rat subarachnoid hemorrhage model, *Stroke,* 26, 2347, 1995.
38. Solomon, R.A. et al., Decrease in cerebral blood flow in rats after experimental subarachnoid hemorrhage: A new animal model, *Stroke,* 16, 58, 1985.
39. Rosenberg, G.A. et al., Collagenase-induced intracerebral hemorrhage in rats, *Stroke,* 21, 801, 1990.
40. Bederson, J.B. et al., Acute vasoconstriction after subarachnoid hemorrhage, *Neurosurgery,* 42, 352, 1998.
41. Schwartz, A.Y. et al., Experimental models of subarachnoid hemorrhage in the rat: A refinement of the endovascular filament model, *J Neurosci Methods,* 96, 161, 2000.
42. Prunell, G.F., Mathiesen, T., and Svendgaard, N.A., Experimental subarachnoid hemorrhage: Cerebral blood flow and brain metabolism during the acute phase in three different models in the rat, *Neurosurgery,* 54, 426, 2004.
43. Bosman, F.T. et al., Basement membrane heterogeneity, *Histochem J,* 21, 629, 1989.
44. Stanley, J.R. et al., Structure and function of basement membrane, *J Invest Dermatol,* 79(Suppl 1), 69s, 1982.
45. Mun-Bryce, S. and Rosenberg, G.A., Matrix metalloproteinases in cerebrovascular disease, *J Cereb Blood Flow Metab,* 18, 1163, 1998.
46. Rosenberg, G.A., Matrix metalloproteinases in brain injury, *J Neurotrauma,* 12, 833, 1995.
47. Clark, A.W. et al., Increased gelatinase A (MMP-2) and gelatinase B (MMP-9) activities in human brain after focal ischemia, *Neurosci Lett,* 238, 53, 1997.
48. Lukes, A. et al., Extracellular matrix degradation by metalloproteinases and central nervous system diseases, *Mol Neurobiol,* 19, 267, 1999.
49. Prunell, G.F., Tiit Mathiesen, T., and Svendgaard, N.A., A new experimental model in rats for study of the pathophysiology of subarachnoid hemorrhage, *NeuroReport,* 13, 2553, 2002.

19 Endpoints for Stroke Studies

Yanlin Wang-Fischer and Lee Koetzner

CONTENTS

INTRODUCTION

Animal models of stroke, like any other disease model, begin with an intervention (to produce a pathological state) and end with an endpoint (to measure the pathological state). These endpoints can address a variety of questions, each important in a different experimental context. As a research group begins work on a model, the most important endpoints measure the group's ability to produce a workable disease model. When this is established, questions turn to understanding the pathological processes in the model or interventions that reduce the pathology of the model. These differing purposes have resulted in the introduction of many different endpoints for stroke models; this chapter serves as a guide to those endpoints.

Endpoints aimed at measuring the model prove useful in the setup process as well as establishing the impact of possible confounding variables. Histochemical techniques can establish the tissue- and cell-level impact of a model; these techniques are described in Chapter 20. Magnetic resonance imaging (MRI), while newer, is also useful; it is described in Chapter 22. In ischemic stroke models, measurements of cerebral blood flow (CBF) provide immediate feedback on whether the model produced sufficient ischemia. Hemorrhage measurements assess the success of intracerebral hemorrhage and can also assess the complications of filament occlusion models. Measurements of body temperature, edema, and intracranial pressure address complications that each confound study results and make data interpretation difficult. These techniques are described in this chapter.

Other endpoints are most useful for understanding the workings of the model. Collection of cerebrospinal fluid (CSF) allows an investigator to directly measure changes in transmitters, hormones, cytokines, or any other factor that might be important. Changes in blood–brain barrier (BBB) function, common in stroke models, contribute to changes in signaling molecules in the brain. Experimental methods for CSF collection and BBB assessment are described in this chapter.

Finally, there are several endpoints that have been very useful for measuring the ability of a drug to prevent or reverse the effects of a stroke model. Histochemical and MRI techniques are useful for these studies. In addition, neurological and behavioral tests provide important tests of functional impairment. The broad range of these tests reflects the broad range of impairments seen in stroke models. These functional assays are described in this chapter.

TECHNIQUES FOR DIFFERENT ENDPOINT MEASUREMENTS

INTRACRANIAL PRESSURE

Intracranial pressure (ICP) is a concern following stroke. Various methods for measuring ICP in rodents have been studied, including methods utilizing epidural monitors[1,2] and subdural,[3] ventricular,[4–8] lumbar, or cisterna magna catheters.[9–15] Each method has its advantages and disadvantages. Mandell[6] developed a method to continuously measure CSF pressure in unrestrained rats through an intraventricular cannula. In 1988, Andrews et al.[1] described an epidural monitor that is an elegant and reliable system for measuring acute changes in ICP; the limitation of this technique is that the hardware is not permanent and cannot be used for later measurements without subjecting the rat to a second operation. In 1989, Rahimifar et al.[3] developed a permanent subdural catheter for ICP monitoring and injection in the cisterna magna of the rat. ICP measurements with a catheter placed directly into the cisterna magna through an atlantooccipital membrane puncture have been presented.[11,13,14] These catheters are very useful for the immediate measurement of ICP but are not designed for permanent implantation.

To develop a reliable and simple system to measure ICP in the rat, Barth et al. (1992)[16] developed a permanent cisterna magna cannula. The rat was allowed to recover from the mechanical effects of catheter placement or spillage of CSF that occurred during insertion of this permanent catheter. This technique is relatively easy and the catheter simple to insert. The catheter is inexpensively constructed from readily available laboratory materials. This technique provides consistently reliable

ICP readings at 48 hours after placement and during experimental procedures. Their data showed that ICP varied between 1.0 and 10.0 cm H_2O, with a mean of 5.6 cm H_2O (about 3.6 mm Hg, where 1 mm Hg = 1.3595 cm H_2O) in 12 normal rats. The limitation of this method is that the rat is anesthetized. The original catheter is not designed for continuous monitoring of ICP in a conscious rat, but it may be modified, for example, using a telemetry system connected to the catheter. Zwienenberg et al. (1999)[17] compared simultaneous ICP measurements, including ventricular, cisterna magna, and intraparenchymal monitoring, during subarachnoid hemorrhage. They found that the ventricular and the intraparenchymal fiber-optic catheters produced reliable and comparable pressure recordings throughout 1 hour of monitoring time (no significant difference; $p = 0.4$). The cisterna magna catheter produced less-reliable and significantly lower readings throughout the monitoring time ($p < .001$) but caused less damage than the other two methods. The intraparenchymal device produced greater cortical damage than the ventricular catheter. Kusaka et al. (2004)[18] reported a new lumbar method for monitoring ICP in rats. They used a PE-10 catheter placed in the subarachnoid space at L5 through the dura mater after laminectomy to measure lumbar CSF pressure. ICP at the cisterna magna was measured simultaneously via a PE-10 catheter in the subarachnoid space at the cisterna magna. They found that the baseline lumbar ICP and cisterna ICP varied between 6 and 8 mm Hg, and respiratory variation could be detected by this method. A similar acute response to subarachnoid hemorrhage (SAH) was measured in both the lumbar ICP and cisterna ICP in rats. However, this technique is very delicate and requires a high degree of training.

Because most of the published techniques are based on Barth et al.'s method,[16] we describe it here as a basic technique.

Procedure: Cannula Construction

1. Remove the hub and point from a 20-gauge needle.
2. Cut a 45° bevel on a 2-cm length of polyethylene tubing (PE-90) and slide the nonbeveled end over the 20-gauge needle.
3. Slide a 2-mm length of tubing from a 25-gauge butterfly needle over the PE-90 tubing to form a supporting ring.
4. Apply a small amount of adhesive material (isobutyl-2-cyanoacrylate cement or Super Glue) to the needle, PE-90 tubing, and ring to secure their positions.
5. Slide a 1-cm length of PE-90 tubing over the other end of the needle to form a cap (Figure 19.1).

Procedure: Cannula Implant

1. Anesthetize a rat; shave the area over the calvarium and dorsal neck and clean it with 2% chlorhexidine.
2. Place the rat in a stereotaxic frame with its head flexed slightly forward.
3. Make a 3-cm midsagittal incision to expose the nuchal ridge and dorsal cervical musculature.
4. The muscle attachments are cut transversely with a sharp scalpel blade at the interparietal-occipital suture, and blunt dissection is carried ventrally along the occipital bone to expose the atlantooccipital membrane.

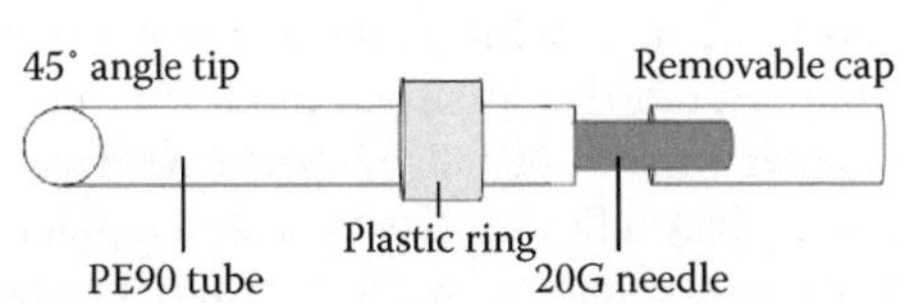

FIGURE 19.1 Diagram showing intracranial pressure (ICP) cannula construction.

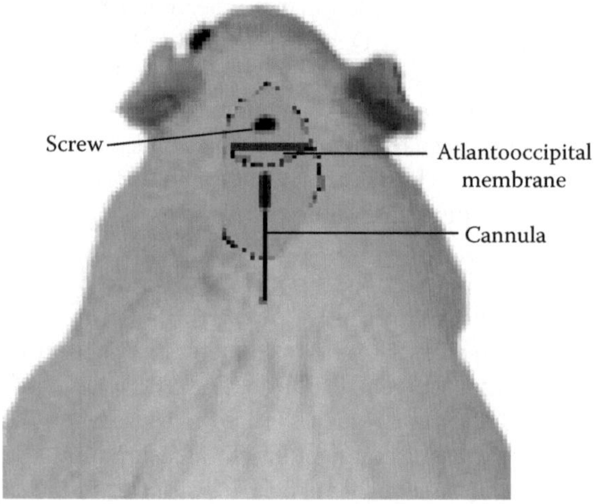

FIGURE 19.2 Schema showing intracranial pressure (ICP) cannula placement in the rat.

5. Drill a small hole through the outer table of the occipital bone; insert a jeweler's screw to provide a suitable surface onto which adhesive can bind (Figure 19.2).
6. Expose the atlantooccipital membrane and pierce it with a 22-gauge needle.
7. Gently twist the beveled tip of the cannula through the hole in the atlantooccipital membrane until CSF is observed pulsating in the catheter.
8. Place a small amount of adhesive material (isobutyl-2-cyanoacrylate cement or Super Glue) around the membrane and cannula.
9. Prepare methyl methacrylate bone cement and pour it into the wound covering the screw, occipital bone, membrane, and cannula, covering the outer ring to prevent slippage.
10. Heat seal the cap with a cautery tool and close the skin with wound clips.

Procedure: Intracranial Pressure (ICP) Measurement

ICP measurements can be recorded at 48 hours after placement of the cannula.

1. The rat is again anesthetized, and the femoral artery is cannulated to monitor blood pressure.
2. The sealed cap is removed from the cannula and connected to a pressure transducer attached to a polygraph.
3. Arterial blood pressure and ICP are recorded simultaneously; compress the abdomen (Valsalva maneuver) to ensure that the system is working.

A sample ICP measurement is shown in Figure 19.3.

Cerebral Blood Flow

Monitoring CBF during stroke surgery allows investigators to reduce experimental variation. Since ischemia is a reduction in blood flow, successful model surgery will generally result in an easily measured reduction in blood flow. Hemorrhagic stroke models produce ischemia by compressing the cerebral vasculature (see Chapter 18); this can also be measured experimentally. While there are numerous techniques for measuring blood flow in tissue, laser Doppler flow measurement provides a good combination of reliability and economy without requiring the harvesting of tissue. This technique uses the Doppler shift of laser light reflected by passing blood to produce a normalized blood flow measurement.[19]

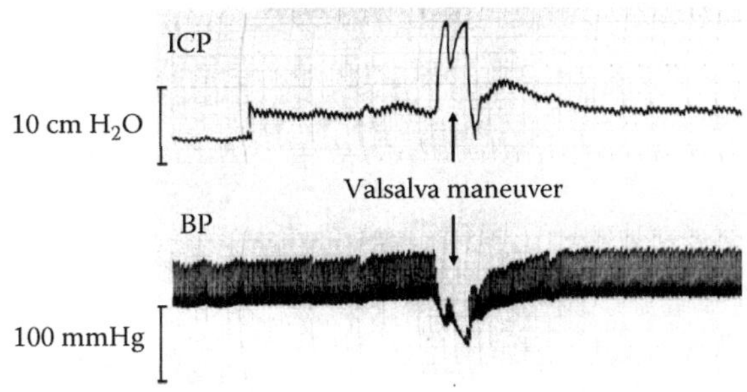

FIGURE 19.3 Simultaneous recording of intracranial pressure (ICP) and femoral arterial blood pressure (BP). Manual compression of the abdomen (Valsalva maneuver) leads to immediate transient rises in ICP. 1 mm Hg = 1.3595 cm H_2O.

Procedure

The *laser Doppler blood flow monitors and equipment* were purchased from Moor Instruments (moorLAB with satellite channels and p10d probes, at www.moor.co.uk). The data acquisition system (PowerLab 16/30, Cat. No. ML880) and system software (Chart 5) were from AD Instruments (www.adinstruments.com).

For SAH studies, do the following:

1. Two 1-mm holes are drilled in the skull (3 mm anterior to the coronal suture and 2 mm lateral to the midline).
2. Optical fibers (probes) are placed and glued bilaterally over the frontal cortices at the epidural surface, away from large vessels. This placement allows the monitoring of a large area of tissue over the site of blood injection.

For ischemic stroke models, the procedure is as follows:

1. A 2-mm hole is made by a drill on the lesioned side through the squamosal bone, 2 to 3 mm in front of the arch rostrum near the eye. The hole is drilled down to dura; at this point, the middle cerebral artery (MCA) can be seen through the dura.
2. A single fiber probe (Moor Instruments, Cat. No. SOF400) is cut flat at the end, placed onto the dura, and fixed with Super Glue liquid (Cat. No. 402628, Office Depot) and cyanoacrylate accelerator (Cat. No. CYA-3, Plastruct, www.plastruct.com) to speed up the bonding process. This probe placement allows the specific measuring of a very small amount of tissue, focused on the MCA.

For both types of study, the signals are monitored by laser Doppler flowmetry (Figure 19.4); connections among the different devices are shown in Figure 19.5. A sample from a filament ischemic stroke experiment shows the CBF changes before MCA occlusion, during occlusion, and after reperfusion (Figure 19.6).

Several criteria indicate successful MCA occlusion in the filament stroke model:

1. CBF decreases to less than 30% during the period of suture occlusion (baseline CBF is set at 100% before insertion of suture into the internal carotid artery [ICA]).
2. CBF remains at this low level during the entire occlusion time (1 to 2 hours).
3. CBF recovers to over 60% of baseline within 30 minutes of filament withdrawal.

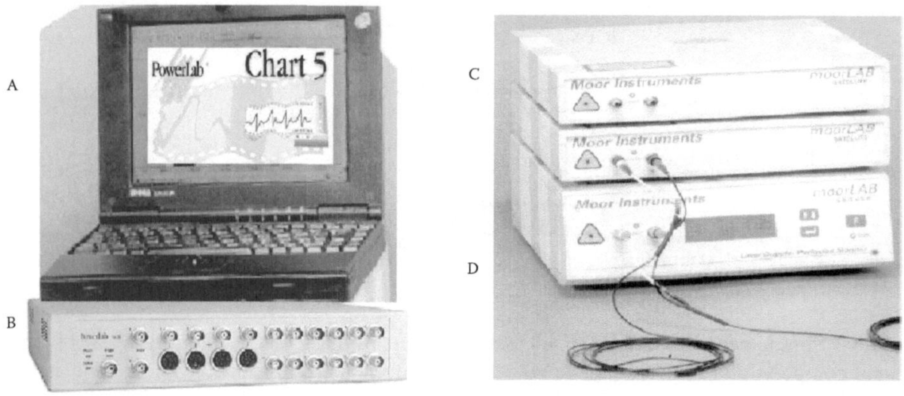

FIGURE 19.4 Devices and software for cerebral blood flow measurement: (A) computer with Chart 5 software; (B) data acquisition unit; (C) moorLAB satellite channels; and (D) laser Doppler monitor. (Courtesy of Moor Instruments.)

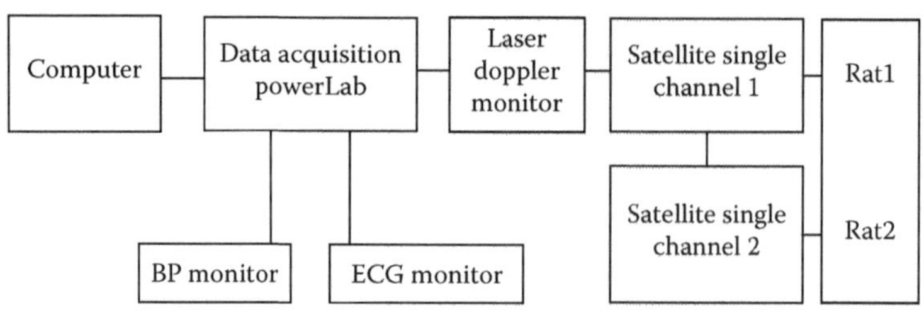

FIGURE 19.5 Schematic diagram of the interconnected devices for measurement of cerebral blood flow, blood pressure (BP), and electrocardiogram (ECG).

Criteria for excluding animals from further study are as follows:

1. CBF does not decrease to 30% of baseline or less.
2. CBF slowly recovers to 60% of baseline during occlusion.
3. CBF does not recover within 30 minutes after removing the filament (because a blood clot may have formed, hence there is no reperfusion).

A slight CBF decrease (>4% of baseline) or recovery during occlusion is associated with no brain infarct or a small brain infarct (see Figure 19.7). In our study, CBF monitoring decreased the variability of brain edema and infarct size in suture models (unpublished data).

Monitoring CBF has pitfalls. Monitoring adds another aspect of surgical damage to the animal. In addition, a small percentage of rats (13%) may show false-positive results; they are not excluded by the CBF criteria but do not show substantial stroke damage. Even though the CBF decreased to about 20% of baseline during 2 hours of MCA occlusion in these rats, only a very small volume of edema or infarct resulted. The cause is not clear but may be related to variations in MCA branching and cortical distribution (for an example of a duplicated MCA, see Chapter 4).

Temperature Measurement

Stroke models can produce both decreases in temperature (hypothermia) and increases in temperature (hyperthermia). Measurement and control of changes in temperature are critical as changes in temperature have profound effects on neuropathology. Hypothermia is neuroprotective, while

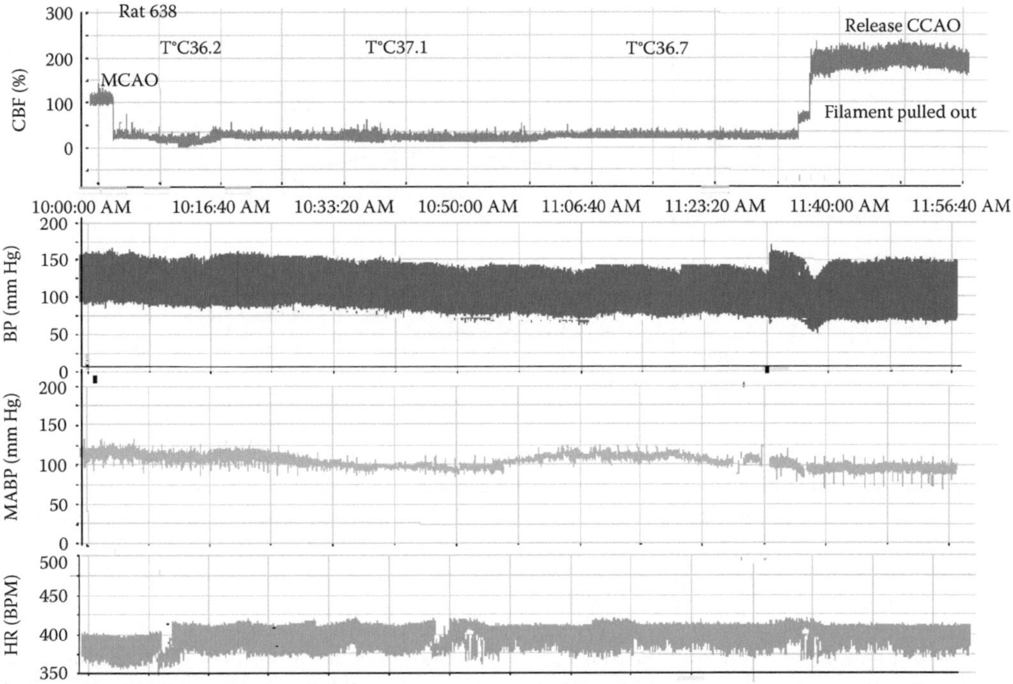

FIGURE 19.6 Cerebral blood flow (CBF) changes before, during middle cerebral artery occlusion (MCAO; suture model), and after reperfusion. CBF baseline is set at 100%. CBF decreased to 20% to 30% of baseline after MCAO and immediately recovered to 100% after withdrawal of the suture from the MCA. Blood pressure (BP), mean arterial blood pressure (MABP), heart rate, and body temperature were measured simultaneously.

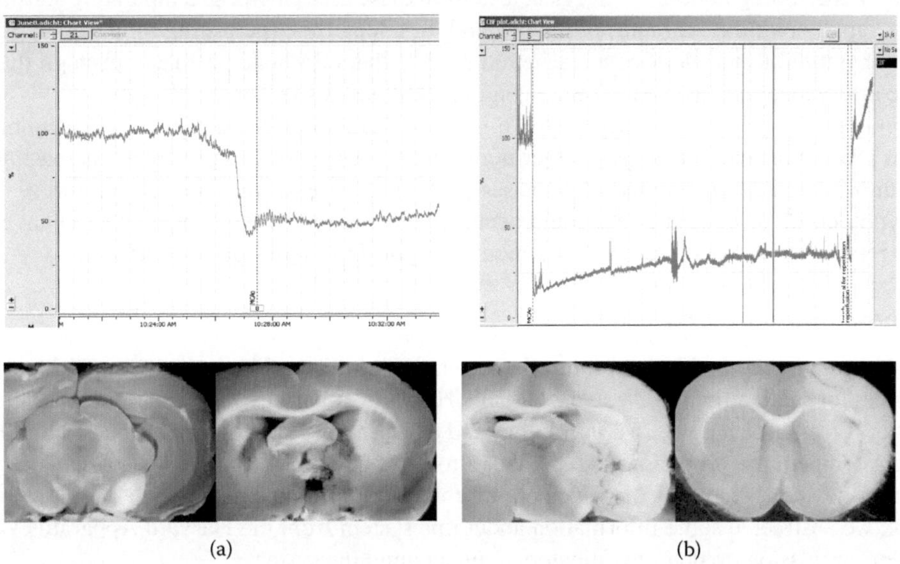

FIGURE 19.7 (a) Incomplete ischemia: Cerebral blood flow (CBF) decreased only 50% after the suture was inserted; this rat had a temperature of 39.4°C and a behavioral score of 2 after 2-hours occlusion. Triphenyltetrazolium chloride (TTC) staining at 24 hours postocclusion showed a very small infarct (3.4%). This animal was excluded on the basis of CBF. (b) Full ischemia: CBF decreased to 20% to 30% of baseline after occlusion; the rat had a high temperature of 39.2°C and a behavioral score of 2. TTC staining showed a large infarct of 50% (white) on the ischemia side. **(See color insert following page 146.)**

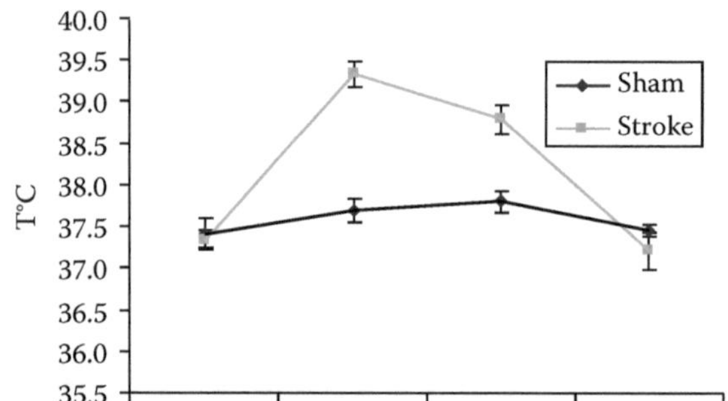

	0'	2 hrs	4–5 hrs	24 hrs
Sham	37.42	37.69	37.80	37.4
Stroke	37.34	39.33	38.79	37.2

Time

FIGURE 19.8 Hyperthermia in a filament-induced stroke model. Male Sprague-Dawley rats were subjected to ischemic stroke by a size 0-3 nylon suture for 2 hours. Rectal temperature was measured at 0, 2, 4, and 24 hours.

hyperthermia exacerbates pathology; this is true for both rodents and humans.[20,21] Temperature must be controlled, regardless of the protective value fever might have in other disease states.[22–30]

Posttraumatic hyperthermia (PTH) has been extensively investigated.[31–35] PTH, also called neurogenic fever or neurocentral fever, is not associated with overt infection but is a fever in the physiological sense: a hyperthermia maintained during changes in ambient temperature. These fevers can be produced by damage to the hypothalamus and preoptic area,[36–38] both of which play prominent roles in thermoregulation.[22,39] Surgical lesions in these areas increase temperature substantially (2.34°C), rapidly (within 2 minutes of injury), and for a long time (persisting for a day).[38] However, PTH is not limited to needle lesions; ischemia models that eliminate perfusion through the hypothalamic artery also increase temperature (Figure 19.8) (see also Chapter 13).

Hypothermia is also a concern in stroke models. Many anesthetic regimens commonly used in rodent procedures cause changes in thermoregulatory physiology, typically by increasing blood flow to uninsulated cooling tissue; this causes profound decreases in core temperature.[23,24] Since either hypothermia or hyperthermia could occur in a stroke model, and since either change would alter the results of the model, temperature monitoring during these studies is critical.

Procedure

For stroke studies, the body temperature is maintained at 37°C. The homeothermic blanket system (system for small animals, Cat. No. 507053F, Harvard Apparatus, www.harvardapparatus.com) consists of three parts: a probe, a control unit, and a heating pad. The animal is placed on the heating pad. A temperature probe is inserted about 2 to 3 cm into the rectum and is attached on the tail with tape. Body temperature is displayed on the control unit's temperature monitor.

Here, we abstracted some information about this system from the Harvard Apparatus Web site (with their permission) to help investigators to understand the system.

> The flexible probe is a precision thermistor encapsulated in a bead of epoxy resin at the top of a flexible hollow plastic tube. The 2 mm diameter probe is 100 mm (4 in) long and has a 2 m (6 ft) long cable with a plug for attachment to the control unit. It is suitable for a wide range of laboratory animals from adult mice to large dogs. The rigid probe is a stainless steel shaft with a diameter of 1.6 mm. The shaft attaches to a 2 m (6 ft) cable with a plug that connects to the control unit. It is suitable for use with mice, rats, rabbits, cats, and dogs. The output of the temperature sensing probe is used by the control unit

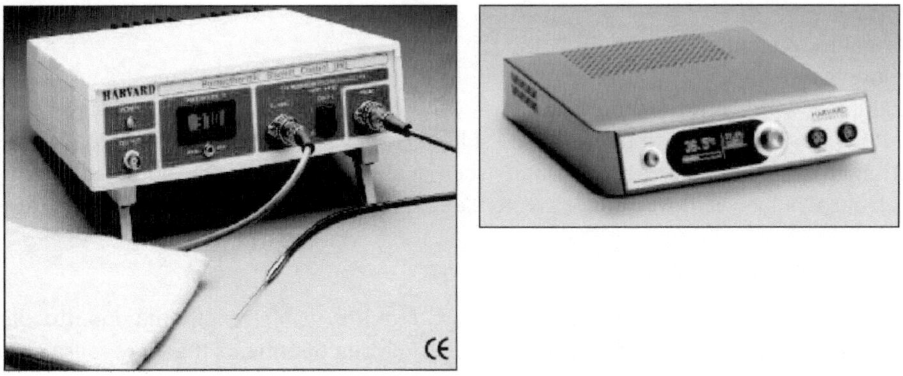

(a) (b)

FIGURE 19.9 Homeothermic blanket system, supplied complete with control unit, temperature probe, blanket, and plastic cover for blanket; (a) the old version, and (b) the new version. (Courtesy of Harvard Apparatus.)

to proportionally control the regulated, low-voltage DC supply to the blanket. This method of control alleviates the interference problems associated with that of non-proportional control, such as switching contacts, thus enabling the system to be used in conjunction with high gain recording systems.

The control temperature is preset at 37°C (98.6°F) at the factory, but can be adjusted within the range of 20° to 50°C (68° to 122°F) by means of a control on the front panel. An LCD [liquid crystal digital] display on the control unit continuously displays the probe temperature. A horizontal LED [light-emitting diode] bar graph indicates the level of power being delivered to the blanket [Figure 19.9]. When core temperature is substantially below the desired temperature the bar graph will indicate maximum power being delivered. The blanket consists of a highly flexible insulated heating element, which can be folded without risking internal damage. It is electrically floating with respect to ground; however, one end of the heating element can be grounded by a switch on the front panel of the control unit. The system operates satisfactorily whether or not the blanket is grounded. Special circuitry eliminates electrical noise thus permitting sensitive recordings. Blankets are available in three sizes:

- *Small Blanket*: Measures 15 × 20 cm (6 × 8 inches) and is suitable for rodents.
- *Medium Blanket*: Measures 45 × 70 cm (18 × 27 inches) and is suitable for rabbits and cats.
- *Large Blanket*: Measures 60 × 90 cm (24 × 36 inches) and is suitable for dogs.

Neurological Behavioral Tests

Behavioral tests may require training prior to surgery in order to measure baseline performance. Consistent methodology and animal handling are very important for these assays. While they are easy to do, it is not as easy to produce consistent data with them.

Neurological Score

Behavioral scoring systems are a trade-off: Thorough exploration of behavior must be weighed against speed and ease of use. While thorough systems like the Irwin screen[25] can be useful, a brief system for scoring stroke-related behaviors is often more practical and more helpful. We have set up a simple assessment system, similar to the Bederson scale,[26] that we find useful for animals with unilateral strokes.

Scoring

0: Normal movement.

1: Failure to extend forearm contralateral to infarct; this level of impairment is usually seen in animals with a mild focal lesion.

2: Walking in a circle toward the infarcted side; this suggests a moderate focal lesion.

3: Falling to the infarct side; this is usually seen in animals with severe (but focal) lesions.

4: Lack of spontaneous movement or greater deficit; this is associated with very severe lesions.

Foot Fault (Wire Screen Test)

The foot fault test, also called the wire screen test, requires grip strength and coordination. Foot fault behavior is used to measure neurological effects of drug treatments that may cause sedation or motor impairment and is a useful endpoint for studies of stroke therapies. [27,28] In this test, the animal is placed on a metal wire mesh screen; each opening in the metal grid is 3 × 3 cm or 6 × 6 cm, and the diameter of the metal wire is 0.5 cm. When animals place forelimbs or hindlimbs inaccurately on the wire screen and any limb falls through an opening in the grid, this behavior is called a foot fault. The number of foot faults (legs falling into the grids) is counted for 2 minutes while the animal is walking on the mesh. The mesh can be modified into parallel bars 3 cm apart (Figure 19.10a). To prod movement across the wire screen, a startling noise (<70 dB) or light touch (for example, near base of tail) may be used. Healthy animals can walk across the top of the grid; paws of lesioned animals will fall into the grids (Figure 19.10b). These faults are usually scored by direct observation. A device can be developed based on this metal mesh to detect and count foot faults automatically.

In some studies, the wire screen may be slowly raised to a vertical position or inverted by rotating it 180°. The time required for the animal to climb on top and latency of slipping or falling may be recorded. The maximum duration of each trial is 5 minutes. The maximum distance of a fall is approximately 6 inches. Pads should be placed under the wire screen.

Procedure

1. Place animal on the wire grid; Figure 19.10a is a modified wire screen.
2. Count the number of faults (paw falling through grid) for 2 minutes.
3. If the animal freezes in place, noise or a light touch to the tail or hindquarters will often get the animal walking.

(a)

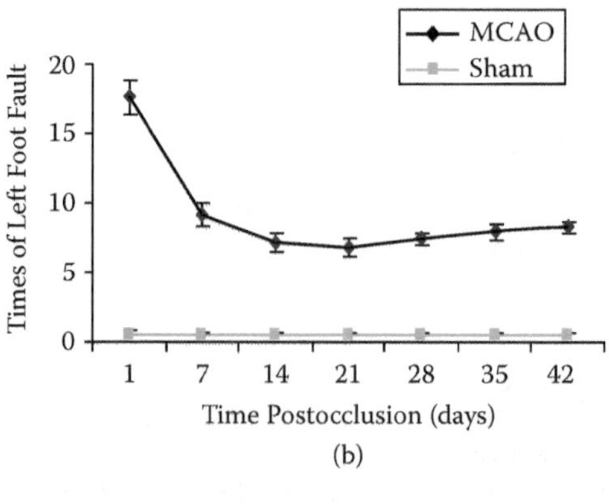

(b)

FIGURE 19.10 Foot fault test: (a) Rat placed on a modified metal grid for testing; (b) group data showing stroke-induced deficit. The metal screen was modified from mesh to parallel bars to help detecting mild neurobehavior damage (a).

Tail Suspension

As used in stroke studies, the tail suspension test is a measure of the lateralization of motor behavior. Since motor behavior requires nigrostriatal function, infarcts and other insults that damage the nigrostriatal system will result in a deficit in motor behavior contralateral to the lesion. Lateral turning during tail suspension was initially described for use in Parkinson's disease models[29]; it was subsequently adapted for use in stroke studies.[30]

Procedure

1. Place the animal in a box for several minutes of habituation.
2. Once the animal has put itself in a neutral position (all four paws on the bottom of the cage), gently grasp the base of the tail and lift so that the animal is completely off the floor of the cage and no more than 10° off vertical.
3. Record the direction of each swing the animal attempts (Figure 19.11a); attempts are ended when the animal's head and forelimbs return to vertical.
4. If the animal grabs its tail or does not return to vertical, place it on the floor of the cage and continue the observation when the animal has put itself in a neutral position.
5. Continue observation until 20 swings have been attempted; if the animal stops before 20 swings, gently pinch the tip of its tail.

Instead of holding the animal by the tail, the investigator may tape the tail to a metal bar or a strain gauge. The strain gauge allows the investigator to automatically measure total motor behavior. Whatever equipment is used, the tail should be secured near the base to reduce the risk of degloving (that is, loss of skin at the tip of the tail).[40]

Data can be presented as the number or percentage of swings to one side.[41] The percentage should be approximately 50% to either side for intact animals, while a lateralized infarct will result in lateralization of behavior. For example, for a rat with ischemia on the right side, in 20 consecutive trials the rat swings to the left 18 times and to the right 2 times; the percentage of swings to left side is 90% or 0.9 (18/20 × 100), and the percentage to the right side is 10% or 0.1 (2/20 × 100). The test also can be directly presented as the swing number in 20 trials (Figure 19.11b).

Paw Tape

Animals will work to remove adhesive tape from their paws; damage to the somatosensory cortex will slow or eliminate removal of tape from the contralateral side. Although this assay was initially

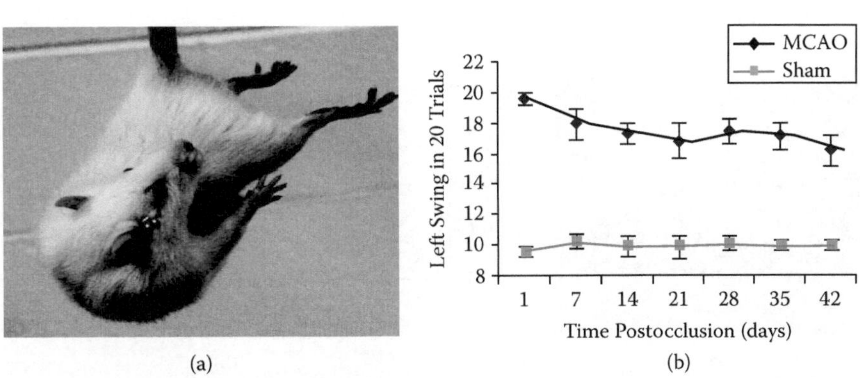

FIGURE 19.11 Elevated body swing test: (a) Rat showing turning toward one side; (b) group data showing dramatic one-sidedness of turning following stroke.

used for mechanistic studies of unilateral sensory neglect, the same behavior is seen after cortical lesions.[42] This assay has been run in two variations; we present both here.

Procedure 1: Traditional Test

Do the following for the traditional test[42,43]:

1. An adhesive dot (typically a commercially available round adhesive label, 0.5 to 1 cm in diameter) is placed on the medial aspect of each forepaw; to eliminate handling bias, the order of placement is randomized or alternated, and both paws are handled immediately before returning the animal to the observation area.
2. The latency to remove each adhesive dot is recorded; the trial ends when either both dots are removed or 10 minutes have passed, whichever comes first.
3. The experimenter removes any dots not removed by the animal.
4. Baseline (prestroke) sessions are usually required, both to train animals and to reduce variability; multiple postinfarct sessions may be averaged to reduce variability.

Procedure 2: Modified Test

The modified test procedure is as follows[44]:

1. One piece of adhesive tape (10 to 13 mm wide) is wrapped all the way around each forepaw, slightly proximal to the toes (Figure 19.12a); to eliminate handling bias, the order of placement is randomized or alternated, and both paws are handled immediately before returning the animal to the observation area.
2. The animal is observed for 5 minutes, and either the number of attempts to remove the tape on each paw or the time spent attempting to remove the tape from each paw is recorded.

Data can be presented as a count, fraction, or percentage. Typically, a stroke model affecting the somatosensory cortex (such as MCA occlusion) will produce unilateral sensory neglect (Figure 19.12b). Some improvement was seen but not complete resolution.

Both procedures will yield data on an animal's ability to respond to somatosensory stimulation. However, the modified procedure appears to produce less-variable data and require less training.[45]

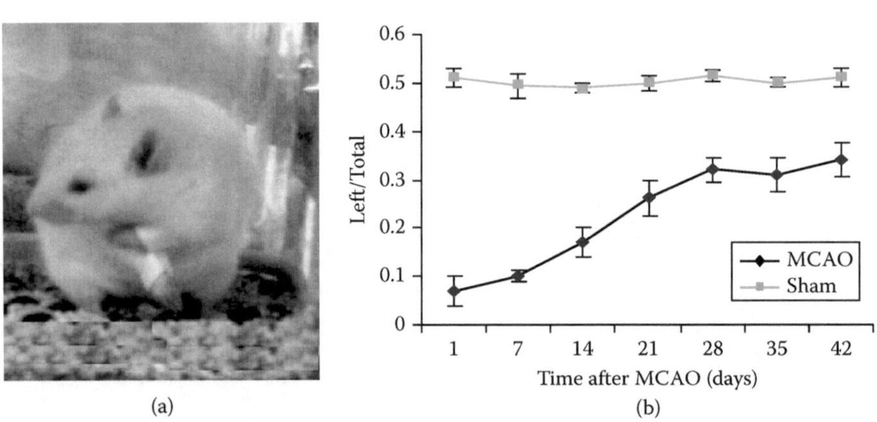

FIGURE 19.12 Modified adhesive tape test: (a) Rat attempting to remove tape from one forelimb; (b) group data showing unilateral sensory neglect after stroke.

Tactile Sensitivity

Like the adhesive tape test, the tactile sensitivity test is a test of somatosensory function. However, in rodents, vibrissae occupy a very large fraction of the primary somatosensory cortex[46]; this may make somatosensory deficits in the vibrissae more sensitive to infarction. This procedure has been used to measure stroke as well as drug effects on recovery.[47]

Procedure

A tactile stimulus (for example, a cotton swab) is applied to the vibrissae for 2 seconds, first on one side and then on the other. To eliminate handling bias, the order of stimulation is randomized or alternated. Latency to orient to the stimulus is recorded for each side; alternately, behavior can be scored.

Scoring

0: No response during 2 seconds of stimulation.
1: Delayed response.
2: Quick response in 1 second.

Rotarod

The original rotarod technique was published[48] in 1957 as a side-effect assay for anticonvulsant development. The original form of the assay placed rats or mice on a rod rotating at a constant speed. Typically, healthy control animals will walk at a pace matching the rod, while animals with deficits will fall off. However, to reduce the variability of the data, a variation was introduced in which animals are placed on a bar rotating at a low speed; the bar accelerates over the course of the observation period.[49] Deficits in rotarod performance have been interpreted as lack of coordination, fatigue, or learning deficits (when repeated trials are used).[48,49]

Rotarod testing devices can be very simple (Figure 19.13). The rotating rod is machined to allow some grip; rods for mice are typically 2 to 3 cm in diameter, while rods for rats are typically about 7 cm in diameter. Often, apparatus is supplied with a mouse-appropriate rod and bolt-on adapters for rats. Large disks separate segments of the rod, so that several animals can be tested at once. Mechanical switches or photobeams detect falling animals. Some investigators consider a "free ride" (clinging to the rod as it rotates instead of walking) to be a deficit[50]; as a result, some photobeam systems include sensors designed to count this behavior.

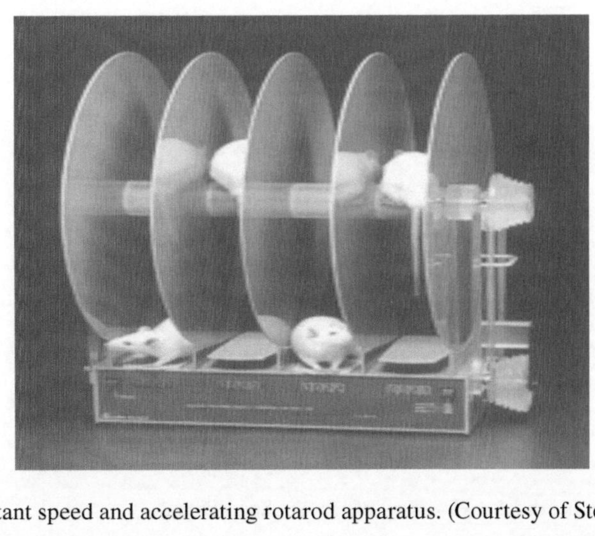

FIGURE 19.13 Constant speed and accelerating rotarod apparatus. (Courtesy of Stoelting.)

Procedure

Rotarod equipment can be purchased from Stoelting (Cat. No. 57620, www.stoeltingco.com). These quality instruments are manufactured by Basile and may be connected to the Basile 52600 computer interface for automated data collection. The data, stored in the 52600's internal memory and shown on its graphic display, can be printed out in real time or routed to a personal computer, for which the 52010 Win-DAS software is required.

1. The trial begins by placing the animal onto a rod that is rotating slowly enough that all animals are able to stay on (typically, 3 to 5 revolutions per minute [rpm] for rats, stationary for mice). Over the course of a minute, the rod accelerates to its final speed (40 to 50 rpm). Alternately, animals are placed onto a rod rotating at a constant speed (30 to 50 rpm). The latency for each animal to fall off the rod is recorded.
2. Animals may be trained before the test (before surgery in stroke study) to get baseline performance levels.
3. The duration of training and testing sessions may range from 1 to 10 trials per day for up to 8 weeks. Time on the rotarod should not exceed 60 minutes per day. If the trial duration is 3 minutes or greater, at least 5 minutes should elapse between trials. Trial duration will typically range between 30 seconds and 6 minutes. With repeated testing, performance can improve (i.e., longer latencies to fall); proper control groups are important.

Balance Beam

The balance beam test measures hind limb grip strength, gait coordination, postural stability, and paw sensory function. An alternate version of this assay uses parallel beams; these assays yield similar information.

Procedure: Balance Beam

For the balance beam procedure,[47,51] the animal is placed on a narrow beam (18- to 25-mm diameter for rats) elevated over the floor. Behavior can be scored either with the rat standing in place or walking across the beam. In the static assay, the time to fall and number of paw slips are recorded; in the walking assay, the time required to cross the beam and number of paw slips are recorded.

Procedure: Parallel Beam

For the parallel beam procedure[47]: The parallel bar apparatus consists of two parallel wooden rods (1-cm diameter, 115 cm long, separated by 2.5 cm) connected to platforms at each end (15 × 50 cm). The animal is placed on one of the platforms (Figure 19.14). The number of times that the animal

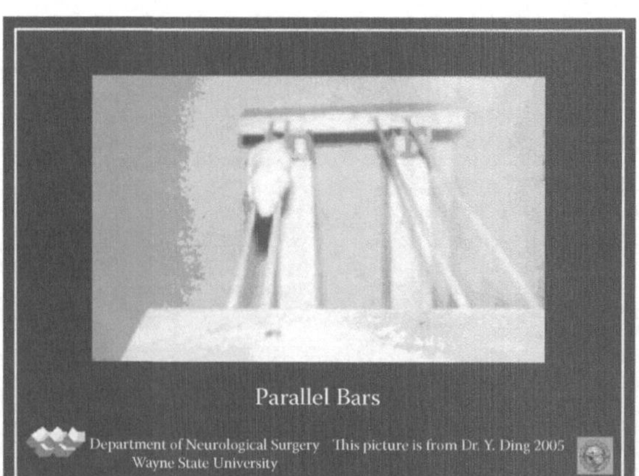

FIGURE 19.14 Balance beam test. (Courtesy of Dr. Yuchuan Ding.)

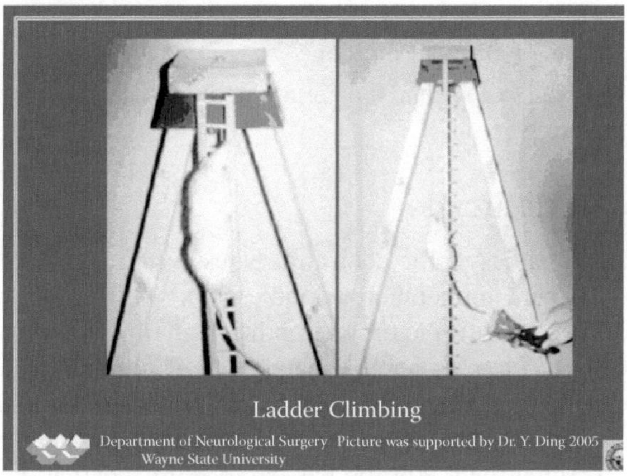

FIGURE 19.15 Climbing test. (Courtesy of Dr. Yuchuan Ding.)

places two hind paws on one rod, drops a hind paw below the rods, or falls or swings under the rods is recorded, as is the time required to cross the beams.

Climbing

The climbing test measures balance and coordination of animals while they climb up a rope (1.5-cm diameter, 1 m high) or ladder (1 m high, with 3 to 4 cm wide × 0.5-cm diameter rungs set at 3-cm intervals; Figure 19.15).[47] Animals may be pretrained before testing, with gentle prodding if needed; normal animals can learn to climb to the top within 15 seconds without prodding. The trial ends when the animal reaches the top or after 3 minutes have elapsed, whichever comes first.

Inclined Plane

Neurological deficits can reduce the ability of a rat or mouse to maintain its position on a tilted surface. This can be tested by placing the animal on a wire cage lid, held flat, and slowly rotating the lid until the animal loses grip. Some investigators build devices to make quantitative measurements easier; the simplest is a pair of plywood sheets connected by a hinge, with a protractor to measure angles.[52,53]

Forelimb Placing

The forelimb placing test measures sensorimotor responses to visual and vibrissal stimulation.[51,52] The assay relies on rodents' tendency to place all four paws securely on a surface and to reach for any surface they can find if they are not standing on their paws. Normal animals place both forelimbs quickly (less than 2 seconds) on a tabletop in response to either visual or tactile cues.

Procedure: Vision-Guided Placing
Cradle the animal with forelimbs dangling free and bring it toward the edge of the table, with the body perpendicular to the edge of the table; animals can also be tested with the body nearly parallel to the edge of the table to assess vision on one side.

Procedure: Touch-Guided Placing
Cradle the rat with forelimbs dangling free and the body angled up at 45°; gently bring the vibrissae in contact with the edge of the table.

Scoring

0: No placing.
1: Incomplete and/or delayed (>2 seconds) placing, including interspersed flailing.
2: Immediate and complete placing.

Cerebrospinal Fluid (CSF) Collection

CSF is extremely useful for studies of the central nervous system (CNS): Since it clears material from the brain, it will contain any unmetabolized molecules.[54] CSF can be assayed for neurotransmitters, hormones, cytokines, growth factors, and other signaling molecules. Rats weighing 300 to 400 g will produce CSF at a rate of approximately 2.2 μL/minute[55]; the total volume of their ventricles will be around 135 μL.[56] However, there is additional fluid in the extracellular space of CNS tissue; this has been estimated to be approximately one-fifth the volume of the tissue.[57,58] This, combined with the CSF surrounding the brain, has led to estimates that the CSF volume of a rat's head is approximately 400 μL.[59] However, not all of this can be recovered in a sample. Excessive CSF drainage is thought to play a role in the remarkable pain seen during postdural puncture headache.[58] In addition, the quality of CSF samples decreases with increasing sample volume.[60] Many labs have developed empirically based maximum sampling volumes and frequencies. Few groups report sampling over 100 μL at one time; intervals of several days between samples are common. However, we encourage the reader to develop schedules on the basis of their own experience.

Many procedures for CSF collection from the cisterna magna have been reported; we describe three practical options. Which of the three is most suitable for a laboratory will be determined by whether the analyses of interest are more affected by anesthetics or handling stress.

Procedure 1: Anesthetized, with Stand

For the following procedure, the animal is anesthetized with stand[40]:

1. The rat is anesthetized (e.g., with pentobarbital 50 mg/kg i.p.).
2. The hair on the back of the neck is shaved, and the skin is cleaned with 0.5% Betadine and 75% ethanol or 2% chlorhexidine.
3. The rat is placed on the stand, prone, with its head out of the stand; the neck will arch slightly (Figure 19.16). The long axis of the body lies at a 40° angle from the horizontal. This position draws the tissue tight over the occipital bone. Directly caudal to this, a depression can be felt. This depression is the space between the occipital bone and the axis (C1 vertebra); the atlantooccipital membrane and cisterna magna lie beneath this space.
4. A 25-gauge needle attached to a polyethylene tube (which is kept 3 cm below the cisternal space) or a needle alone is slowly pushed into the center of this depression; when it enters the cistern magna, a sudden decrease in resistance should be felt. The CSF will flow out of the needle hub into the polyethylene tube. If a needle is used without a tube, when CSF slowly flows out of the top of the needle hub, the needle is

FIGURE 19.16 Diagram of stand for performing cisternal puncture and injection in rats.

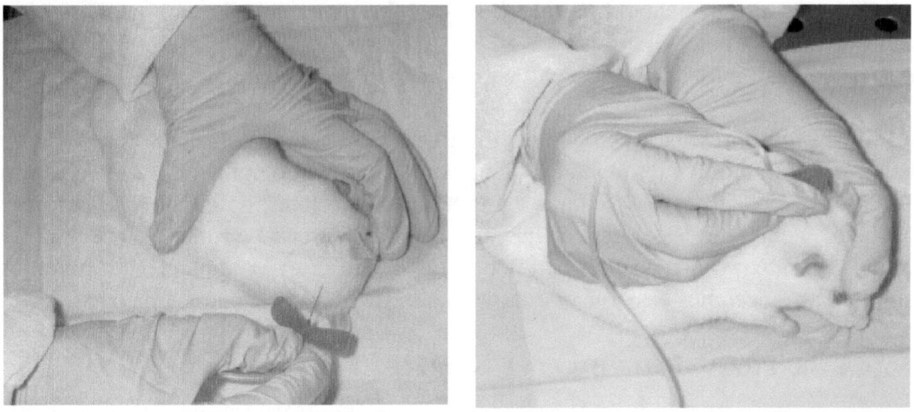

(a) (b)

FIGURE 19.17 Position for cisternal puncture in rats: (a) position for an anesthetized rat; (b) position for a conscious rat. A 25-gauge butterfly needle is used for cisterna magna puncture. This needle is connected to a 1-cc syringe.

removed and placed into an Eppendorf tube and centrifuged for 2 to 3 seconds to transfer the CSF to the tube.

5. In some cases, it may be necessary to withdraw the needle slightly or provide gentle suction to start the flow.

Procedure 2: Anesthetized, Lateral Recumbent

1. The rat is anesthetized (e.g., with pentobarbital 50 mg/kg i.p.).
2. The hair on the back of the neck is shaved, and the skin is cleaned with 0.5% Betadine and 75% ethanol or 2% chlorhexidine.
3. The rat is placed in a lateral recumbent position, and the head is bent toward the chest, drawing the tissue tight over the occipital bone (see Figure 19.17a). Directly caudal to this, a depression can be felt. This depression is the space between the occipital bone and the axis (C1 vertebra); the atlantooccipital membrane and cisterna magna lie beneath this space.
4. A 25-gauge needle attached to a polyethylene tube connected to a 1-mL syringe is slowly pushed into the center of this depression; when it enters the cisterna magna, a sudden decrease in resistance should be felt. CSF will flow immediately with gentle aspiration (using the syringe).
5. In some cases, it may be necessary to withdraw the needle slightly.

Procedure 3: Awake

In the third procedure, the rat is awake[59]:

1. Rats are trained to remain immobile by being held and stroked for 15 minutes each day for 2 weeks.
2. About 10 minutes before the CSF puncture, 0.2 mL lidocaine (1%) is injected into the posterior cervical skin.
3. The rat is placed in a prone position or lateral recumbent position with manual restraint (see Figure 19.17b). The neck is flexed to reveal the space between the occipital bone and the axis (C1 vertebra); the atlantooccipital membrane and cisterna magna lie beneath this space.
4. A 25-gauge needle attached to a polyethylene tube connected to a 1-mL syringe is slowly pushed into the center of this depression; when it enters the cisterna magna, a sudden

decrease in resistance should be felt. CSF will flow immediately with gentle aspiration (using the syringe).

5. In some cases, it may be necessary to withdraw the needle slightly.

In all three procedures, between 50 and 100 μL of CSF can be collected. Potential complications include contamination with blood and behavioral suppression in the rat. Blood contamination can be reduced by collecting less volume, using less vacuum to aspirate CSF, and puncturing the atlantooccipital membrane less often. Behavioral suppression can be reduced by sampling small volumes and less often; sharp needles and noncoring needles (for example, Huber point) can also help. Sampling also becomes easier and less traumatic with several months' practice.

Brain Edema and Blood–Brain Barrier Function

Stroke models produce BBB deficits in rats. Biphasic opening BBB after stroke was shown using transient MCA occlusion in rats (Figure 19.18).[61] This has been a consistent finding. BBB changes after ischemia were studied using a modified radiotracer method. The greatest BBB deficits (measured as permeation of [^{3}H] sucrose) were located in the ipsilateral cerebral cortex, which was dissected to include both the core and edge of the region perfused by the MCA. Biphasic opening was clearly demonstrated by an initial 10-fold increase in K_i during the first half hour of reperfusion, followed by partial closing, and then a delayed but progressive opening between 22 and 46 hours after reperfusion.[62]

This profile of BBB deficit is also seen using Evans blue extravasation in the cat.[63] One hour of temporary MCA occlusion caused staining of brain parenchyma. This was followed by a refractory period and then a delayed opening, which was visible in cats sacrificed 5 hours or 3 days after stroke. The initial acute opening has been described as a "hemodynamic" BBB opening.[64,65] Reperfusion caused excessive blood flow or "luxury perfusion" owing to acidosis, loss of autoregulation, and vasodilation of the cerebral vasculature. It was found that high intraluminal blood pressure in the cerebral microvasculature induced abnormal pinocytotic transport across endothelial cells and opened interendothelial tight junctions.[66,67]

Following brain ischemia or traumatic damage, BBB dysfunction results in the extravascular leakage of plasma proteins and other solutes, leading to an imbalance of osmotic forces, such that excess water is drawn into the tissue (vasogenic edema).[68] A large hemispheric infarct can cause cerebral edema,[69–71] which can result in asymmetric shifting of the contents of the cranium, leading to transtentorial herniation.[72,73] Use of MRI or axial computed tomography (CT) can detect subtle signs of elevated tissue water content within hours of stroke onset,[74] but the clinical symptoms of

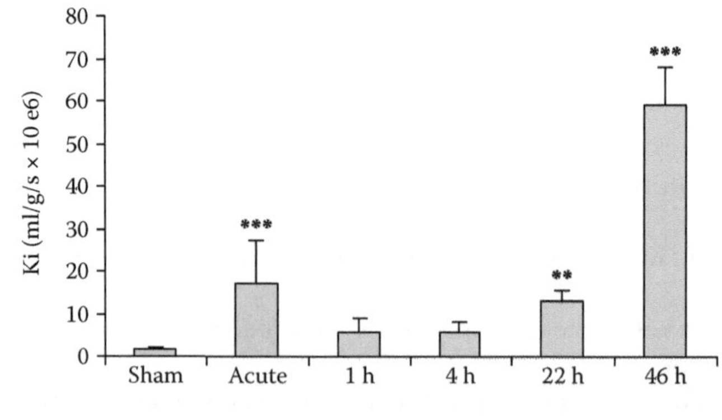

FIGURE 19.18 Biphasic opening of the blood–brain barrier (BBB) in right cerebral cortex of rats after 2 hours of occlusion and 46 hours of reperfusion: Opening of BBB is indicated by an increase in transfer constant K_i for blood-to-brain diffusion of ^{3}H-sucrose. Mean ± standard deviation (n = 5) for each time period. $^{**}p < .01$; $^{***}p < .001$, significantly increased from control values.

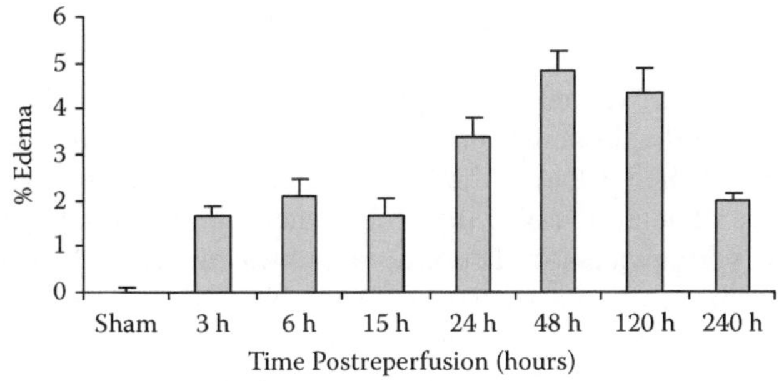

FIGURE 19.19 Time course of edema (wet:dry weight method) following middle cerebral artery occlusion in rats.

cerebral edema typically reach a peak after a delay of 1 to 5 days.[75] In our study on a rat suture model of stroke, edema peaked at 48 hours after the ictus (Figure 19.19). Tissue swelling in the confined skull increases ICP, ensuring secondary ischemia owing to compression of microvasculature and ultimately brain herniation.[64]

Investigators must use caution during their experiments as procedural details can influence estimates of BBB function. We compared the effect of CO_2 and pentobarbital euthanasia on BBB function in rats (unpublished data). Animals were sacrificed by either carbon dioxide suffocation or by pentobarbital injection (100 mg/kg). CO_2 resulted in a breakdown of the BBB, allowing triphenyltetrazolium chloride (TTC) dye or immunoglobulin G (IgG) serum proteins into the brain parenchyma (Figure 19.20). If BBB function is to be studied, animals should be anesthetized with a barbiturate such as Nembutal and then perfused.

We present two methods for determining BBB permeability ([^{3}H] sucrose permeation and Evans blue permeation) and one method for determining edema (wet:dry weight ratios). These techniques have different material and labor costs and produce different data. We summarize them here to help investigators choose between them.

The radiotracer method is based on a two-compartment (plasma/brain) simple diffusion model. This model assumes that the amount of [^{3}H] sucrose that permeates the microvasculature into brain parenchyma is proportional to the time integral of plasma tracer concentration. Normal BBB permeability and opening are indexed in the ratio of parenchymal uptake relative to the plasma integral and are calculated in terms of the transfer constant K_i.[76,77] This assay can easily be adapted to produce regional measurements and quantifies even minor degrees of BBB opening with a high degree

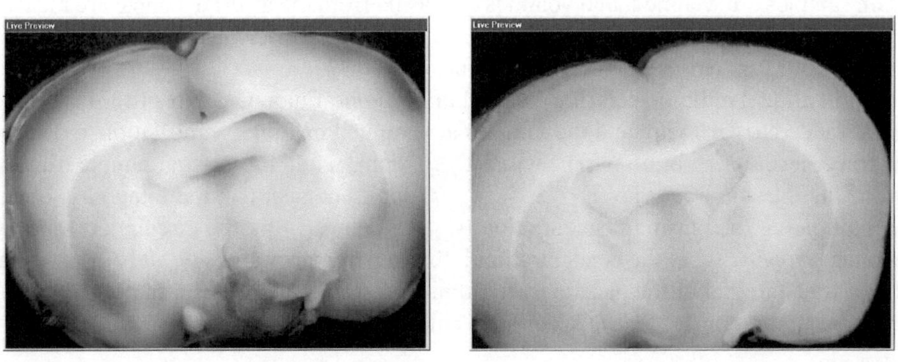

(a) 70%CO_2 + 30%O_2 (b) Pentobarbital

FIGURE 19.20 Triphenyltetrazolium chloride (TTC; 2%, 35 minutes) solution perfused into rats: (a) rats euthanized with CO_2 (5 minutes); (b) rats euthanized with pentobarbital. **(See color insert.)**

of sensitivity.[76,77] However, there are substantial costs associated with the purchase, storage, and disposal of radioisotopes.

Evans blue is an anionic dye and is a large molecule closely related to trypan blue. It has been used to measure blood volume because it binds to serum proteins and stays in the circulation for a few hours.[78] When it leaves the blood, some of it binds to collagen (its elongated structure favors this), and some is taken into cells, including macrophages and neurons. The dye–protein complex is fluorescent (red emission) and was the first fluorescent tracer of neuronal uptake and retrograde axonal transport. Evans blue can be eliminated from tissue by washing in slightly alkaline water. Evans blue and trypan blue both can be used to determine cell vitality; live cells exclude the dye, dead cells take it up, forming the basis of the trypan blue exclusion test. Evans blue dye has been used for quantitative evaluation of BBB permeability in stroke and in myocardial ischemia.[79,80] Infarcted tissue stains blue, but normal tissue does not stain. This technique avoids the costs associated with radioisotope use and is simpler than the [^{3}H] sucrose method since it does not require an integral sample. The only equipment required is a spectrophotometer. However, the Evans blue technique can only produce total permanent values; rate constants cannot be estimated this way.

Finally, we introduce the wet:dry weight method. This method does not measure BBB permeability but has been used to measure brain water content for many years.[81,82] The technique is simple and requires only an oven to dry the tissue.

Procedure: [^{3}H] Sucrose

1. Rats are anesthetized with pentobarbital (50 mg/kg i.p.).
2. The femoral artery and vein and carotid artery are cannulated; [^{3}H] sucrose (Cat. No. NET331, PerkinElmer, las.perkinelmer.com) is injected intravenously (20 μCi/100 g in 0.5 mL saline). There has been a tremendous reshuffling of vendors in radiochemical study supplies and equipment; PE and Amersham now own essentially everything. These materials can been found at PE, Amersham, or other companies.
3. Immediately on tracer injection, syringe-pump sampling of femoral arterial blood is begun at a constant rate (39 μL/minute) and continued for 30 minutes (1800 seconds total; 1.17 mL blood).
4. Sampling is stopped and the brain immediately cleared of intravascular tracer[62] by perfusing 25 mL saline at 100 to 130 mm Hg pressure through the carotid artery cannula.[83]
5. The rat is decapitated, and the brain is removed and dissected bilaterally into the cortex (about 180 mm^3, representing the complete MCA supply territory), striatum, and hippocampus.
6. Brain samples are weighed, and volumes of plasma from the arterial sample are measured; samples are placed in scintillation vials and solubilized overnight at 37.5°C in 1.3 mL Soluene 350 (Cat. No. 6003038, PerkinElmer).
7. A 10-mL aliquot of liquid scintillation cocktail (Hionic-Fluor, Cat. No. 6013311, PerkinElmer) is added to each vial, and the samples are counted by liquid scintillation to determine the tracer level in the brain parenchyma (C_{paren}, dpm/g) and the plasma sample (time integral of the plasma tracer level, $\alpha^{1800} C_{plasma}\, dt$, dpm.seconds/mL); the integral is obtained by multiplying the plasma concentration (C_{plasma}, dpm/mL) by the circulation time (1800 seconds). *Note*: C_{paren} is the concentration in brain parenchyma; $C_{plasma}\, dt$ is the concentration of time integral of the plasma tracer; dpm is disintegrations per minute.
8. The transfer constant (K_i, mL/g.seconds) is calculated from the relationship $K_i = C_{paren}/\alpha^{1800} C_{plasma}\, dt$.[76]

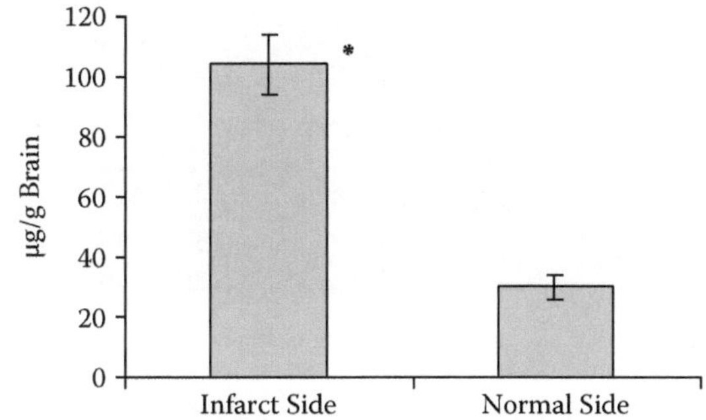

FIGURE 19.21 Evans blue assay shows blood–brain barrier (BBB) damage in suture-stroked rats at 24 hours (mean ± standard deviation, n = 4). $*p < .01$.

Procedure: Evans Blue

The following is the Evans blue procedure[80]:

1. Animals are given an intravenous dose of 4 mL/kg Evans blue (Cat. No. T-1824, Sigma-Aldrich; 2% weight/volume in saline or phosphate-buffered saline [PBS]); the dye is allowed to circulate for at least 60 minutes (or better, 2 hours).
2. Animals are anesthetized and perfused with saline (or PBS) through the left ventricle until the perfusate runs clear. *Note*: Because the free dye is washed out during perfusion, the dye in tissue is bound to protein.
3. Brains are removed; the hemispheres are separated, weighed, and placed in tubes. Add 1 mL of 50% trichloroacetic acid (weight/volume) solution to each sample. Tissue samples are mechanically homogenized and centrifuged (21,000 g for 20 minutes), and the supernatant (containing the dye) is diluted 1:3 with ethanol.
4. A fluorescence spectrophotometer or plate reader is used to determine the fluorescence intensity of the dye (620-nm excitation, 680-nm emission, with background subtraction).
5. Calculations are based on external standards (100 to 1000 ng/mL) in solvent (1:3 mixture of 50% trichloroacetic acid and ethanol).
6. Fluorescence is linear with respect to dye concentration over the range of the standard curve, so linear regression can be used to construct a standard curve and estimate dye concentrations in tissue. The results are then expressed in terms of concentration per gram of wet weight of each hemisphere (see our unpublished data in the rat model of stroke, Figure 19.21).

Procedure: Wet:Dry Weight

1. Animals are killed by decapitation, the brains are collected, and the cerebral hemispheres are exposed. The olfactory projections and cerebellum of the brain are removed and discarded.
2. The brain is placed in a brain blocking matrix (with blade positions for 1-mm coronal cuts; for example, Kent Scientific, Cat. No. RBMA-300C). In the suture MCA occlusion model, most of the infarct is in the central part of the hemisphere (Figure 19.22). The brain is blocked over the region spanning 4 mm from the frontal pole and 4 mm from the occipital pole of the cortex and is separated at the midline. Each part is then separately weighed and the weight recorded to within 0.01 mg with a precision balance (for example, Sartorius 2462, Sartorius Werke).
3. The hemisphere brain tissues are dried in a vacuum oven (Cat. No. 52201-504, VWR, www.vwrsp.com) at 100°C under suction at –20°C for 24 to 48 hours (Figure 19.23). The dried brain tissue is weighed and the weight recorded to within 0.01 mg.

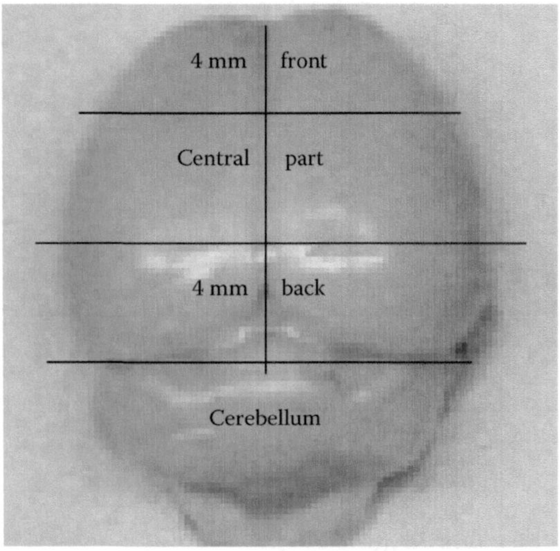

FIGURE 19.22 Blocking cut lines to prepare samples for wet:dry weight measurement.

FIGURE 19.23 Vacuum oven apparatus. (Courtesy of VWR, Cat. No. 52201-504.)

4. The percentage H_2O in each tissue sample is then calculated according to the following equation: $\%H_2O$ = ((Wet Weight – Dry Weight)/Wet Weight) × 100.81. The left and right hemispheres are compared. For example:

$$\text{Left hemisphere:} \quad ((0.5555 - 0.2112)/0.5555) \times 100 = 61.9\%$$

$$\text{Right hemisphere:} \quad ((0.5567 - 0.1701)/0.5567) \times 100 = 69.4\%$$

The difference between left and right hemisphere is 69.4% – 61.9% = 7.54% (see Figure 19.19, our unpublished data from a rat model of stroke).

Intracerebral Hemorrhage

Intracerebral hemorrhage is both a complication of some ischemia models (see Chapter 13) and the goal of some stroke models (see Chapter 18). Therefore, it is important to have a measure of hemorrhage available.

Procedure: Solutions

Drabkin's Solution

1. Reconstitute one vial of Drabkin's reagent (Sigma Diagnostics, Cat. No. D5941, K3Fe(CN)6 200 mg/L, KCN 50 mg/L, $NaHCO_3$ 1 g/L, pH 8.6) with 1000 mL of water.
2. Add 0.5 mL of a 30% Brij 35 solution (Sigma, Product Code No. B 4184) to each liter of reconstituted Drabkin's reagent. Mix well and filter if insoluble particles remain. With protection from light, this solution is stable for at least 6 months at room temperature.

Hemoglobin Standard

For the hemoglobin standard,[84] add 1.8 g of the appropriate species hemoglobin powder to 10 mL Drabkin's reagent; store tightly capped, protected from light, and refrigerated at 2°C to 8°C.

Procedure: Harvest of Experimental Samples

Several publications indicated that TTC staining does not interfere with hemoglobin assay,[84] so tissue can be used for hemoglobin assay after TTC staining (Figure 19.24).

1. Rats are anesthetized and perfused with normal saline (250 mL; efflux should run clear).
2. Brains are collected, and the cerebellum and olfactory tracts are removed.
3. Hemispheres are separated, taking great care to identify ipsilateral and contralateral samples.
4. Each hemisphere is transferred to a 14-mL tube (VWR, Cat. No. 60818-725, disposable round bottom, plastic culture tube with closure) with 1 mL of PBS, and the contents are mechanically homogenized (Ultra-Turrax T8; IKA Works, www.ika.net).
5. The sample is sonicated on ice for 1 minute (ultrasonic probe disruptor, 100 W or greater, such as VWR, Cat. No. 40000-608).
6. The sample is centrifuged at 27,000 g for 30 minutes.
7. The supernatant is collected, and 400 μL is placed into a reaction tube; each sample must be run in duplicate.

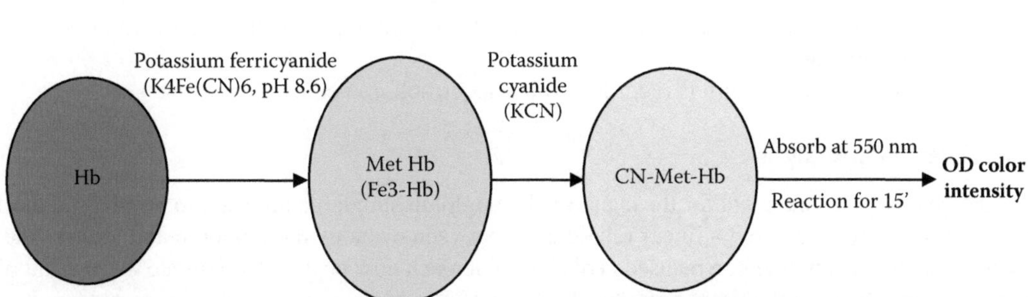

FIGURE 19.24 Schematic mechanism of hemoglobin assay.

TABLE 19.1
Preparation of Standard Working Solutions

Tube	Diluted CN-met Hb	Drabkin Solution (mL)	CN-F-Hb Concentrate (mg/mL)
1	0	1	Blank (0)
2	0.125	2.875	15
3	0.25	2.75	30
4	0.5	2.5	60
5	1	2	180
6	2	1	360

Procedure: Preparation of Tissue Standards

The following procedure is used for preparation of tissue standards[85]:

1. About 2 mL of fresh heparinized blood is sampled from a rat by cardiac puncture.
2. Normal rats are anesthetized and perfused with normal saline (250 mL; efflux should run clear) to eliminate intravascular blood.
3. Brains are collected, and the cerebellum and olfactory tracts are removed; hemispheres are separated.
4. Homologous blood, in amounts ranging from 0.5 to 200 μL, is added to each hemisphere (Table 19.1).
5. PBS is added to reach a total volume of 1 mL for each hemisphere (Table 19.1).
6. Hemispheres are mechanically homogenized for 1 minute (Ultra-Turrax T8).
7. Samples are sonicated on ice for 1 minute.
8. The sample is centrifuged at 27,000 g for 30 minutes.
9. The supernatant is collected, and 400 μL is placed in a reaction cuvette.

Procedure: *In Vitro* Assay Work

1. To prepare a dilute CN-met Hb standard work solution, 40 μL stock cyanmethemoglobin (CN-met Hb) standard solution is added to 10 mL of Drabkin's solution. This standard working solution should be used within 4 hours.
2. Prepare working standards by mixing thoroughly the solutions as shown in Table 19.2. *Note*: The concentration range should not start too high. For instance, if one starts at 60 mg/mL and intracerebral hemorrhage is small, samples may fall below the range of the standard curve (for example, a 100-μL bleed will result in an assay concentration of only 14 mg/mL cyanomethemoglobin).
3. For tissue samples, the reaction is started by adding 1600 μL of Drabkin's solution to 400 μL of the supernatant prepared in the previous steps. Samples are incubated for 15 minutes at room temperature.
4. The absorbance at 540 nm is recorded with a spectrophotometer.

Procedure: Data Analysis

The absorbance measurements for the cyanomethemoglobin standards are used to build a standard curve. These procedures yield a linear relationship between cyanomethemoglobin and absorbance; linear regression techniques can be used. This regression can now be used to estimate the amount of cyanomethemoglobin in the tissue samples (both tissue standards and experimental samples).

Alternately, the absorbance measurements for the tissue standards can be used to build a standard curve, with the amount of blood added to the tissue as the independent variable.

TABLE 19.2
Intracerebral Hemoglobin Standard Assay Curve

Hemisphere	Blood ((L)	PBS ((L)	Supernatant ((L)	Drabkin Solution ((L)
1	0	1000	400	1600
2	0.5	999.5	400	1600
3	1	999	400	1600
4	2	998	400	1600
5	4	996	400	1600
6	8	992	400	1600
7	16	984	400	1600
8	32	968	400	1600
9	50	950	400	1600
10	100	900	400	1600
11	200	800	400	1600

Note: PBS, phosphate-buffered saline.

These procedures yield a relationship between hemoglobin concentration in perfused brain and the volume of blood added (Figure 19.25), so linear regression techniques can be used. This regression can now be used to estimate the amount of blood in the tissue samples. These data can also be plotted into an exponential curve (Figure 19.26).

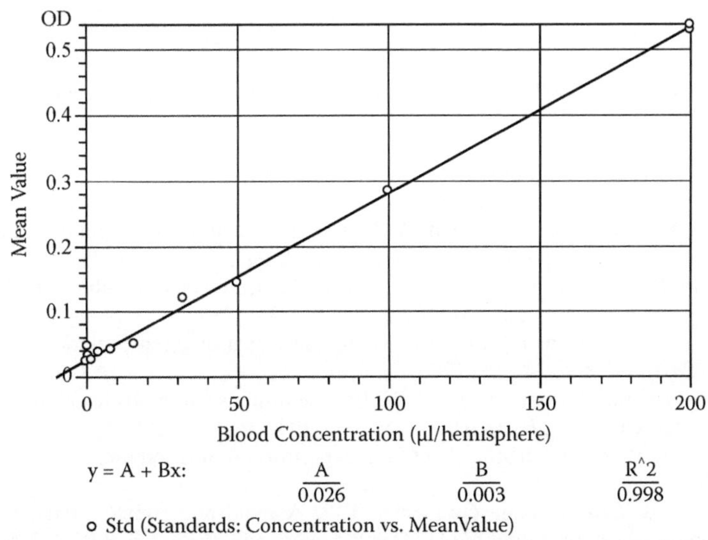

FIGURE 19.25 Standard curve for intracerebral hemorrhage. The data from Table 19.1 were plotted linearly.

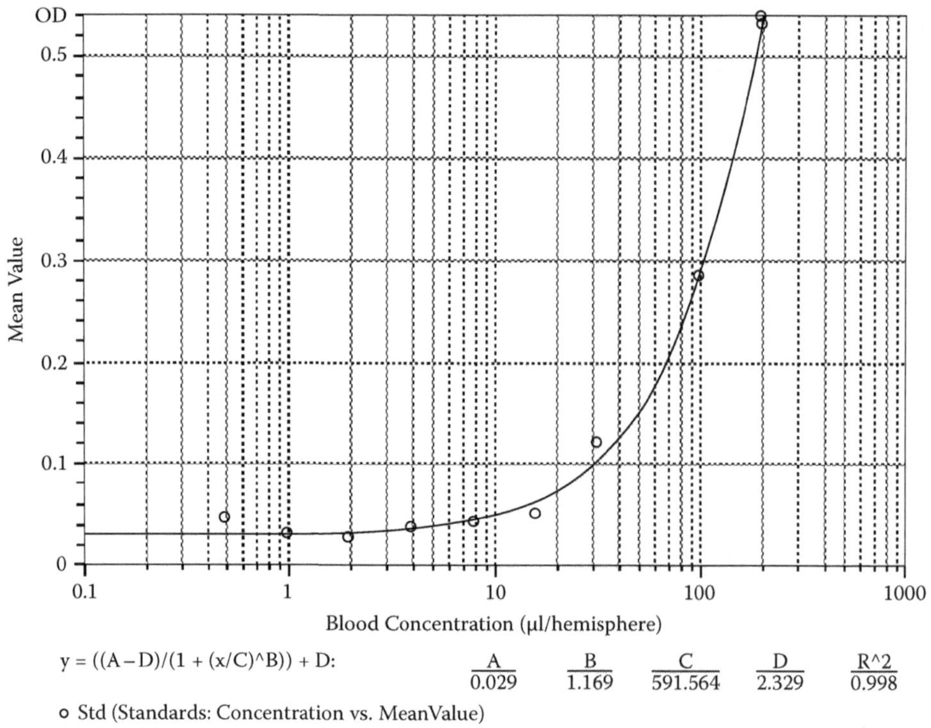

FIGURE 19.26 A standard curve for intracerebral hemorrhage. The data from Table 19.1 were plotted by an exponential curve.

ACKNOWLEDGMENTS

We thank Drs. Zhihong Haung and Jianya Ma from Johnson & Johnson for their technical assistance and advice.

REFERENCES

1. Andrews, B.T. et al., An epidural intracranial pressure monitor for experimental use in the rat, *Neurol Res,* 10, 123, 1988.
2. Giulioni, M. et al., Epidural pressure measurement in the rat, *J Neurosurg Sci,* 30, 177, 1986.
3. Rahimifar, M. et al., Effect of blood transfusion, dopamine, or normal saline on neurogenic shock secondary to acutely raised intracranial pressure, *J Neurosurg,* 70, 932, 1989.
4. Brakkee, J.H., Wiegant, V.W., and Gispen, W.H., A simple technique for rapid implantation of a permanent cannula into the rat brain ventricular system, *Lab Anim Sci,* 29, 78, 1979.
5. Doczi, T. et al., Blood–brain barrier damage during the acute stage of subarachnoid hemorrhage, as exemplified by a new animal model, *Neurosurgery,* 18, 733, 1986.
6. Mandell, E.C. and Zimmerman, E., Continuous measurement of cerebrospinal fluid pressure in unrestrained rats, *Physiol Behav,* 24, 399, 1980.
7. Nakamura, K., Osborn, J.W., and Cowley, A.W., Pressor response to small elevations of cerebroventricular pressure in conscious rats, *Hypertension,* 10, 635, 1987.
8. Nath, F.P. et al., Early hemodynamic changes in experimental intracerebral hemorrhage, *J Neurosurg,* 65, 697, 1986.
9. Bouman, H.T. and VanWimersma Greidanus, T.B., A rapid and simple cannulation technique for repeated sampling of cerebrospinal fluid in freely moving rats, *Brain Res Bull,* 4, 575, 1979.
10. Davis, R.P. et al., The protective effect of experimental subarachnoid hemorrhage on sodium dehydrocholate-induced blood-brain barrier disruption, *Acta Neurochir (Wien),* 83, 138, 1986.

11. Hayes, K.C. and Corey, J., Measurement of cerebrospinal fluid pressure in the rat, *J Appl Physiol,* 28, 872, 1970.
12. Hollis, P.H. et al., Effects of etoposide-induced blood-brain barrier disruption on brain water, intracranial pressure, and cerebral vasomotor tone, *Exp Neurol,* 99, 428, 1988.
13. Jones, H.C. and Bucknall, R.M., Changes in cerebrospinal fluid pressure and outflow from the lateral ventricles during development of congenital hydrocephalus in the H-Tx rat, *Exp Neurol,* 98, 573, 1987.
14. Melton, J.E. and Nattie, E.E., Intracranial volume adjustments and cerebrospinal fluid pressure in the osmotically swollen rat brain, *Am J Physiol,* 246, R533, 1984.
15. Solomon, R.A. et al., Decrease in cerebral blood flow in rats after experimental subarachnoid hemorrhage: a new animal model, *Stroke,* 16, 58, 1985.
16. Barth, K.N. et al., A simple and reliable technique to monitor intracranial pressure in the rat: technical note, *Neurosurgery,* 30, 138, 1992.
17. Zwienenberg, M. et al., ICP monitoring in the rat: comparison of monitoring in the ventricle, brain parenchyma, and cisterna magna, *J Neurotrauma,* 16, 1095, 1999.
18. Kusaka, G. et al., New lumbar method for monitoring cerebrospinal fluid pressure in rats, *J. Neurosci Meth*, 135(May 30), 121, 2004.
19. Dirnagl, U. et al., Continuous measurement of cerebral cortical blood flow by laser-Doppler flowmetry in a rat stroke model, *J Cereb Blood Flow Metab,* 9, 589, 1989.
20. Busto, R. et al., Small differences in intraischemic brain temperature critically determine the extent of ischemic neuronal injury, *J Cereb Blood Flow Metab,* 7, 729, 1987.
21. Kabon, B. et al., Therapeutic hypothermia, *Best Pract Res Clin Anaesthesiol,* 17, 551, 2003.
22. Cooper, K.E., *Fever and Antipyresis, The Role of the Nervous System*, Cambridge University Press, Cambridge, U.K., 1995.
23. Wixson, S.K. et al., The effects of pentobarbital, fentanyl-droperidol, ketamine-xylazine and ketamine-diazepam on core and surface body temperature regulation in adult male rats, *Lab Anim Sci*, 37, 743, 1987.
24. Boswell, M.V. and Collins, V.J., Halothane. In: *Physiologic and Pharmacologic Bases of Anesthesia*, Collins, V.J., ed., Williams & Wilkins, Baltimore, Maryland, 1996, pp. 663–686.
25. Irwin, S., Comprehensive observational assessment: Ia. A systematic, quantitative procedure for assessing the behavioral and physiologic state of the mouse, *Psychopharmacology*, 13, 222, 1968.
26. Bederson, J.B. et al., Rat middle cerebral artery occlusion: evaluation of the model and development of a neurologic examination, *Stroke*, 17, 472, 1986.
27. Hernandez, T.D. and Schallert, T., Seizures and recovery from experimental brain damage, *Exp Neurol*, 102, 318, 1988.
28. Stroemer, R.P., Kent, T.A., and Hulsebosch, C.E., Enhanced neocortical neural sprouting, synaptogenesis, and behavioral recovery with *d*-amphetamine therapy after neocortical infarction in rats, *Stroke*, 29, 2381, 1998.
29. Borlongan, C.V. and Sanberg, P.R., Elevated body swing test: a new behavioral parameter for rats with 6-hydroxydopamine-induced hemiparkinsonism, *J Neurosci,* 15, 5372, 1995.
30. Borlongan, C.V. et al., Transplantation of cryopreserved human embryonal carcinoma-derived neurons (NT2N cells) promotes functional recovery in ischemic rats, *Exp Neurol,* 149, 310, 1998.
31. Thompson, H.J. et al., Hyperthermia following traumatic brain injury: a critical evaluation, *Neurobiol Dis,* 12, 163, 2003.
32. Thompson, H.J. et al., Development of posttraumatic hyperthermia after traumatic brain injury in rats is associated with increased periventricular inflammation, *J Cereb Blood Flow Metab*, 25, 163, 2005.
33. Dietrich, W.D., The importance of brain temperature in cerebral injury, *J Neurotrauma,* 9, S475, 1992.
34. Dietrich, W.D. et al., Delayed post-traumatic brain hyperthermia worsens outcome after fluid percussion brain injury: A light and electron microscopic study in rats, *Neurosurgery,* 38, 533, 1996.
35. Kakuda, W. et al., Body temperature increases following embolic stroke correlating with size of infarction, *J Cereb Blood Flow Metab,* 17, S43, 1997.
36. Rossi, S. et al., Brain temperature, body core temperature, and intracranial pressure in acute cerebral damage, *J Neurol Neurosurg Psychiatry,* 71, 448, 2001.
37. Rudy, T.A., Pathogenesis of fever associated with cerebral trauma and intracranial hemorrhage. In: *Thermoregulatory Mechanisms and Their Therapeutic Implications, Fourth International Symposium on the Pharmacology of Thermoregulation*, Cox, B., Lomax, P., Milton, A.S., and Schönbaum, E., eds., Karger, Basel, 1980, pp. 75–81.

38. Rudy, T.A., Williams, J.W., and Yaksh, T.L., Antagonism by indomethacin of neurogenic hyperthermia produced by unilateral puncture of the anterior hypothalamic/preoptic region, *J Physiol*, 272, 721, 1977.
39. Saper, C.B. and Breder, C.D., The neurologic basis of fever, *N Engl J Med,* 330, 1880, 1994.
40. Waynforth, H.B. and Flecknell, P.A., *Experimental and Surgical Technique in the Rat*, 2nd ed., Academic Press, San Diego, California, 1992.
41. Motulsky, H., *Intuitive Biostatistics*, Oxford University Press, New York, 1995, pp. 11–21.
42. Schallert, T. et al., Tactile extinction: Distinguishing between sensorimotor and motor asymmetries in rats with unilateral nigrostriatal damage, *Pharmacol Biochem Behav*, 16, 455, 1982.
43. Schallert, T. and Whishaw, I.Q., Bilateral cutaneous stimulation of the somatosensory system in hemidecorticate rats, *Behav Neurosci*, 98, 518, 1984.
44. Sughrue, M.E. et al., An improved test of neurological dysfunction following transient focal cerebral ischemia in rats, *J Neurosci Meth*, 151, 83, 2006.
45. Hermer-Vazquez, L. et al., Somatosensation. In: Whishaw, I.Q. and Kolb, B.K., eds., *The Behavior of the Laboratory Rat: A Handbook with Tests*, Oxford University Press, New York, 2005, pp. 60–68.
46. Barth, T.M., Grant, M.L., and Schallert, T., Effects of MK-801 on recovery from sensorimotor cortex lesions, *Stroke*, 21, SIII153, 1990.
47. Ding, Y. et al., Motor balance and coordination training enhances functional outcome in rat with transient middle cerebral artery occlusion, *Neuroscience*, 123, 667, 2004.
48. Dunham, N.W. and Miya, T.S., A note on a simple apparatus for detecting neurological deficit in rats and mice, *J Am Pharm. Assn., Sci. Ed.*, 46, 208, 1957.
49. Jones, B.J. and Roberts, D.J., The quantitative measurement of motor incoordination in naïve mice using an accelerating rotarod, *J Pharm Pharmacol*, 20, 302, 1968.
50. Watzman, N. et al., Influence of certain parameters on the performance of mice on the rotarod, *Arch Int Pharmacodyn*, 169, 362, 1967.
51. Alexis, N.E. et al., Nonocclusive common carotid artery thrombosis in the rat results in reversible sensorimotor and cognitive behavioral deficits, *Stroke*, 26, 2338, 1995.
52. De Ryck, M. et al., Photochemical stroke model: Flunarizine prevents sensorimotor deficits after neocortical infarcts in rats, *Stroke*, 20, 1383, 1989.
53. Rivlin, A.S. and Tator, C.H., Objective clinical assessment of motor function after experimental spinal cord injury in the rat, *J Neurosurg*, 47, 577, 1977.
54. Cserr, H., Potassium exchange between cerebrospinal fluid, plasma and brain, *Am J Physiol*, 209, 1219, 1965.
55. Levinger, I.M., The cerebral ventricles of the rat, *J Anat*, 108, 447, 1971.
56. Rapoport, S.I. et al., Drug entry into and distribution within brain and cerebrospinal fluid: [^{14}C] urea pharmacokinetics, *Am J Physiol Reg Integr Comp Physiol*, 242, R339, 1982.
57. Höistad, M. et al., Quantitative dual-probe microdialysis: Evaluation of [^{3}H]mannitol diffusion in agar and rat striatum, *J Neurochem*, 81, 80, 2002.
58. Jankowski, C.J., Neuraxial anesthetic techniques. In: *Textbook of Regional Anesthesia*, Raj, P.P., ed., Churchill Livingstone, New York, 2002, pp. 829–852.
59. Takasugi, Y. et al., Transcutaneous cisternal puncture for sampling of cerebrospinal fluid in awake rat, *Exp Anim,* 54, 193, 2005.
60. Suckling, A.J. and Reiber, H., Cerebrospinal fluid sampling from guinea pigs: sample volume-related changes in protein concentration in control animals and animals in the relapsing phase of chronic relapsing experimental allergic encephalomyelitis, *Lab Anim*, 18, 36, 1984.
61. Huang, Z.G. et al., Biphasic opening of the blood-brain barrier following transient focal ischemia: Effects of hypothermia, *Can J Neurol Sci*, 26, 298, 1999.
62. Preston, E., Allen, M., and Haas, N., A modified method for measurement of radiotracer permeation across the rat blood-brain barrier: The problem of correcting brain uptake for intravascular tracer, *J Neurosci Meth*, 9, 45, 1983.
63. Kuroiwa, T., Ting, P., and Klatzo, I., The biphasic opening of the blood brain barrier to proteins following temporary middle cerebral artery occlusion, *Acta Neuropathol*, 68, 122, 1985.
64. Klatzo, I., Concept of ischemic injury associated with brain edema. In: *Brain Edema*, Inaba, Y., Klatzo, I., and Spatz, M., eds., Springer, Tokyo, 1984, pp. 1–5.
65. Westergaard, E., van Deurs, B., and Brondsted, H.E., Increased vesicular transfer of horseradish peroxidase across cerebral endothelium, evoked by acute hypertension, *Acta Neuropathol*, 37, 141, 1977.

66. Nagy, Z., Mathieson, G., and Huttner, I., Blood–brain barrier opening to horseradish peroxidase in acute arterial hypertension, *Acta Neuropathol*, 48, 45, 1979.
67. Cole, D.J. et al., Time- and pressure-dependent changes in blood-brain barrier permeability after temporary middle cerebral artery occlusion in rats, *Acta Neuropathol*, 82, 266, 1991.
68. Betz, A.L. et al., Blood-brain barrier permeability and brain content of sodium, potassium and chloride during focal ischemia, *J Cereb Blood Flow Metab*, 14, 29, 1994.
69. Ng, L.K.Y. and Nimmannitya, J., Massive cerebral infarction with severe brain swelling: A clinicopathological study, *Stroke*, 1, 158, 1970.
70. Halsey, J.H., Jr. and Capra, N.F., The course of experimental cerebral infarction: the development of increased intracranial pressure, *Stroke*, 3, 268, 1972.
71. Sharma, H.S., Hyperthermia induced brain edema: current status and future perspectives, *Ind J Med Res*,123, 629, 2006.
72. Hacke, W. et al., "Malignant" middle cerebral artery territory infarction: Clinical course and prognostic signs, *Arch Neurol*, 53, 309, 1996.
73. Berrouschot, J. et al., Mortality of space-occupying ("malignant") middle cerebral artery infarction under conservative intensive care, *Intensive Care Med*, 24, 620, 1998.
74. Moulin, T. et al., Early CT signs in acute middle cerebral artery infarction: predictive value for subsequent infarct locations and outcome, *Neurology*, 47, 366, 1996.
75. Yoshimoto, T. et al., Clinical course of acute middle cerebral artery occlusion, *J Neurosurg*, 65, 326, 1986.
76. Ohno, K., Pettigrew, K.D., and Rapoport, S.I., Lower limits of cerebrovascular permeability to nonelectrocytes in the conscious rat, *Am J Physiol*, 235, H299, 1978.
77. Preston, E. and Haas, N., Defining the lower limits of blood–brain barrier permeability: Factors affecting the magnitude and interpretation of permeability-area products, *J Neurosci Res*, 6, 709, 1986.
78. Sarelius, I.H., Plasma volume shifts during moderate exercise in splenectomized greyhounds, *J Physiol*, 292, 391, 1979.
79. Yepes, M. et al., Tissue-type plasminogen activator induces opening of the blood–brain barrier via the LDL receptor-related protein, *J Clin Invest,* 112, 1533, 2003.
80. Kyoi, S. et al., Loss of intracellular dystrophin: a potential mechanism for myocardial reperfusion injury, *Circ. J.,* 67, 725, 2003.
81. Lin, T.N., Effect of brain edema on infarct volume in a focal cerebral ischemia model in rats, *Stroke*, 24, 117, 1993.
82. Elliott, K.A. and Jasper, H., Measurement of experimentally induced brain swelling and shrinkage, *Am J Physiol*, 157, 122, 1949.
83. Ishimaru, S. and Hossman, K.A., Relationship between cerebral blood flow and blood–brain barrier permeability of sodium and albumin in cerebral infarcts of rats, *Acta Neurochir*, 51, 216, 1990.
84. Choudhri, T.F. et al., Use of a spectrophotometric hemoglobin assay to objectively quantify intracerebral hemorrhage in mice, *Stroke*, 28, 2296, 1997.
85. Sumii, T. and Lo, E.H., Involvement of matrix metalloproteinase in thrombolysis-associated hemorrhagic transformation after embolic focal ischemia in rats, *Stroke*, 33, 831, 2002.

20 Tissue Staining Techniques for Stroke Studies

Yanlin Wang-Fischer and Lee Koetzner

CONTENTS

INTRODUCTION

Histological and immunochemical stains are important tools for stroke research. Here, we introduce some basic techniques (mainly immunohistochemical stains) that we trust will benefit beginners in central nervous system (CNS) studies.

TISSUE PREPARATION

Tissue Preparation for Immunohistochemical Staining

Saline/Formaldehyde Perfusion

The saline/formaldehyde perfusion is a standard brain perfusion for rats at 24 to 48 hours poststroke.

Materials

1. Saline: 0.9% sodium chloride (approximately 300 mL/rat).
2. 4% paraformaldehyde (PFA) in freshly prepared $NaPO_4$ buffer solution (PB) (approximately 300 mL/rat).
3. Perfusion is performed with a pump system (Figure 20.1), which can be placed in a hood to evacuate the formaldehyde vapor released by PFA. The system can be purchased from VWR: Varistaltic Pump Systems, Vera, Barnant, Cat. No. BR72-315-000 (for 115 V), BR72-317-230 (for 220 V), BR72-317-000 (115 V, with remote capabilities), BR72-315-230 (220 V, with remote capabilities).

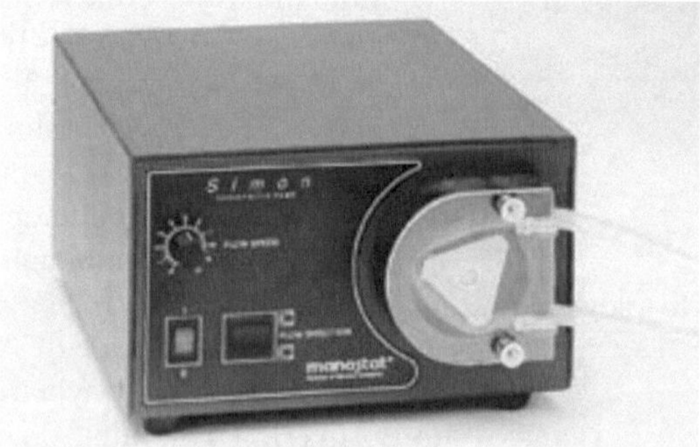

FIGURE 20.1 A Varistaltic Pump System from VWR. These self-priming, tubing pumps deliver precise movement of liquids or gases. The tubing can be ordered from VWR. (Reproduced courtesy of VWR.)

Here, we abstract some information from the VWR Web site (with permission) to help investigators better understand the system:

> This self-priming tubing pump delivers precise movement of liquids or gases. It is ideal for a wide variety of laboratory applications, including sterile pumping, positive pressure filtration, circulation of buffers, running columns, and handling of caustics or corrosives in a hood. The pumping head utilizes a squeezing action of rollers on flexible tubing to create flow of liquids or gases. The pump can be set digitally from 0 to 99% of full speed and allows resetting and duplication of flow rates, and forward or reverse operation. The head accepts up to six different sizes of tubing. The unit is compact and stackable, with the following characteristics: flow rate: 1 to 3,400 mL/min, maximum speed: 720 rpm, dimensions: 26L × 22.9D × 13H cm ($10^1/_4 \times 9 \times 5^1/_8$"), and weight: 5.6 kg (12 lbs.).

Alternatively, a 60-mL syringe can be used to pump perfusate.

4. Tubing system: Tubes connect to the pump, two-way switch, and a G18 round-tip needle (a gavage needle for mice).
5. Because PFA is a chemical hazard, a waste container is needed for the used PFA solution.

Procedure for Transcardiac Perfusion

1. The rat is anesthetized with pentobarbital and placed on a wire mesh screen over a sink.
2. The chest is opened. Blood (1 to 2 mL) can be collected directly from the left ventricle if needed.
3. A small hole is cut at the cardiac apex, and a feeding needle, which is connected to the pump by a tube, is inserted into the left ventricle up to the aorta; blunt needles can also be used; however, using standard needles increases the risk of puncture to the opposite wall of the heart or aorta.
4. Two ventricles are blocked with a hemostatic forceps (see Figure 20.2). Cooley-Derra anastomosis clamps such as Roboz, Product No. RS-7660 will make this easier. The descending aorta can be blocked to shorten the perfusion time (only perfusing the upper body).
5. The right atrium is cut open to allow the perfusate to drain.
6. About 200 to 300 mL normal saline is infused to displace the blood in the vessels. Continue this perfusion until the draining perfusate is clear (free of blood). Delays in reaching this step will allow blood to begin coagulating *in situ*, which will reduce the quality of the perfusion.

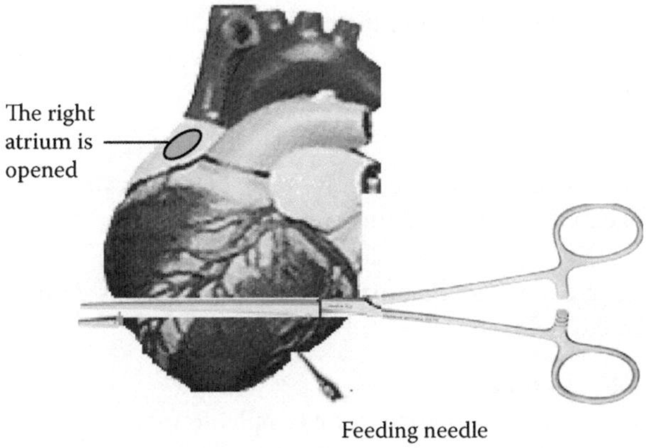

FIGURE 20.2 A procedure for transcardiac perfusion. A feeding needle with a rounded tip is inserted into the left ventricle up to the aorta. The ventricles are blocked with a hemostatic forceps. The right atrium is opened for drainage of the perfusate.

7. This is followed by infusing 4% PFA to fix the tissue. As the perfusate displaces the saline, the change in tonicity will cause spontaneous muscle contractions. When these contractions cease, perfusion is complete. Typically, a 350-g rat can be adequately perfused with about 300 mL.
8. The brain is removed into 30 to 50 mL of 4% PFA in $NaPO_4$ buffer solution and postfixed for 1 to 2 hours.
9. The brain is placed in 20% sucrose in 0.1*M* $NaPO_4$ buffer, pH 7.4, 50 mL/brain at 4°C for 18 to 24 hours. Then, the brains are cut into sections for staining or frozen quickly and stored at −80°C for future use. Rapid freezing is important; slower rates of freezing lead to the formation of larger ice crystals, which produce artifacts during processing.[1] With adequate cryoprotection, tissue may be frozen by covering it with dry ice. However, the fastest technique for freezing tissue is immersion in dry ice-chilled 2-methylbutane; this is even faster than immersion in liquid nitrogen.[2]

Perfusion with 1% Gelatin

Perfusion with 1% gelatin is used for preparation of animal brain tissue after more than 3 days poststroke, because the infarcted areas tend to fall away when the free-floating sections are collected, or later when the sections are stained. We developed a technique with Dr. Robert C. Switzer III to overcome this problem by perfusion of gelatin into the brain before brain sectioning. (Dr. Switzer is affiliated with NeuroScience Associates and is an expert in neurohistological staining procedures. Company information can be found at www.NSALabs.com, 10915 Lake Ridge Drive, Knoxville, Tennessee 37922, 865-675-2245.)

1. An adult rat is anesthetized with pentobarbital (Nembutal, 50 mg/kg i.p.). Then, 0.1 to 0.2 mL 1% sodium nitrite (Sigma, Cat. No. S2252) is injected into the heart just before perfusion to dilate the entire vascular system and thus allow the ingress of more perfusate. However, if the purpose is to study brain vessel response to stroke, this procedure should be avoided.
2. Phosphate-buffered saline (PBS) or normal saline is perfused transcardially until the outflow solution is clear (see Figure 20.2).
3. 1% gelatin in PBS (about 100 to 120 mL) is then perfused. The solution is kept at about 45°C to 50°C so that it stays fluid.
4. 4% PFA in PBS (about 100 to 200 mL) is then perfused.
5. The brain is left in the skull overnight for postfixing. Next morning, the brain is removed from the skull and postfixed in 4% PFA for at least 36 hours (but less than 48 hours) and then transferred to PBS.

This is a standard perfusion with the addition of 1% gelatin at the midpoint of the procedure. The main concern is to keep the gelatin solution warm so it does not thicken before perfusion. We suggest keeping the reservoir of gelatin in a double boiler set up to control the temperature, and if the perfusion tubing is long enough so that the gelatin solution cools en route, then immersing the tubing in a warm bath would help.

Preparation of 1% Gelatin Solution

Regarding preparation of 1% gelatin solution, for example, to make up 2 g gelatin in 200 mL PBS, the PBS is heated to 55°C to 60°C and with a strong magnetic stirrer, the PBS is stirred so there is a deep vortex. Gelatin powder is gradually poured on the sides of the vortex, not directly down the middle. The gelatin powder will seem to disappear as it enters the solution from the surface of the vortex if added slowly enough. On the contrary, if added too quickly, the gelatin will clump and take longer to go into solution; patience will be needed if that happens. The solution can be made ahead of time and stored in the refrigerator at 4°C. The solution is reheated using a double-boiler setup in which the gelatin container is placed in a hot water bath. *Note*: Do not microwave the solution as there can be "hot spots" in the solution that could exceed 90°C or so, whereupon a "metal gel" will form that will not "cure" when exposed to formalin. The gelatin can be purchased from Fisher Scientific (Cat. No. G8-500; -### refers to the number of grams).

Tissue Preparation for Biochemical Assay

Perfusion fixation is not necessary for some biochemical assays and actively interferes with others (for example, enzyme activity or *in situ* hybridization). For these experiments, tissue can be collected after rapid decapitation. This procedure produces rapid loss of consciousness, even without the use of anesthetics.[3] Readers unfamiliar with this procedure are encouraged to seek a demonstration from more experienced colleagues before attempting the procedure themselves as mistakes can be dangerous to both animals and experimenters.

Procedure

1. Gently but firmly grasp the rat so the head is free; quickly place the rat's head in the lower blade of a guillotine (e.g., Harvard Apparatus Product No. 550012) and sever the rat's neck. For mice, a sturdy pair of shears with good handles can be substituted (e.g., Roboz Product No. RS-7097).
2. Rapidly remove the brain from the skull and freeze the brain in 2-methylbutane (formerly called isopentane) chilled on dry ice. The tissue is cooled to about –20°C; when it turns white (typically 20 to 30 seconds), the tissue can be stored at –80°C.

Tissue Preparation for Histological Stains

Brains can be prepared in 10% neutral buffered formalin, then dehydrated, embedded in paraffin, and cut on a rotary microtome for histological examination (for a detailed method, see the section "Hematoxylin-Eosin [H&E] Staining" later in this chapter).

Tissue Sectioning

The choice of section thickness represents a compromise between ease of section and cell count; large neurons can have bodies 50 μm wide, while the smallest neurons are 10 μm wide or less. Therefore, a thicker section (25 to 50 μm) will contain most of the soma of the largest cells and may superimpose several smaller cells. These sections perform well with antibody stains (for example, immunochemical stains) and are relatively easy to cut. Thinner sections (10 μm) generally do not superimpose cell bodies and therefore are much less likely to give false-positive results for colocalization studies; however, they are much more difficult to cut. Sections can be prepared for cutting either by embedding in matrix or freezing; while freezing is simpler, many staining techniques have been developed around paraffin embedding. Sections either embedded or frozen can be cut on a sliding microtome; frozen sections require the controlled cryostat environment.

The next section describes cutting the frozen tissues. Embedded tissue cutting is described in the H&E staining method for paraffin-embedded sections.

HOW TO USE A MICROTOME

A microtome is a device used for cutting sections. This device is usually combined with a temperature controller and a water tank. A microtome that slides the specimen across the block is called a *sliding microtome*. The microtome may be different in each laboratory; rotary microtomes use

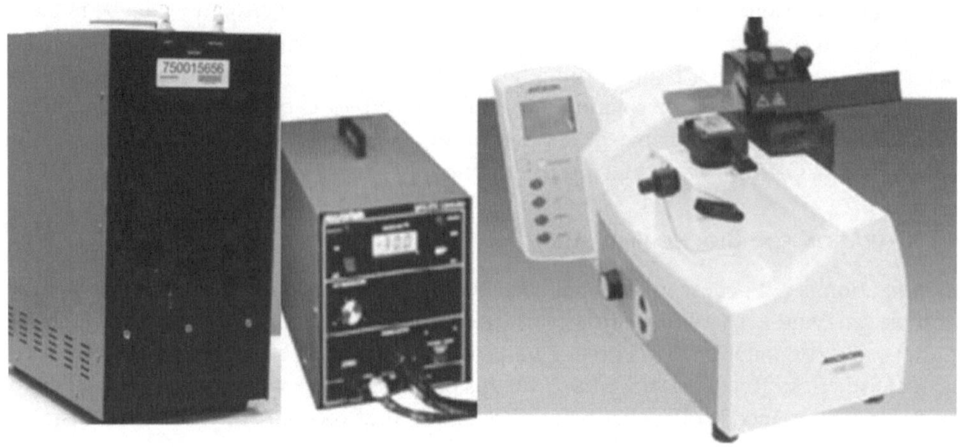

(a) (b) (c)

FIGURE 20.3 The microtome system includes a pump and tank unit (a), Physitemp BFS-30 TC controller (b), and Microm HM 450 microtome (c). (Courtesy of Physitemp Instruments.)

a rotary actuator to advance the specimen across the blade. The basic theories are the same. We introduce the system from our laboratory as a sample; it includes a water pump and tank unit (PTU-3), temperature controller (Physitemp BFS-30 TC), and microtome (HM 450) (Figure 20.3). This system was purchased from Physitemp Instruments, and with their permission we present some information about it from their Web site (Figure 20.3) (154 Huron Avenue, Clifton, New Jersey 07013, 973-779-5577, 800-452-8510, physitemp@aol.com).

PTU-3 Pump and Tank Unit

The PTU-3 pump and tank unit is designed to be attached to a sliding system for controlling temperature. A closed-loop water system is desirable for installations where an external water source is not available.

Temperature Controller (Physitemp BFS-30TC)

The BFS-30TC freezing stage fits easily onto most microtomes. These thermoelectric devices have no moving parts and need only alternating current and a trickle of water for operation. Freezing starts immediately and can be maintained indefinitely. The stage temperature is adjustable so that optimum cutting temperatures can be set for different kinds of tissue.

Microtome HM 450

This is a unique sliding microtome with an automatic fine/trim-feed system.

> This microtome can be used for sectioning paraffin, frozen specimens, and specialty applications in biological, botany, and materials science. Specimen sizes up to 155 × 90 mm can be accommodated. The extraordinarily smooth knife carrier movement guarantees fast, comfortable and nontiring operation. Intelligent and modern electronic control systems coupled with high-precision mechanics result in optimal section quality and reproducibility of even difficult-to-cut embedded specimens. Section thickness can range from 1 to 100 microns.

How to Operate the System to Cut Frozen Tissue

1. Before starting the system, make sure the pump and tank unit are filled with distilled water. Turn the power on at the Physitemp unit (red button). Set temperature at –30°C using the white button and the adjusting switch.
2. Turn on the microtome power switch, located in the back.
3. Spread water as a flat layer onto the surface of the stage and wait until the water freezes.
4. Place a rat brain flat onto the plate in preparation for making coronal sections. From personal experience gives preference for brains prepared in cryoprotectant solution before the section. This eases fixation of brain tissue on the frozen water surface.
5. The brain is frozen by placing pulverized dry ice around it.
6. A microtome knife is placed into its holder, and the screws are tightened on the top. *Be careful*: This knife is very sharp; do not touch the edge.
7. Adjust the feed to 40 μm (dependent on your need) and trim by pressing the round white button and black switch on the left side of the microtome.
8. Adjust the stage up and down by the up-and-down button on the left side of the microtome. Move the plate to the left or right by the smaller black switch and move the plate forward or backward by the larger black switch. The plate is fixed in place or loosened by turning the front black handle up or down.
9. Always work smoothly when sectioning tissue; erratic movements produce poor sections. A noise will be heard while moving the handle back; the plate is raised by 40 μm and is ready to cut the next section.
10. When sectioning is finished, turn the microtome temperature control back to 23°C and clean the microtome.
11. The sectioned slides are placed in 50 m*M* TBS buffer (50 m*M* Tris buffer plus normal 0.9% NaCl, pH 7.4, containing 1 m*M* sodium azide) with a fine paintbrush and are stored free floating in the solution at 4°C.

Preparation of Gelatin/Chrom Alum Subbed Slides

The purpose of coating (also called *subbing*) slides is to make them sticky so sections stay on the slides.

1. Load slide racks.
2. If the slides will be used for receptor-binding studies, they should be precleaned in acid alcohol (a mixture of 3 mL 12*N* HCL plus 92 mL ethanol and 5 mL water) and rinsed in distilled water prior to coating.
   Alternately, slides can be precleaned in Chromerge (chromic-sulfuric acid mixture, a cleaning solution for laboratory glassware). Immerse each rack of slides in Chromerge for 5 minutes. Rinse slides five times for 2 minutes each in distilled water. The purpose of precleaning is to remove protein/DNA contamination and bleach any color on the slides.
3. Prepare coating (or subbing) solution (500 mL for rat slides):
   a. Heat 500 mL dH_2O (distilled water) to 45°C to 55°C (*do not heat above 60°C*).
   b. Add 5 g gelatin and dissolve with gentle stirring; allow the solution to cool to 20°C to 21°C.
   c. Add 0.5 g chromium potassium sulfate and dissolve with gentle stirring.
4. Dip slides into this coating solution for 30 seconds and place them into an oven (37°C to 50°C) overnight to dry.
5. Carefully tap slides to remove from slide racks. (*If acid washed, handle slides by the edges.* Otherwise, you defeat the purpose of the acid wash.) Store slides in slide boxes.

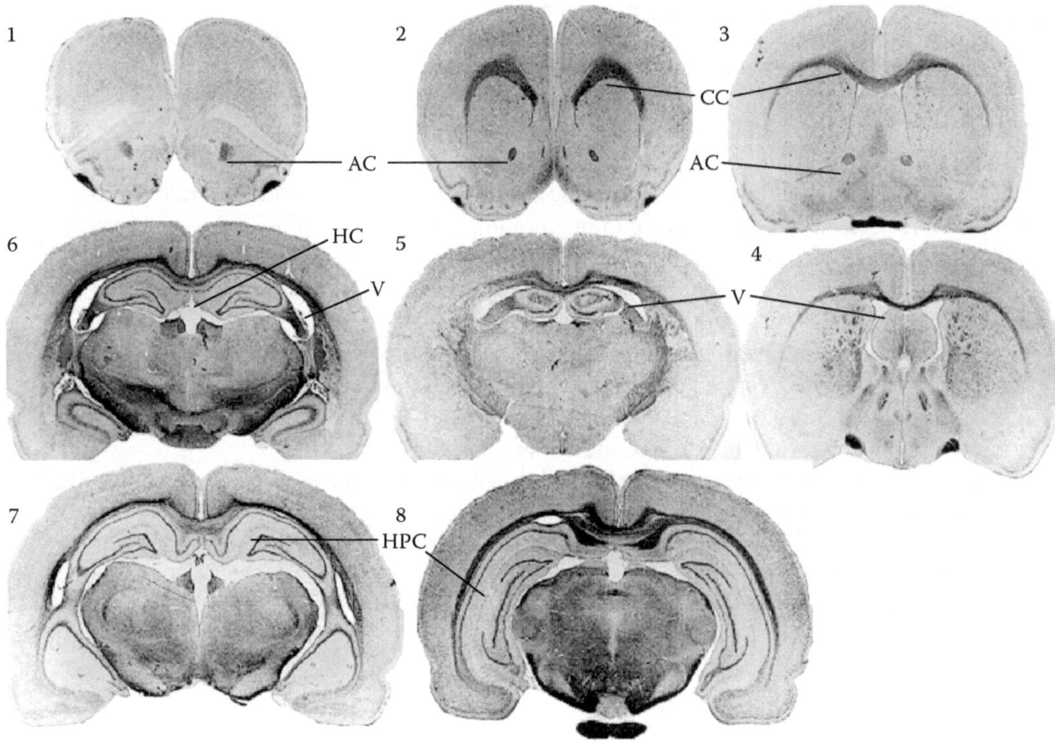

FIGURE 20.4 The order from anterior to posterior (slices 1 to 8) is based on the anatomical structure characteristics of white matter, ventricles, and hippocampus. AC, anterior commissure; CC, corpus callosum; HC, hippocampal commissure; HPC, hippocampus; V, ventricle.

How to Place Brain Slices on Subbed Slides in Order

Brain slices that are free floating in solution are mounted one by one on the subbed slides using a fine paintbrush. Ventricles and fiber tracts make the best landmarks on gross inspection of a frozen brain; these landmarks can be used to organize sections from rostral to caudal (see Figure 20.4).

1. At the rostral pole, slices with the anterior commissure (AC; two black spots on the slices) start from far apart and become closer (see Figure 20.4 from 1 to 3).
2. Subcortex and cortex are delineated by the corpus callosum (a bridge between two hemispheres; see Figure 20.4 from 2 to 4).
3. The lateral ventricles grow larger in more caudal sections (see Figure 20.4 from 4 to 6).
4. The hippocampus becomes larger and finally occupies the entire subcortex (see Figure 20.4 from 6 to 8).

PREPARATION OF SOLUTIONS

Sodium Phosphate Buffer, 0.1*M*

Preparation of 0.2*M* Sodium Phosphate Buffer

Solution A: 27.6 g $NaH_2PO_4{\cdot}H_2O$ per liter (0.2*M*). Molecular weight is 137.99.
Solution B: 53.65 g $Na_2HPO_4{\cdot}7H_2O$ per liter (0.2*M*). Molecular weight is 268.07.

TABLE 20.1
Preparation of 0.1*M* Sodium Phosphate Buffers at Different pH*

Desired pH	Solution A (ml)	Solution B (ml)	Desired pH	Solution A (ml)	Solution B (ml)
5.7	93.5	6.5	6.9	45	55
5.8	92	8	7	39	61
5.9	90	10	7.1	33	67
6	87.7	12.3	7.2	28	72
6.1	85	15	7.3	23	77
6.2	81.5	18.5	7.4	19	81
6.3	77.5	22.5	7.5	16	84
6.4	73.5	26.5	7.6	13	87
6.5	68.5	31.5	7.7	10.5	90.5
6.6	62.5	37.5	7.8	8.5	91.5
6.7	56.5	43.5	7.9	7	93
6.8	51	49	8	5.3	94.7

If using Na_2HPO_4.anhydrous, the molecular weight is 141.96. Refer to Table 20.1 for the desired pH* obtained by mixing the indicated volumes of 0.2*M* solutions A and B, then diluting with distilled water to 200 mL, which equals 0.1*M*.

Preparation of 4% Paraformaldehyde Solution in 0.1*M* Phosphate Buffer for Perfusion

Caution: Preparation of 4% PFA in 0.1*M* phosphate buffer (PB) should be done in a chemical hood. An example of using 0.2*M* PB to prepare 4% PFA in 0.1*M* PB is as follows:

1. PFA (8 g) is dissolved in 100 ml dH_2O.
2. The solution is heated (on a heated stirrer while stirring with A magnetic bead) to 55°C to 65°C to partially dissolve PFA.
3. 1*M* NaOH (one or two drops) is added dropwise until the solution clears.
4. Add 100 mL 0.2*M* PB and mix the solution, filter with Whatman filter paper (grade no. 4), and cool to 4°C before use. The pH of this solution is adjusted around to 7.4. The solution is stored at 4°C. Prolonged storage (>1 week) may cause the pH to change. If the PFA powder is directly added into 0.1*M* PB solution, there is no need to readjust pH.

Note: Do not autoclave. To achieve a better fixture result, PFA must be completely dissolved into its stock solution and cooled to 4°C before use (hot PFA may damage the cells).

Cryoprotectant Solution with 0.1*M* Phosphate Buffer

The most commonly used cryoprotectant solution is 20% sucrose. This solution contains 800 mL PB (0.1*M*, pH 7.4) and 200 g sucrose. Bring to 1000 mL with 0.1*M* PB at pH 7.4.

If cryopreservation with 20% sucrose is not sufficient to prevent the appearance of freezing artifacts (this often happens with larger blocks of tissue), a glycerol-DMSO (dimethyl sulfoxide) solution can be used. This solution contains the following:

Phosphate buffer (0.12*M*, pH 7.4): 780 mL
Glycerol: 200 mL
DMSO: 20 mL

Adjust to pH 7.4 with concentrated HCl or NaOH; store in refrigerator.

TBS (1*M* Tris Stock Solution with 0.9% NaCl, pH 7.4) for Immunohistochemical Staining

For TBS, use 1*M* Tris base (Sigma), molecular weight 121.14. For example, to make 2000 mL of 1*M* Tris base stock solution, use the following:

Tris base: 242.28 g
Double-distilled water: 1800 mL
NaCl: 18 g

Bring to total volume of 2000 mL with double-distilled water. Mixed with a stirrer. Adjust to pH 7.4 with 12*N* HCl.

TBS (50 m*M*, pH 7.4) and 0.5% Triton X-100 for Immunohistological Staining

For example, to make 400 mL of the 50 m*M* TBS and 0.5% Triton X-100 solution, 1*M* stock solution from the TBS (1*M* Tris stock solution with 0.9% NaCl, pH 7.4) is diluted 20 times to yield 50 m*M* solution, then 398 mL of 50 m*M* TBS plus 2 mL Triton X-100 are mixed in a warm bath. Store at 4°C.

TBS (50 m*M*, pH 7.4) and 0.1% Triton X-100 for Immunohistological Staining

For example, to make 400 mL of TBS (50 m*M*, pH 7.4) and 0.1% Triton X-100 solution, use 399.6 mL 50 m*M* TBS and 0.4 mL Triton X-100. Mix in a warm bath (37°C). Store at 4°C.

TBS 50 m*M*, 3.3% Normal Serum, and 0.1% Triton X-100 for Immunohistological Staining

For example, to make 400 mL of TBS 50 m*M*, 3.3% normal serum, and 0.1% Triton X-100, use the following:

0.4 mL Triton X-100
13.2 mL goat serum
386.4 mL TBS

Mix in a warm bath to yield 400 mL. Store at 4°C.

50 m*M* Tris, pH 7.6, for Immunohistological Staining

To make 50 m*M* Tris, pH 7.6, first make a 1*M* Tris base stock solution at pH 7.6. Then, for example, to make 2000 mL of this solution, add 242.28 g Tris base to 1800 mL ddH_2O and adjust to pH 7.6 with 12*N* HCl, then bring to 2000 mL with double-distilled water. The stock solution is diluted 20 times to yield 50 m*M*; store at 4°C.

Developing Solution for Immunohistochemical Stains

Developing solution for immunohistochemical stains consists of Tris buffer containing 0.04% DAB (3,3´-Diaminobenzidine tetrahydrochloride tablets, 10 mg/tablet, Sigma, Cat. No. D5905-100 tab) with 0.3% nickel ammonium sulfate, which is then vacuum filtered over Whatman No. 4 filter paper, after which H_2O_2 is added to 0.003% concentration. The solution has to be prepared just before use.

For example, to make 25 mL of this solution,

Tris buffer: 25 mL
DAB tablets: 10 mg
Nickel ammonium sulfate: 75 mg

Add 30% 2.5 μL hydrogen peroxide solution, and Filter over a Whatman No. 4 filter paper.

HISTOLOGICAL STAINING

2,3,5-Triphenyltetrazolium Chloride (TTC) Stain

Staining with TTC is a common method to assess lesion size in rat brains after stroke.[4–6] TTC is itself a white or faint yellow powder and is colorless in solution. When TTC diffuses into actively respiring tissues, it accepts electrons from the mitochondrial electron transport chain, and the stain is reduced to yield a deep pink compound, formazan. Accumulation of the pink formazan stains the tissues red, and the intensity of the red color is proportional to the rate of respiration in those tissues.[7,8] Infarcted brain regions do not convert TTC and remain unstained. Therefore, this method can distinguish live (stained red) from infarcted (unstained white) tissues. TTC stain demonstrates ischemic lesions that can be appreciated visually even without microscopic examination. Infarct areas measured with TTC correspond closely with those measured with other histological methods.[4,9,10] This technique enables assessment of lesion size with minimal tissue preparation. Quantitative measurements of infarct volume determined by this method have proven useful in determining the extent of brain injury in experimental stroke models and in assessing potential neuroprotective agents for cerebral ischemia.[4–6]

Procedures for 2,3,5-Triphenyltetrazolium Chloride Stain

1. 2% TTC (Sigma-Aldrich, Cat. No. T8877): For example, to make 50 mL 2% TTC, 1 g TTC is solved in 50 mL normal saline (0.9%) at room temperature to yield 2% solution. *Note*: Based on the data safety sheet, TTC may irritate eyes, respiratory system, and skin.
2. Prepare the brain tissue: Rat brains have to be prepared freshly on wet ice. A brain is sliced in a brain matrix to 2-mm-thick sections by means of a sharp blade (Accu-Edge High Profile Microtome blades). Usually, the brain of an adult rat should be cut into eight pieces. The brain matrix has many grids; each grid is 1 mm thick (Figure 20.5). The brain matrix is made of either stainless steel or acrylic. The grids are sturdy and are able to withstand the rigors of daily use; they can be heated, chilled, autoclaved, and scrubbed. The acrylic coronal brain matrix has the additional feature of a midline sagittal cut to facilitate the splitting of the left and right hemispheres. The brain matrix can be purchased from Kent Scientific (1116 Litchfield Street, Orrington, Connecticut 06790, 860-626-1172 or 888-5RATTUS, www.kentscientific.com) or other companies.
3. The brain tissue slices are soaked in TTC solution in a plastic dish (a cell culture dish) and stained for 25 to 30 minutes at room temperature (the time is dependent on the color of the tissue; when the tissue turns pinkish, the tissue is ready to fix).
4. To fix the brain tissue, gently aspirate the TTC solution around the brain tissue and add 10% neutral buffered formalin into the dish. This step is repeated twice to rinse away the TTC solution. The fixed brain tissue can be scanned or photographed for analysis of infarct size.

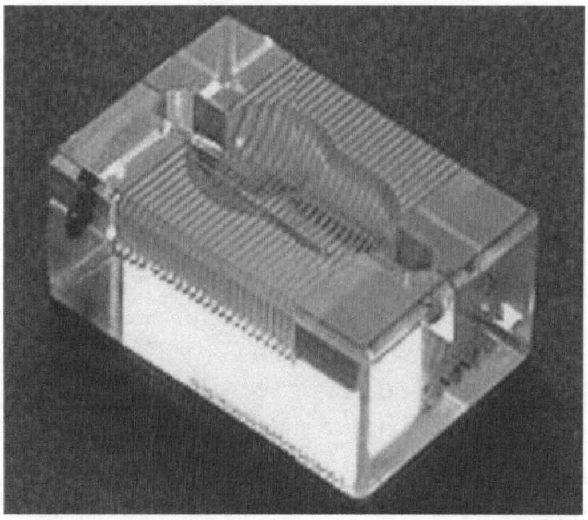

FIGURE 20.5 A brain matrix for adult rats (product RBMA-300C) from Kent Scientific. (Courtesy of Kent Scientific.)

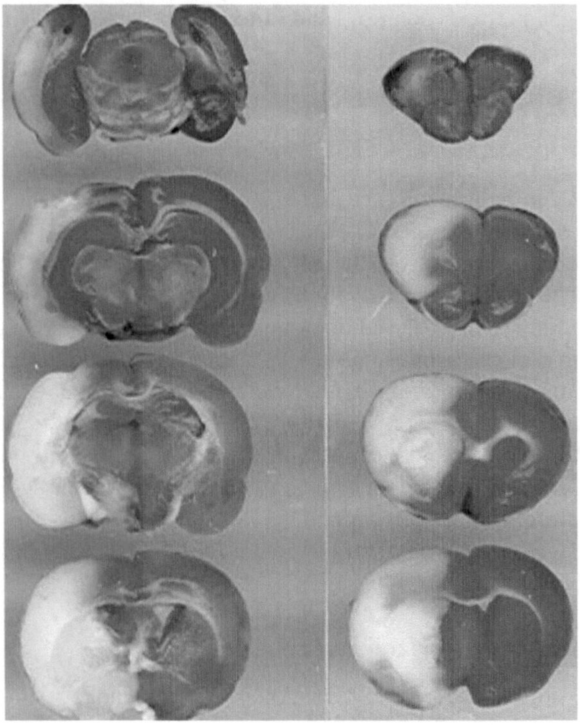

FIGURE 20.6 A rat brain stained with 2% triphenyltetrazolium chloride (TTC) for 30 minutes at room temperature. White areas represent infarct, and red areas represent living brain tissue (24 hours postsuture model of stroke in an adult rat).

5. Analysis of infarct size: The 2-mm-thick sections are placed under a microscope with an attached camera. Images are captured using a software package (Image ProPlus), and the size of the infarct is measured using tools present in the software. Infarct volume is determined by adding the infarct areas of individual sections and is represented as a percentage of total volume of the hemisphere. Figure 20.6 is a sample from our studies.

Hematoxylin-Eosin (H&E) Staining

H&E Staining for Frozen Specimens (Cryostat Sections)

Note: Brain sections must be defatted (see step 3) before H&E staining.

1. The 4% PFA-perfused brains are sliced in a cryostat microtome at 40-μm thickness.
2. The slices are mounted on subbed glass slides in order and air dried overnight.
3. Defatting the brain sections:
   a. Sections are soaked for at least 1 hour in a mixture of chloroform and 100% ethanol (1:1).
   b. Sections are treated by successive 5-minute washes in 100%, 95%, 70%, and 50% EtOH and are rinsed under tap water (the successive washes are to rehydrate the sections). *Note*: Wait until the sections dry a little bit (no water is dripping). If the slides had been prepared several days before, the slides must be put into double-distilled water for 5 minutes before staining.

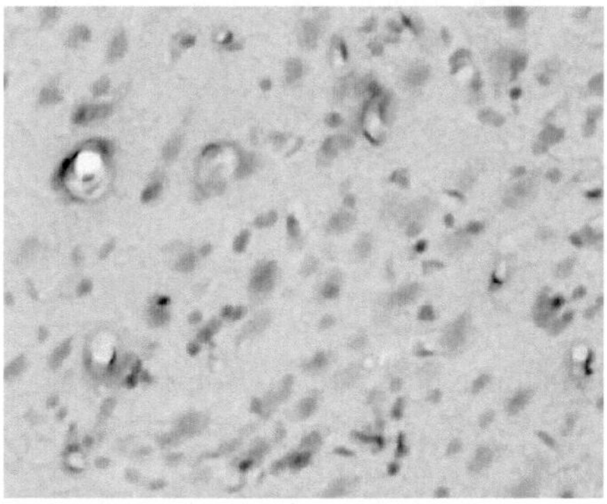

(a) Infarct side

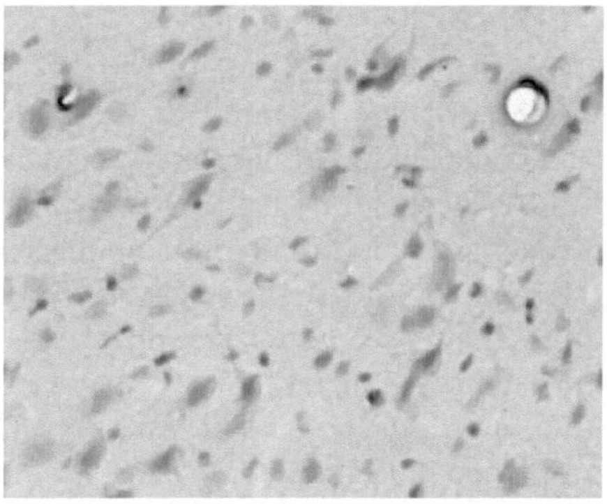

(b) Contralateral (normal)

FIGURE 20.7 Hematoxylin and eosin Y (H&E) stain on rat brain at 3 weeks after stroke, ipsilateral infarct side (a) and contralateral normal side (b); ×20. **(See color insert following page 146.)**

4. Staining procedure:
   a. Acid hematoxylin solution (eosin-hematoxylin solution, Sigma-Aldrich, Cat. No. 2852) is applied over the tissue slides and stained for 20 to 25 minutes. The process is monitored visually, by eye or under a microscope, and is sufficient when the cell nuclei turn blue.
   b. Wash with tap water twice.
   c. The slides are washed gently (to prevent tissue dislodgement) in 0.25% ammonia solution for about 10 seconds.
   d. Wash with tap water twice.
   e. Eosin Y solution is applied over the tissue slides (solution alcoholic, Sigma-Aldrich, Cat. No. HT110116) for 10 seconds.
   f. Wash with tap water twice.
   g. Dehydrate with 95% alcohol, EtOH, twice for 5 minutes each.
   h. Continue to dehydrate with 100% EtOH for 5 minutes.
   i. The tissue slides are placed into xylene or xylene substitute solution for 5 minutes (Thermo Electron, Cat. No. 9990505).

The slides are covered and fixed with Permount adhesive (Fisher Scientific, Cat. No. S70104). The resulting nuclei are blue; the cytosol is pink (see Figure 20.7).

H&E Staining for Paraffin-Embedded Sections

For H&E staining for paraffin-embedded sections, do the following:

1. Trimming the tissue: If the brain is prepared in 10% formalin, the procedure for trimming the brain tissue has to be performed in a chemical hood. The cerebellum and olfactory sections are removed. The brain is put in water to clean it and keep it wet and cut into three coronal sections. The sections are placed in a small plastic basket (4 × 6 × 2 cm) and kept in water. This small basket along with the sections is moved into a big metal basket (30 × 20 × 10 cm) for processing.
2. Processing and paraffin embedding: This procedure is carried out automatically in a processor machine (Tissue-Tex VIP). The metal basket is placed in this machine overnight. The procedure gradually dehydrates the tissue and fills any water-containing spaces with paraffin. These steps can also be done manually; the time for each step can be shorted to 15 minutes for processing small biopsy specimens (e.g., needle biopsies).

a. 10% formalin for 2 hours
b 70% alcohol for 1.5 hours
c. 80% alcohol for 1 hour
d. 95% alcohol for 1 hour for two changes
e. 100% alcohol for 1 hour for two changes
f. Alcohol/xylene 50/50 for 45 minutes
g. Xylene for 1 hour for two changes
h. Paraffin for 1 hour for four changes

The next day, the metal basket is removed, and the machine is cleaned automatically.

3. Tissue embedding process (Tissue-Tex): This dispensing console consists of
   a. Thermal or heating console at 59°C to 61°C
   b. Dispensing console at 59°C to 61°C
   c. Cryo or cold console plate at 5°C

   The tissue in a plastic basket is transferred into the thermal console and soaked into the paraffin (the tissue becomes a little wrinkled). Then, the tissue sections are transferred into a small metal plate (3 × 4 × 2 cm) on the dispensing console.

   Note: Be very careful not to flip tissue sections onto their opposite side; they also must be kept in the same order. Place the plastic basket bottom over the tissue sections. The plastic basket top is removed.

   Warm paraffin is added over the plate to just cover the basket.

   Leave this metal-plastic plate on the cold console for 20 minutes to solidify the paraffin.

   Separate the metal plate and plastic basket. The brain tissue is embedded in the paraffin along with the plastic basket plate (tissue-paraffin block).
4. Cutting tissue by microtome at room temperature:
   a. Set up a water bath at 45°C to 50°C (flotation bath).
   b. The tissue-paraffin block is placed in an ice box with ice and water (to cool and moisten the tissue before cutting).
   c. Set the microtome (HM355 Microm) at room temperature.
   d. Use a slide warmer to dry the glass slides for 20 minutes at 45°C.
   e. The tissue paraffin-plastic plate is fixed on the microtome, and the surface is trimmed until the brain tissue appears.
   f. The brain is cut into sections 5 μm thick for three to five sections (depending on your need).

The sections are transferred into the water bath at 45°C to 50°C and mounted on glass slides with a delicate paintbrush.

Prepare labels for the glass slides either by hand or by a label machine (TBS Innovation for Science, Shur/Mark). The tissue section slides are labeled for staining.

5. Prestaining: The slides are transferred into a slide cassette and placed in an oven at 60°C for 15 minutes to dry and melt the paraffin and to evaporate the remaining alcohol. *Note*: Use a cassette holder to hold the cassette and slides.
6. H&E staining: The following protocol can be performed manually or automatically. Automatic staining is carried out in a stainer (Tissue-Tex DRS, Sakura, Japan).
   a. Xylene, three changes
   b. 100% alcohol, two changes
   c. 95% alcohol, two changes
   d. Hematoxylin (Acid Hematoxylin solution, Eosin-Hematoxylin solution, Sigma-Aldrich, Cat. No. 2852)
   e. 95% alcohol, two changes
   f. 70% alcohol

g. Eosin Y (Eosin Y solution alcoholic, Sigma-Aldrich, Cat. No. HT110116, Certified Eosin Y, 0.5% weight/volume in acidified 90% ethanol)
h. 70% alcohol
i. Xylene, three changes, for the final brain slide staining

The whole process takes about 25 to 30 minutes. The slides are transferred into a hood. Two to three drops of Permount glue are placed on the slides over the tissue and then covered with a very thin glass coverslip. *Note*:

a. Do not wait for the slides dry before placing Permount glue on the slides. Permount will permanently mount the tissue.
b. Keep the slides flat until totally dry.
c. Check the slides under a microscope at ×10 to see if there are any procedural artifacts. For example, if the slides are not soaked long enough on the icebox, the tissue will look wrinkled. Poor flotation also causes wrinkles on the slice. If the floating temperature is too high, the slice falls into pieces; if the temperature is too low, the tissue slices will not float.

An alternative H&E staining procedure (manual) is as follows:

a. Xylene, three changes, 2 minutes each to remove fat, wax
b. Absolute EtOH (100%), 10 dips each
c. Alcohol, 95% EtOH, two changes, 10 dips each
d. Tap water; rinse evenly
e. Acid hematoxylin solution (Sigma-Aldrich, Cat. No. 2852) for 15 minutes
f. Tap water, two changes; rinse evenly
g. Ammonia water, 0.25%, until blue
h. Tap water, two changes; rinse evenly
i. Eosin Y solution 0.5% (alcoholic, Sigma-Aldrich, Cat. No. HT110116), 10 to 20 dips each
j. Alcohol, 95% EtOH, two changes, 10 to 15 dips each
k. Absolute EtOH (100%), 10 to 15 dips each
l. Xylene, three changes, 10 to 15 dips each

Let slides remain in last container until a coverslip is applied.

Protocol for Thionine Stain

Mechanism

The thionine stain method stains soma (cell bodies) purple by binding with acidic components of cells.

Prepare Solution

To make 1 L of 0.05% thionine, use the following:

1. Distilled water: 940 mL
2. Sodium acetate: 37 g
3. Thionine acetate (Sigma-Aldrich, Cat. No. T3387): 0.5 g
4. Glacial acetic acid (pH between 4.2 and 4.5 is acceptable): 30 to 35 mL

Mix and vacuum filter over Whatman No. 4 filter paper. *Note*: Thionine solution can be kept for months; however, the solution should be filtered before use because crystals form readily, and the pH should be checked to make sure it is between 4.2 and 4.5. Brain slices are mounted on glass slides in order.

Procedure

1. Defatting: *Note*: Sections must be defatted prior to staining with thionine if the sections are from fresh brains sliced on a cryostat. Sections are soaked for at least 1 hour in a mixture of chloroform and 100% ethanol (1:1).
2. Sections are hydrated by successive 5-minute washes in 100%, 95%, 70%, and 50% EtOH and distilled water. Wait for them to dry a little bit (no water drips). If the slides were prepared some days before, the slides have to be placed in double-distilled water for 5 minutes before staining.
3. Staining:
   a. Tissue slices are stained in thionine solution for about 2.5 to 3 minutes. Optimal staining times will vary between brains depending on the fixative used and the section thickness. Trials should be run using spare sections to determine the optimal staining time (*note*: unfixed cryostat sections require a minimum of 10 minutes in thionine).
   b. Rinse sections in distilled water for 30 seconds.
   c. Differentiate in 70%, 95%, and 100% EtOH; times will vary and need to be judged "by eye." For fixed tissue, 30 seconds to 1 minute in each ethanol bath is usually sufficient. If one has trouble achieving satisfactory differentiation (unstained white matter), try adding 5 to 10 drops of acetic acid to the 95% ethanol bath. *Note*: For unfixed cryostat sections, the differentiation steps are very short (<20 seconds each).
   d Sections are soaked in xylene solution (twice for 5 minutes).
   e. Place one or two drips of Permount glue on the slice and cover slide with a coverglass.

Result

The thionine staining leaves the nucleus blue and the cytosol purple (see Figure 20.8).

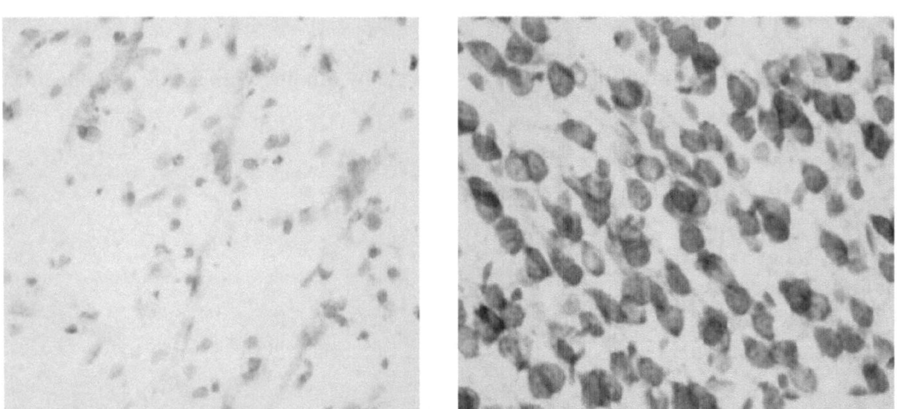

(a) Infarct area (b) Contralateral

FIGURE 20.8 Thionine stain on rat brain at day 2 after suture stroke (×40). (a) Infarct area shows that most neurons have disappeared; (b) the contralateral side shows normal neurons. **(See color insert.)**

IMMUNOHISTOCHEMICAL STAINS

Tissue Preparation

1. Animals are anesthetized and perfused through the ascending aorta with saline followed by fixative containing 4% PFA in 0.1*M* $NaPO_4$ buffer.
2. Brains are removed from the skull and immersed for 24 to 48 hours in cryoprotectant solution containing 20% sucrose in 0.1*M* $NaPO_4$ buffer, or 20% glycerol, 2% DMSO in 0.1*M* $NaPO_4$ buffer, pH 7.4.
3. Brains are frozen in dry ice and cut into 40-μm sections on a sliding microtome.
4. Collected sections are "free floated" in 50 m*M* TBS buffer (pH 7.4) containing 1 m*M* sodium azide and stored at 4°C.

Mechanism of the Assay

Briefly, immunohistochemistry uses antibodies (Ab) to detect antigens (Ag) in tissue via a chain of binding events analyzed under microscopy: Ag (tissue) + Primary Ab + Biotin-conjugated secondary Ab + Avidin-conjugated horseradish peroxidase (HRP), which is magnified as an enzymatic reaction producing a color precipitate staining on the tissue slices.[11] See the scheme in Figure 20.9.

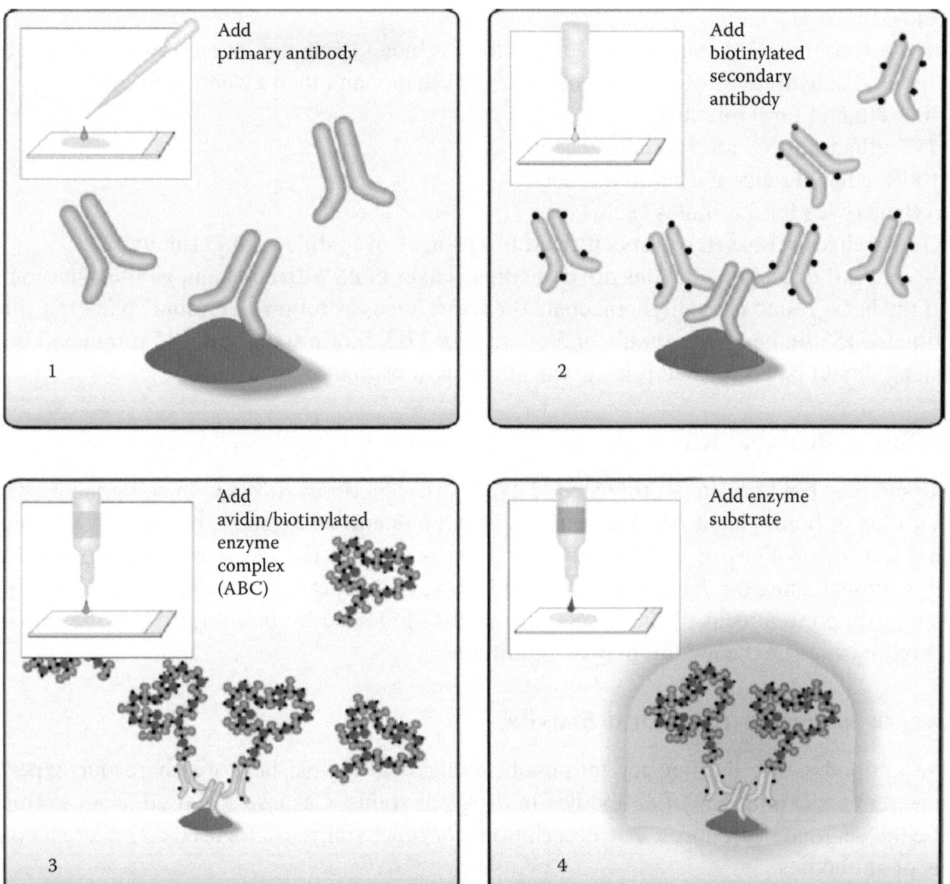

FIGURE 20.9 This scheme shows the mechanism of immunohistochemical staining. (Courtesy of Vector laboratories, www.vectorlabs.com.)

Immunohistochemical Procedure

1. Rinse the tissue twice for 5 minutes each in TBS (50 m*M* Tris buffer plus 0.9% NaCl, pH 7.4).
2. Soak tissue in TBS containing 0.5% Triton X-100 for 45 minutes at 4°C.
3. Rinse tissue in TBS twice for 5 minutes each.
4. Soak tissue in antibody "vehicle" solution for 45 minutes: Vehicle contains TBS, 3.3% normal serum (from same species as host of second antibody), and 0.1% Triton X-100.
5. Rinse tissue in TBS twice for 5 minutes each.
6. Soak overnight at 4°C in vehicle containing primary antibody, diluted as appropriate; e.g., 1:50 to 1:100 for ED1 (CD68).
7. Rinse tissue in TBS five times for 5 minutes each.
8. Soak tissue at 4°C in vehicle containing biotinylated second antibody at 1:200 (provided as Vectastain Elite peroxidase kit) for 1 hour.
9. Rinse tissue in TBS five times for 5 minutes each.
10. Soak tissue at 4°C in vehicle containing ABC Elite reagent (streptavidin-HRP complex) for 1 hour (purchased from www.vectorlabs.com, Cat. No. PK-7200).
11. Rinse in TBS twice for 5 minutes each.
12. Rinse in 50 m*M* Tris, pH 7.6, three times for 3 minutes each.
13. Develop in Tris buffer (pH 7.6) containing 0.04% DAB, 0.3% nickel ammonium sulfate, and 0.003% H_2O_2.
14. Rinse sections in Tris buffer, three times for 5 minutes each, mount onto subbed slides, air dry, and dehydrate through a graded series of ethanol and then xylene as follows:
    70% ethanol for 5 minutes
    95% ethanol twice for 5 minutes each
    100% ethanol twice for 5 minutes each
    Xylene twice for 5 minutes each
    Or use citrus oil-based solvents instead of xylene. Cover slide with Permount.
    Note: Washes and incubations are done on a shaker table with constant, gentle agitation.
    If the background is too high, incubate the tissue slices as follows: Avidin D blocking solution for 15 minutes, rinse tissue for 5 minutes in TBS, biotin solution for 15 minutes. These steps should be performed prior to the addition of primary Ab.

Avidin/Biotin Blocking Kit

Some tissue may bind avidin, biotinylated HRP, or other biotin/avidin system components without prior addition of biotinylated Ab. This binding may be due to endogenous biotin or biotin-binding proteins, lectins, or nonspecific binding substances present in the tissue section. If a high background is present using the ABC reagents (or other avidin conjugate) in the absence of biotinylated secondary Ab, pretreatment of the tissue with avidin, followed by biotin (to block the remaining biotin-binding sites on the avidin) may be required.

Examples of Immunohistochemical Staining

Because antibodies are the key for immunohistochemical stains, here we share our experience with different concentrations of antibodies in different stains. Caspase-3 is used as an example to describe the detailed procedures. The descriptions for other stains are focused on different concentrations of antibodies.

Caspase-3 Staining

Objective

Caspase-3 staining is done to detect apoptosis in rat brain after suture-induced stroke.

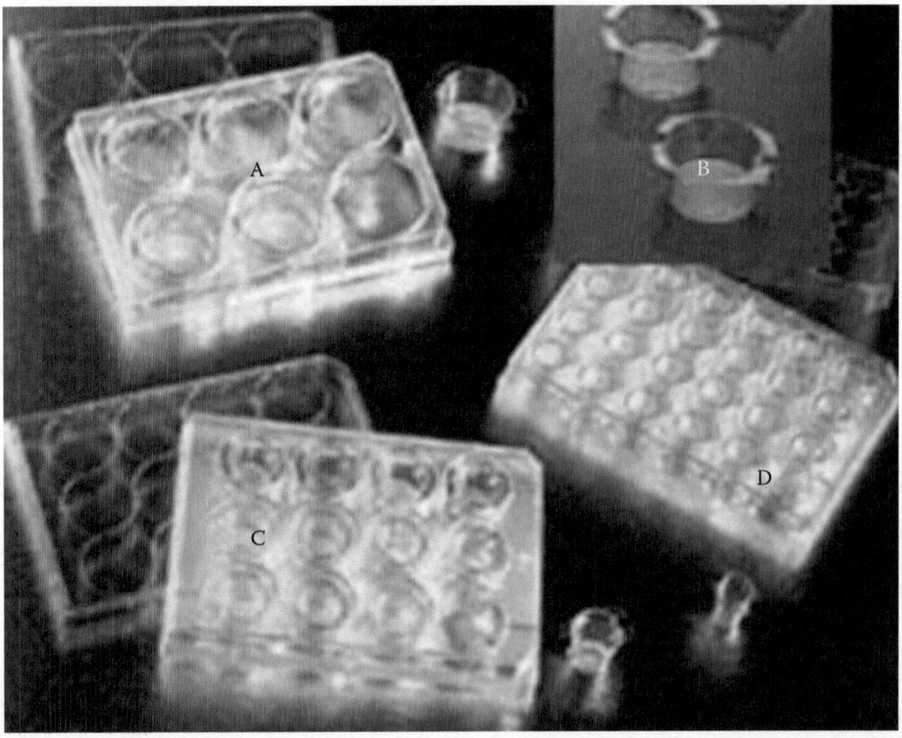

FIGURE 20.10 Cell culture dish plates with 6 wells (A) and nets (B), 12 wells (C), or 24 wells (D). (Courtesy of VWR.)

Materials

1. Primary antibody from Chemicon (Cat. No. AB3623) or Cell Signaling Technologies (Cat. No. G661) recognizing the cleaved form (p17) of caspase-3 Ab; both are polyclonal rabbit antirat sera that do not recognize the uncleaved enzyme.
2. Normal goat serum purchased from Chemicon (Cat. No. S26); the species is chosen to match the same host of the second antibody.
3. Biotinylated second antibody (goat antirabbit immunoglobulin [Ig] G), Chemicon (Cat. No. AP132B).
4. RTU Vectastain® Elite ABC reagent (containing vehicle solution, avidin–HRP complex), Vector Laboratories (Cat. No. PK7100).
5. BD Falcon cell culture dishes and nets from VWR, 6 wells (Cat. No. 62406-163 or BD 353090) or 12 wells (Cat. No. 62406-174 or BD 353180) (see Figure 20.10). The purpose of using a dish net is to transfer tissue easily from different solutions during the procedures.

Immunohistochemical Procedure

1. About three to five brain slices are placed in a dish net and then soaked in a well with TBS (50 m*M* Tris buffer plus 0.9% NaCl, pH 7.4) twice for 5 minutes each.
2. Soak tissue for 45 minutes at 4°C in TBS containing 0.5% Triton X-100.
3. Rinse tissue twice for 5 minutes each in TBS.
4. Soak tissue for 45 minutes in antibody vehicle solution. Vehicle contains 3.3% TBS, normal serum (from goat antimouse: same species as host of second antibody), and 0.1% Triton X-100.

5. Rinse tissue twice for 5 minutes each in TBS.
6. Soak tissue overnight at 4°C in vehicle containing primary antibody: rabbit antirat active caspase-3 IgG Ab, diluted as appropriate, recommended 1:100 (50 μL/5 mL). When starting any new work with a new antibody, a pilot experiment with a dilution series is useful; in this case, one could try 1:500 (10 μL in 5 mL), 1:1,000 (5 μL in 5 mL), and one well without caspase-3 Ab as control.
7. Rinse tissue five times for 5 minutes each in TBS.
8. Soak tissue for 1 hour at 4°C in vehicle containing biotinylated second antibody (goat antirabbit IgG) at 1:250, 20 μL in 5 mL or 28 μL in 7 mL (manufacturer's recommendation 1:500 to 1:5,000; again, a pilot experiment with a dilution series is recommended).
9. Rinse tissue five times for 5 minutes each in TBS.
10. Soak tissue for 1 hour at 4°C in RTU Vectastain Elite ABC reagent (avidin–HRP complex), 3 mL ABC for each well.
11. Rinse twice for 5 minutes each in TBS.
12. Rinse three times for 3 minutes each in 50 m*M* Tris, pH 7.6.
13. Develop in freshly made Tris buffer containing 0.04% DAB (10 mg in 25 mL), 0.3% nickel ammonium sulfate (75 mg in 25 mL), and 0.003% H_2O_2 (2.5 μL of 30% H_2O_2 in 25 mL). *Note*: DAB is extremely poisonous; the waste must be collected in a chemical waste container. Weighing must be done in a hood. H_2O_2 is added into the solution just before the stain. Closely watch the tissue; when the staining is complete, the tissue will change to a deep purple color.
14. Stop the reaction by rinsing sections three times for 5 minutes each in Tris buffer (50 m*M*, pH 7.6).
15. Mount the slices on subbed slides and air dry.
16. Dehydrate through a graded series of ethanol and then xylene as follows: 70% ethanol, 5 minutes; 95% ethanol, twice for 5 minutes each; 100% ethanol, twice 5 minutes each; xylene twice for 5 minutes each or xylene substitute (Thermo Electron, Cat. No. 9990505). Cover the slide with Permount or DPX; be careful that no air bubbles appear under the coverslip. *Note*: Washes and incubations are done on a shaker table with constant, gentle agitation.

Result

For the result, see Figure 20.11.

Background

We abstracted some information about caspase-3 from the Chemicon Web site (with permission) to help investigators better understand the method.

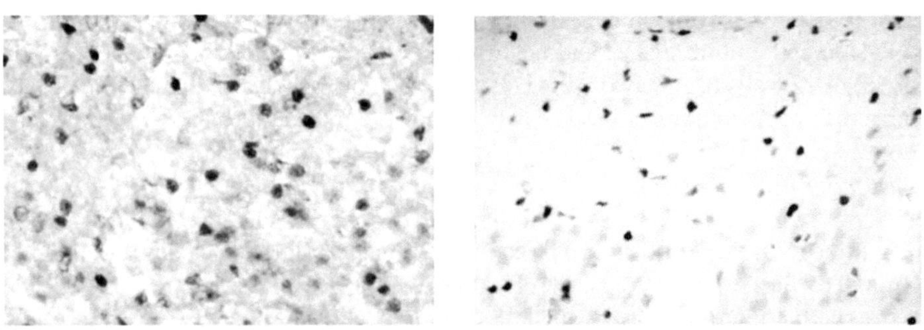

(a) Infarct side (b) Sham surgery

FIGURE 20.11 Immunohistochemical staining shows that caspase-3, a marker for apoptosis, is upregulated in a stroked rat. (a) Infarct side at 48 hours after stroke; (b) sham-operated rat (×40). **(See color insert.)**

Caspase-3 (CPP32, Apopain) is the most extensively studied apoptotic protein. Caspase-3 is synthesized as an inactive proenzyme (32 kDa) that is processed in cells undergoing apoptosis by self-proteolysis and/or cleavage by another upstream protease. The processed form of caspase-3 consists of large (17 kDa) and small (12 kDa) subunits which associate to form an active enzyme. The active caspase-3 proteolytically cleaves and activates other caspases, as well as relevant targets in the cells such as PARP and DFF. AB3623 detects only the cleaved p17 fragment and does not detect the precursor form in cells undergoing apoptosis.

ED1 Stain

Objective

The objective of the ED1 stain is to recognize monocyte-derived cells (macrophages) in the brain.

Materials

1. Primary antibody: Mouse monoclonal ED1 (CD68, antirat) (Serotec, Cat. No. MCA341R)
2. Normal goat serum: Goat antimouse (same species as host of second antibody)
3. Secondary antibody: Biotinylated goat antimouse IgG (Chemicon International, Cat. No. AP124B)

Immunohistochemical Procedure

Immunohistochemical procedures are performed in a six-well dish with a dish net (see Figure 20.10) with the following antibodies:

Primary antibody: ED1 (CD68, mouse antirat), diluted as appropriate (manufacturer's recommendation 1:50 to 1:100). We tried 1:100 (70 μL in 7 mL), 1:500 (14 μL in 7 mL), and 1:1000 (7 μL in 7 mL). The 1:1000 dilution worked well (see Figure 20.12).

Secondary antibody: Biotinylated goat antimouse IgG worked well at 1:250 (manufacturer's recommendation 1:500 to 1:5000).

Background

ED1 antibody (Cat. No. MCA341R) recognizes a protein that appears to be the rat homolog of human CD68 and mouse macrophages. The ED1 antigen is a 110-kDa glycoprotein that is expressed predominantly on the lysosomal membrane of myeloid cells; its expression on the cell surface is weak. The antigen is expressed by the majority of tissue macrophages and weakly by peripheral blood granulocytes. ED1 is also expressed by many (but not all) microglia.[12] Stroke leads to profound cellular responses both in the vicinity of the infarct and remote regions (see Figure 20.12).

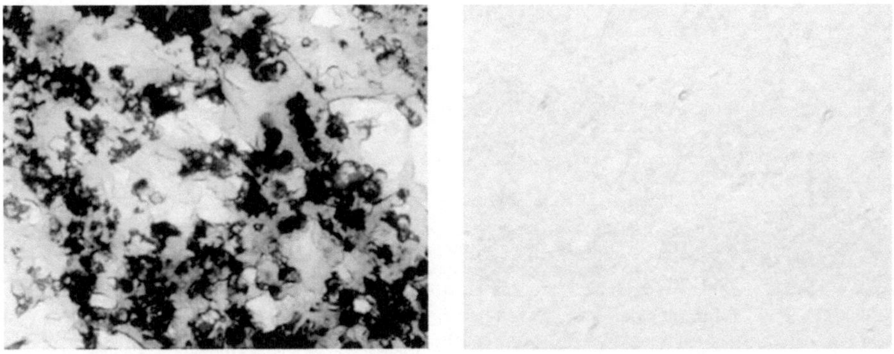

(a) Infarct area (b) Contralateral (normal side)

FIGURE 20.12 ED1 immunohistochemical stain shows macrophages in the infarct area (a) but not on the contralateral side (b) (×40; 3 weeks postsuture stroke rats). **(See color insert.)**

The response includes activation and proliferation of resident glia10 and attraction of blood-derived leukocytes to the region of ischemic brain damage. After stroke, the initial phagocytic response is mainly performed by resident microglia. The majority of hematogenous macrophages are recruited later to facilitate the removal of necrotic tissue.[13]

Glial Fibrillary Acid Protein Stain

Objective

The objective of this stain is to recognize the glial fibrillary acid protein (GFAP) in the brain.

Materials

1. Primary antibody: Murine monoclonal antirat GFAP (Chemicon, Cat. No. MAB360).
2. Normal goat serum: Goat antimouse (same species as host of second antibody) (Chemicon, Cat. No. S26, 100 mL).
3. Secondary antibody: Biotinylated goat antimouse antibody IgG (Chemicon, Cat. No. AP124B).

Immunohistochemical Procedure

The immunohistochemical procedure is performed in a six-well dish with a dish net.

1. Primary antibody: Mouse antirat GFAP (manufacturer's recommendation 1:400 to 1:800). We tried 1:800 (8.75 μL in 7 mL), 1:8,000 (0.875 μL in 7 mL), and 1:10,000 (0.7 μL in 7 mL). Of these three, 1:8,000 was the most effective (see Figure 20.13).
2. Secondary antibody: Biotinylated goat antimouse IgG worked successfully at 1:250 (28 μL in 7 mL) (manufacturer's recommendation 1:500 to 1:5,000).

Background

Information from Chemicon indicates that "GFAP (Glial fibrillary acidic protein), MW ~50 kDa, recognizes astrocytes and Bergmann glia cells, glioma and glial cell-derived tumors. It shows cross-reactivity with human, rat and porcine but no cross-reactivity with vimentin." Studies on humans[14] and animals[15] showed that GFAP was increased in acute stroke patients, indicating that GFAP is expressed in brain ischemia.

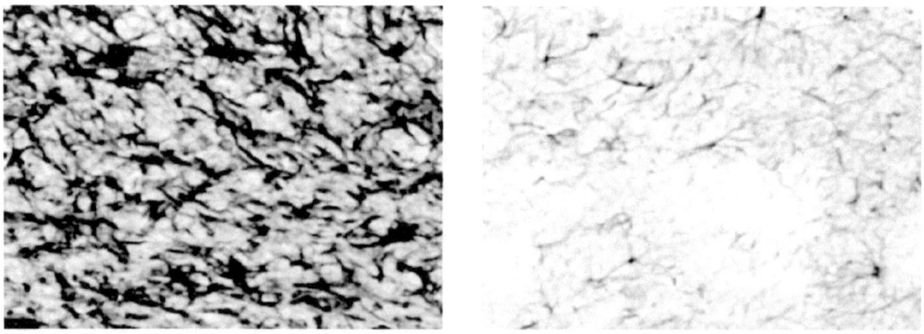

(a) Infarct side (b) Contralateral

FIGURE 20.13 Immunohistological staining shows increased GFAP (glial fibrillary acid protein) on the infarcted side of a stroked rat (a) but no increase on the contralateral side (b) (×40; 3 weeks postsuture stroke). Note that the morphology of astroglia changes from small bodies (b) to larger bodies and thicker processes (a). **(See color insert.)**

Microtubule-Associated Protein Staining

Objective

The objective of microtubule-associated protein (Map2) staining is to recognize neurons and dendrites in brain.

Materials

1. Primary antibody: Murine monoclonal antirat Map2 (Chemicon, Cat. No. MAB3418).
2. Normal goat serum: Goat antimouse (same species as host of second antibody) (Chemicon, Cat. No. S26, 100 mL).
3. Secondary antibody: Biotinylated goat antimouse IgG (Chemicon, Cat. No. AP124B).

Immunohistochemical Procedure

The immunohistochemical procedure performed in dishes with a dish net.

1. Primary antibody: Map2 mouse antirat monoclonal Ab (manufacturer's recommendation 1:200 [5 μg/mL]). We tried 1:400 (17.5 μL in 7 mL), 1:2000 (3.5 μL in 7 mL), 1:4,000 (1.75 μL in 7 mL), and 1:5000, with one well without primary Ab as a control. Dilutions of 1:4000 and 1:5000 showed good staining (Figure 20.14).
2. Biotinylated second antibody (goat antimouse IgG) works well at 1:250 (28 μL in 7 mL) (manufacturer's recommendation 1:500 to 1:5000).

Background

Information from Chemicon (with permission) indicates that

> Map-2 is one of several high molecular weight proteins that play an important role in brain microtubule assembly. In addition to its association with microtubules, Map2 associates with neurofilaments and actin filaments suggesting that it may guide interaction among microtubules, other cytoskeletal elements, and cytoplasmic organelles. Map2 is a stringent marker for neurons; in addition, Map2 displays intracellular specificity. In the central nervous system, Map2 is confined to neuronal cell bodies and dendrites. There are exceptions, however, where some axons stain positive for small amounts of Map2. Map2 is uniformly distributed throughout the cell when first expressed in cultured neurons but becomes selectively localized as dendritic development proceeds.

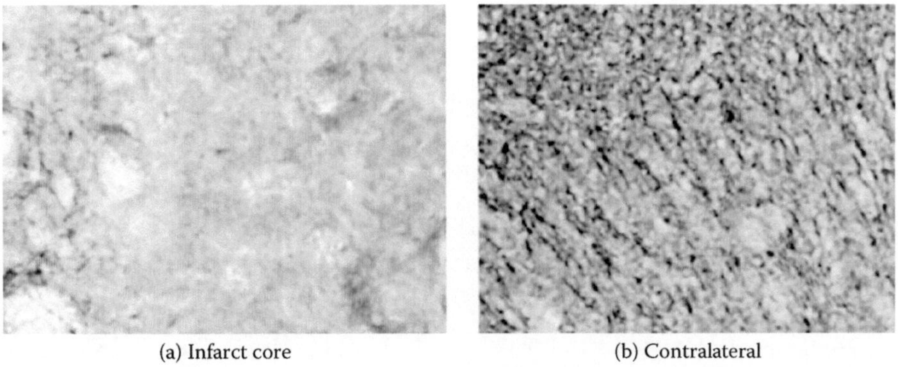

(a) Infarct core (b) Contralateral

FIGURE 20.14 Map2 (microtubule-associated protein) staining shows disappearance of neuronal dendrites in the infarct core (a) compared to the contralateral side of a stroked rat at 3 weeks poststroke (b); ×40. **(See color insert.)**

Neuronal Nuclear Stain

Objective

The neuronal nuclear (NeuN) stain recognizes neuronal nuclei in the brain.[16]

Materials

1. Primary antibody: NeuN mouse antirat monoclonal Ab (Chemicon, Cat. No. MAB377).
2. Normal goat serum: Goat antimouse (same species as host of second antibody) (Chemicon, Cat. No. S26, 100 mL).
3. Secondary antibody: Biotinylated goat antimouse IgG (Chemicon, Cat. No. AP124B).

Immunohistochemical Procedure

The immunohistochemical procedure is performed in dishes with a dish net.

1. Primary antibody: NeuN mouse antirat monoclonal Ab (manufacturer's recommendation at 1:100 to 1:1000). We tried 1:1000 (5 μL in 5 mL) and 1:2000 (2.5 μL in 5 mL), with one well without NeuN Ab as control. The Ab was too concentrated (strong background) and had to be diluted by 1:5000 to 1:10,000 (see Figure 20.15).
2. Secondary antibody: Biotinylated goat antimouse IgG worked well at 1:250 (25 μL in 5 mL) (manufacturer's recommendation 1:500 to 1:5000).

Background

Information from Chemicon indicates that

> Vertebrate neuron-specific nuclear protein called NeuN (Neuronal Nuclei) MAB377 reacts with most neuronal cell types throughout the nervous system of mice including cerebellum, cerebral cortex, hippocampus, thalamus, spinal cord and neurons in the peripheral nervous system including dorsal root ganglia, and enteric ganglia. The immunohistochemical staining is primarily in the nucleus of the neurons with lighter staining in the cytoplasm. The few cell types not reactive with MAB377 include Purkinje, mitral and photoreceptor cells.

Platelet Endothelial Cell Adhesion Molecule 1

Objective

Platelet endothelial cell adhesion molecule 1 (PECAM-1; also called CD31) is a marker for blood vessels and can recognize vessel repair.

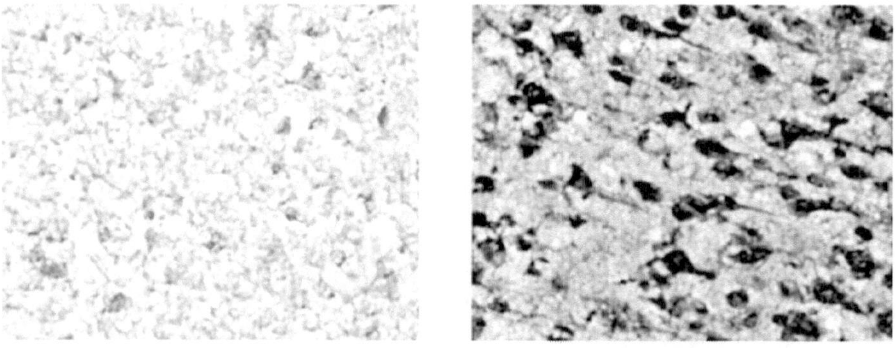

(a) Infarct core (b) Contralateral

FIGURE 20.15 Neuron-specific nuclear protein (NeuN). The stain shows disappearance of neuronal nuclei in the infarct core (a) compared to the contralateral side (b) (48 hours poststroke rat; ×40). **(See color insert.)**

Materials

1. Primary antibody: Murine monoclonal antirat CD31 (Serotec, Cat. No. MCA1334G).
2. Normal goat serum: Goat antimouse (same species as host of second antibody) (Chemicon, Cat. No. S26, 100 mL).
3. Secondary antibody: Biotinylated goat antimouse IgG (Chemicon, Cat. No. AP124B).

Immunohistochemical Procedure

The immunohistochemical procedure is performed in dishes with a dish net.

1. Primary antibody: Mouse antirat CD31 IgG (manufacturer's recommendation 1:100). We tried 1:500 (14 μL in 7 mL), 1:1000 (7 μL in 7 mL), and 1:2000 (3.5 μL in 7 mL), with one well without CD31 Ab as control. A dilution of 1:500 showed good staining.
2. Secondary antibody: Biotinylated goat antimouse IgG worked well at 1:250 (28 μL in 7 mL) (manufacturer's recommendation 1:500 1:5000).

Background

PECAM-1, also called CD31 and EndoCAM, is a member of the immunoglobulin superfamily. PECAM-1 is a transmembrane glycoprotein. The molecular weight of this protein is approximately 130 kDa, depending on the degree of glycosylation. PECAM-1 is constitutively expressed on all vascular cells and has provided a useful immunohistochemical marker for blood vessels,[17] particularly in the setting of angiogenesis.

Matrix Metalloproteinase 9 Antibody Stain

Objective

The objective of this stain is to recognize the MMP-9 (matrix metalloproteinase) in the brain.

Materials

1. Primary antibody: Rabbit polyclonal antirat MMP-9 catalytic domain (Chemicon, Cat. No. AB19016).
2. Normal goat serum: Goat antimouse (same species as host of second antibody) (Chemicon, Cat. No. S26, 100 mL).
3. Secondary antibody: Biotinylated goat antimouse IgG (Chemicon, Cat. No. AP124B).

Immunohistochemical Procedure

The immunohistochemical procedure is performed in dishes with a dish net.

1. Primary antibody: Rabbit antirat MMP-9 (manufacturer's recommendation 1:100). We tried 1:500 (14 μL in 7 mL), 1:1000 (7 μL in 7 mL), 1:2000 (3.5 μL in 7 mL), 1:5000 (1.4 μL in 7 mL), and 1:10,000 (0.7 μL in 7 mL); 1:5000 and 1:10,000 showed good staining (see Figure 20.16).
2. Secondary antibody: Biotinylated goat antirabbit IgG worked well at 1:250 (25 μL in 5 mL) (manufacturer's recommendation 1:500 to 1:5000).

Background

Information from Chemicon indicates that

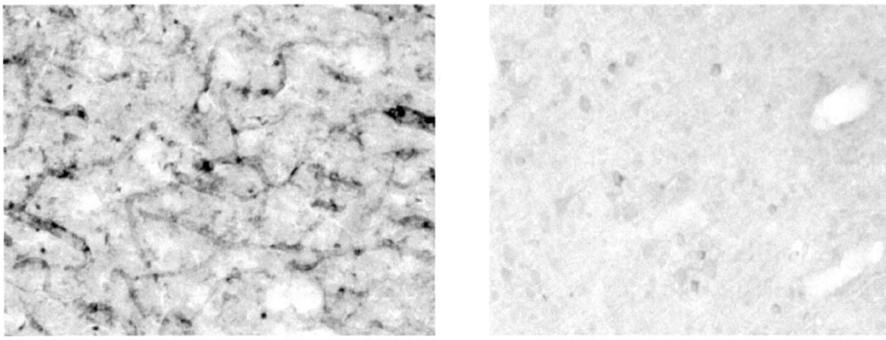

(a) Infarct side (b) Contralateral

FIGURE 20.16 Matrix metalloproteinase (MMP-9) staining of rat brain 24 hours poststroke. (a) Infarct side. (b) Contralateral normal brain. MMP-9 is upregulated in the vascular tree on the stroke side. MMP-9 rabbit antirat antibody 1:5000 diluted at ×20. **(See color insert.)**

MMP-9 rabbit anti-rat Ab recognizes rat MMP-9 pro and active forms. The antibody also reacts with mouse MMP-9 and human MMP-9, and exhibits no cross-reactivity with other MMP family members. The antibody was generated using *E. coli*-expressed active rat 92 kDa type IV collagenase (catalytic domain) as the immunogen.

MMPs (mammalian matrix metalloproteinases) are a family of enzymes. Their major functions are involved in the degradation of extracellular matrix components such as collagen, laminins, and proteoglycans in physiological and pathological processes.[18] In addition to sequence homology, all MMPs share the following characteristics: "1) they are secreted as zymogens, which are activated by removal of an approximately 10 kDa segment from the N-terminus; 2) they cleave one or more extracellular matrix components; 3) the catalytic mechanism is dependent on a zinc ion at the active site and 4) they are inhibited by tissue inhibitor of metalloproteinases (TIMP). These enzymes are involved in normal physiological processes such as embryogenesis and tissue remodeling. They may play an important role in angiogenesis, arthritis, periodontitis, and metastasis. MMP-9 is also called gelatinase B; it is a 92 kDa gelatinase/type IV collagenase (EC3.4.24.35). It has a broad substrate specificity for native collagens including type IV, V, VII and X as well as gelatin, proteoglycans and elastin. After cleavage of a single peptide domain of about 20 amino acids, the MMPs are secreted in latent forms. MMP-9 is secreted as a 92 kDa proenzyme and can be activated *in vitro* by organomercurials (e.g., 4-aminophenylmercuric acetate, APMA), trypsin, and α-chymotypsin and *in vivo* by cathepsin G and MMP-3 to its 83 kDa active form. The TIMPs (in a 1:1 molar ratio) can inhibit the activity of MMP-9. TIMP-1 is known to bind exclusively to pro-MMP-9 (kd approximately 35 n*M*) whereas TIMP-2 binds to pro-MMP-2 with a kd approximately 5 n*M*. Upon activation, the N-terminal propeptide domain is cleaved to generate the active forms of MMP. MMP-9 (92 kDa type IV collagenase, Gelatinase-B) contains the basic structure of propeptide, catalytic and hemopexin domains. It is an important proteinase in tissue remodeling. MMP-9 antibodies recognize both latent (92 and 88 kDa) and active forms (68 kDa and lower) of matrix metalloproteinase-9 (MMP-9). This Ab has no cross reactivity with other MMP family members. MMP-9 is expressed by different cell types, including neurons, glia, dendritic cells, cerebral vessels, trophoblasts, fibroblasts, and some epithelial cells.[18]

MMP-9 has been measured in human plasma from normal individuals and has been shown to be elevated in sera of patients with hepatocellular carcinomas and stroke.[19]

Cytokine/Chemokine/Growth Factor Stain

Objective

The objective of this stain is to recognize the GRO/KC (cytokine/chemokine/growth factor) in the brain.

Materials

1. Primary antibody: Rabbit antirat antibody GRO/KC (CXCL1) (PeproTech, Cat. No. 400-10).
2. Normal goat serum: Goat antimouse (same species as host of second antibody) (Chemicon, Cat. No. S26, 100 mL).
3. Secondary antibody: Biotinylated goat antirabbit IgG (Chemicon, Cat. No. AP124B).

Immunohistochemical Procedure

The immunohistochemical procedure is performed in dishes with a dish net.

1. Primary antibody: Rabbit antirat GRO/KC. We found that 1:1000 (7 μL in 7 mL) showed good staining. Negative control was without Ab.
2. Secondary antibody: Biotinylated goat antirabbit IgG worked well at 1:250 (25 μL in 5 mL) (manufacturer's recommendation 1:500 to 1:5000).

Background

Information from PeproTech indicates that

> Rat GRO (Rat KC, CINC) is a combination of Cytokines/Chemokines/Growth Factors and Neurotrophin that promotes neutrophil chemotaxis and degranulation. All three isoforms of GRO are CXC chemokines that can signal through the CXCR1 or CXCR2 receptors. Rat KC is a 7.8 kDa protein consisting of 72 amino acids. The GRO proteins chemoattract and activate neutrophils and basophils. Cytokine-induced neutrophil chemoattractant (CINC) is related to human 'gro' and murine 'KC.'

Brain chemokines play an important role in neutrophil accumulation in cerebral ischemia, which can contribute to the extent of tissue injury in stroke.[20]

ACKNOWLEDGMENTS

We thank Dr. Kenneth Rhodes for his support, instruction, and wonderful leadership and Dr. Robert Scannevin for his technical assistance.

REFERENCES

1. Rosene, D.L., Roy, N.J., and Danis, B.J., A cryoprotection method that facilitates cutting frozen sections of whole monkey brains for histological and histochemical processing without freezing artifact, *J Histochem Cytochem*, 34, 1301, 1986.
2. Faupel, R.P. et al., The problem of tissue sampling from experimental animals with respect to freezing technique, anoxia, stress and narcosis, *Arch Biochem Biophys*, 148, 509, 1972.
3. Holsen, R.R., Euthanasia by decapitation: Evidence that this technique produces prompt, painless unconsciousness in laboratory rodents, *Neurotoxical Teratol*, 14, 253, 1992.
4. Bederson, J.B. et al., Evaluation of 2,3,5-triphenyltetrazolium chloride as a stain for detection and quantification of experimental cerebral infarction in rats, *Stroke*, 17, 1304, 1986.
5. Park, C.K. et al., Correlation of triphenyltetrazolium chloride perfusion staining with conventional neurohistology in the detection of early brain ischaemia, *Neuropathol Appl Neurobiol*, 14, 289, 1988.
6. Goldlust, E.J. et al., Automated measurement of infarct size with scanned images of triphenyltetrazolium chloride-stained rat brains, *Stroke*, 27, 1657, 1996.
7. Nachlas, M.M. et al., Cytochemical demonstration of succinic dehydrogenase by the use of a new p-nitrophenyl substituted ditetrazole, *J Histochem Cytochem*, 5, 420, 1957.
8. Altman, F.P., Tetrazolium salts and formazans, *Prog Histochem Cytochem*, 9, 1, 1976.
9. Lundy, E.F. et al., Morphometric evaluation of brain infarcts in rats and gerbils, *J Pharmacol Methods*, 16, 201, 1986.

10. Schroeter, M. et al., Astroglial responses in photochemically induced focal ischemia of the rat cortex, *Exp Brain Res,* 106, 1, 1995.
11. Harlow, E. and Lane, D., *Antibodies: A Laboratory Manual*, Cold Spring Harbor Press, Cold Spring Harbor, New York, 1988, chapter 10.
12. Provis, J.M., Diaz, C.M., and Penfold, P.L., Microglia in human retina: A heterogeneous population with distinct ontogenies, *Perspect Dev Neurobiol*, 3(3), 213, 1996.
13. Schroeter, M., et al., Phagocytic response in photochemically induced infarction of rat cerebral cortex. The role of resident microglia, *Stroke*, 28(2), 382, 1997.
14. Lynch, J.R. et al., Novel diagnostic test for acute stroke, *Stroke*, 35, 57, 2004.
15. Chang, C.F., Hoffer, B., et al., Bone morphogenetic proteins are involved in fetal kidney tissue transplantation-induced neuroprotection in stroke rats, *Neuropharmacology*, 43(3), 418, 2002.
16. Mullen, R.J., Buck, C.R., and Smith, A.M., NeuN, a neuronal specific nuclear protein in vertebrates, Development, 116(1), 201, 1992.
17. DeLisser, H.M., Newman, P.J., and Albelda, S.M., Molecular and functional aspects of PECAM-1/CD31, *Immunol Today*, 15(10), 490, 1994.
18. Romanic, A.M. et al., MMP expression increases after cerebral focal ischemia in rats: inhibition of MMP-9 reduces infarct size, *Stroke,* 29, 1020, 1998.
19. Gursoy-Ozdemir, Y. et al., Cortical spreading depression activates and upregulates MMP-9, *J Clin Invest*, 113(10), 1447, 2004.
20. Liu, T. et al., Cytokine-induced neutrophil chemo attractant mRNA expressed in cerebral ischemia, *Neurosci Lett*, 24, 164(1–2), 125, 1993.

21 Protocol for Brain Vessel Corrosion Casting and Embedding

Afshin A. Divani and Yanlin Wang-Fischer

CONTENTS

OVERVIEW

Plastic models of specimens provide true three-dimensional models of blood vessels, tissues, and organs for education and research. Various corrosion casting techniques have been utilized to produce replicas of vascular systems. A resin model can be used for visualizing vascular morphology, measuring vessel lengths and diameter, and volumetric measurements. By knowing the density of the resin used, the volume of a cavity can be easily determined by the cast weight. Resin replicas of blood vessels are frequently used in scanning electron microscopy examinations.[1–5] Resin casts can be used in investment casting techniques for reproduction of the samples. Elastic models of blood vessels or organs can be made via this technique.[6] Different corrosion casting kits, such as Mercox™ Embedding Resin Kit (Structure Probel, West Chester, Pennsylvania, www.2spi.com/agntdist/us.html) and Batson's No. 17 Plastic Replica and Corrosion Kit (Polysciences, Warrington, Pennsylvania, www.polysciences.com) are commercially available. Most of the kits contain a monomer solution (methyl acrylate) that will be polymerized in the presence of a catalyst. Extreme care should be taken when handling and disposing of the materials. Protective gear (goggles and gloves) should be used while handling the solution. During mixing and application, all the materials should be kept under a fume hood. Our laboratory uses Batson's No. 17 for corrosion casting procedures. A technical data sheet (No. 105) is readily available at the Polysciences Web site (www.polysciences.com). Always take the necessary precautions as corrosion casting material is highly toxic and corrosive.

TABLE 21.1
Materials for Vessel Casting and Embedding

Catalog No.	Description	Quantity
07349	**Batson's No. 17 Corrosion Kit**	1 kit
	Monomer Base Solution	940 ml
	Catalyst	100 ml
	Promoter	50 ml
	Red Pigment	10 g
	Blue Pigment	10 g
	Additional Products	
02599	Monomer Base Solution (Methyl Methacrylate)	940 ml
02608	Catalyst	100 ml
02610	Promoter	100 ml
07359	Maceration Solution	940 ml
07350	Red Pigment	100 g
07352	Blue Pigment	100 g
07351	White Pigment	100 g
07353	Green Pigment	100 g
07354	Yellow Pigment	100 g
03573	Methyl Methacrylate - Butyl Methacrylate Embedding Kit	940 ml

In an animal model of stroke, it is important to know the accurate diameter of the targeted brain artery for occlusion. This can be ascertained either by inserting a suture or through deposition of an *in vitro* generated thrombus. In many cases, the targeted vessel is a middle cerebral artery (MCA) or an internal carotid artery (ICA). The plastic models of brain vessels obtained from corrosion casting provide a relatively accurate assessment of vascular anatomy and diameter of the targeted vessel and its distance with respect to the carotid bifurcation. This technique has been used in recent corrosion casting studies, including those for ocular vasoproliferation, arterial proliferation, the peribiliary portal system in the rat, the temporal branches of the MCA, the rabbit eye, aortic endothelium, and dermal microcirculation.[1,3,7]

MATERIALS AND ORDERING INFORMATION

Batson's No. 17 Plastic Replica and Corrosion Kit are commercially available and can be purchased online from Polysciences (www.polysciences.com). However, each component of the kit can also be ordered separately and in larger quantities. All materials needed for this procedure are summarized in Table 21.1.

VESSEL CORROSION CASTING PROCEDURE

Animal Preparation

The animal should be prepared and perfused prior to mixing the solutions. Under heavy anesthesia, the animal will be transcardially perfused with heparinized normal saline (400 to 500 mL) to remove the blood from the vessels. Prior to use, the solution should be brought to body temperature to prevent any vasospasm. It is very important to fully remove the blood from the vessels to ensure

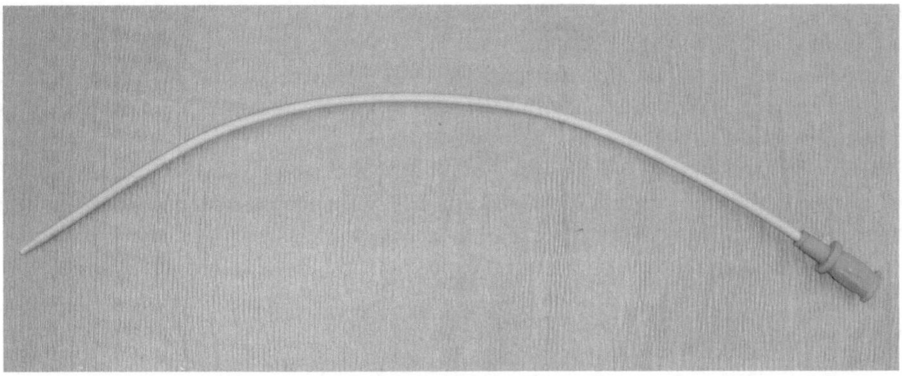

FIGURE 21.1 A 5-French dilator used for injection of the resin into the vascular system.

the patency of the vessels and complete penetration of the resin into the arterial, capillary, and venous system. During the perfusion, the aorta is cannulated with a heavy-gauge needle or a dilator, and the descending aorta is ligated to perfuse only the upper extremity. In our laboratory, the animal is perfused under heavy anesthesia by cannulating the aorta with a 4- or 5-French (3 French = 1 mm) plastic dilator. A 5-French dilator is shown in Figure 21.1. Plastic dilators have a soft tapered tip that can be inserted into the aorta without rupturing the vessel. The proximal end of the dilator has a Luer-Lok that can be used to fasten to a syringe. The advantage of using such dilators is that they are relatively long (about 6 to 10 inches) and flexible. This makes them easy to use during the injection of the resin.

Procedure for Injection and Curing

The monomer base solution, catalyst, and the promoter (from Batson's No. 17 Corrosion Kit) should be mixed in appropriate proportions to initiate the polymerization process prior to injection. Polymerization time can be altered by adjusting the amount of catalyst and promoter. This can also alter the stiffness of the polymerized plastic. Less catalyst and promoter may result in a less-brittle resin model at the cost of longer polymerization time. The mixing and injection of the resin should be performed under a ventilated hood to avoid inhalation of hazardous fumes. The material should be handled with protective gear (that is, gloves, goggles, and a fume mask). The monomer base solution must be kept at 4°C for storage and should be brought to room temperature prior to use. High temperature can influence and initiate the polymerization of the base solution without the addition of the catalyst or the promoter. Therefore, only the proportion of the base solution that will be used for the casting procedure should be brought to room temperature prior to use. All other solutions can be stored at room temperature in the original containers, tightly capped. Protect the containers from light and heat. The following steps are recommended for a corrosion casting procedure:

- The setting time of the mixture can be varied by adjusting the amount of the promoter or catalyst. To ensure proper mixing of the solutions, the base solution can be divided into two equal parts (A and B). The catalyst and promoter can be added to each part. The ratios of the catalyst and promoter can be varied with respect to the base solution to obtain the desired polymerization time. One suggested ratio is 200 mL of the base solution (divided into two equal parts A and B), 24 to 40 mL of catalyst added to part A, and 24 drops of promoter added to part B. Each part should be fully mixed with a spatula. Once each part is ready, the two parts can be mixed together. The approximate working time for the suggested ratio is about 30 to 45 minutes. For a color model, one of the pigments (2% to 10% depending on the desired color depth) can be added to part A or B. For further instructions, please refer to the manufacturer's technical note at www.polysciences.com.

- Injections can be made with a heavy-duty veterinary or disposable polyethylene syringe. If more than one injection is required to fill the vascular systems, care should be taken not to introduce any air bubble into the line. Once the injection is completed, remove the cannula and ligate the aorta permanently. Different injection sites (other than the aorta) can be chosen depending on the targeted cast region.
- The injected specimen should be fully cured in 2 to 3 hours. The polymerization of the resin is an exothermic process. It is recommended to place the specimen in cold water or an ice bath during the curing process to help with the dissipation of the generated heat. Controlling the temperature will prevent expansion and distortion of the specimen.

Maceration Process

Maceration solution (Polysciences, Cat. No. 07359, purchased separately) is used to remove excess tissue from the resin-perfused specimen. Please refer to Technical Data Sheet 105 from Polysciences for recommendations and use instructions. You can also prepare your own maceration solution by making a concentrated alkali solution (5% to 30% KOH or NAOH)[2,4,5,8,9] using NaOH pellets. The pellets can be re-solved in warm water in a proper plastic container. The mixing process will generate heat and hazardous fumes, and it must be performed in a ventilated hood.

Once the maceration solution is ready, the specimen should be placed in a laboratory-grade plastic container with a seal-tight lid. The container should be large enough to accommodate the size of the specimen; otherwise, deformation of the resin model may result after the removal of the tissue. Gently pour the solution over the tissue. Make sure the specimen is completely covered by the solution. The plastic container should be kept in an oven at 40°C to 50°C during the maceration process. The time duration for keeping the specimen warm varies based on the size of the specimen and concentration of the maceration solution. Our laboratory usually keeps the specimen in the solution for 2 weeks. This is followed by another 2 weeks in fresh solution. The specimen should be rinsed carefully with warm water when changing the solution and at the end of the procedure. This usually results in complete removal of the tissue as well as the boney structure. A pair of microforceps and microscissors can be used to clean and prune the resin model as needed. The result of the resin model of a rat's vascular system made in our laboratory is shown in Figure 13.3 in Chapter 13. Prior to maceration, it is recommended to remove any excess skin and tissue from the specimen to speed up the process.

EMBEDDING THE CLEANED SPECIMENS

The cleaned specimens can be subsequently embedded in clear methyl methacrylate resin to further protect any delicate structures. The Methyl Methacrylate Embedding and Casting Kit can be purchased from Polysciences (Cat. No. 03573).

Background

Blending methyl methacrylate (MMA) and butyl methacrylate (BMA) will yield a clear and firm cast. The plastic can be removed from the sections before staining with xylene or acetone treatment. Polymerization can be obtained with the addition of benzoyl peroxide at 60°C or with the addition of benzoin methyl ether, a photocatalyst, and ultraviolet illumination.

The MMA and BMA embedding media can be used for electron microscopy with the addition of a cross-linking agent such as divinylbenzene (not included in this kit). By changing the amount of BMA, the hardness of the block can be varied. A larger ratio of BMA will produce a softer block.

Procedure

1. Prepare a clear plastic or glass container for the polymerization procedure. Note that 40-mL scintillation tubes are sufficient for the rat brain vessels.
2. After fixation, washing, and dehydration, the sample is ready to use.
3. Methacrylates are soluble in ethanol and acetone. Intermediate clearing solvents are not required before infiltration of the tissue with the resin.
4. Solution preparation:
    - 1:1 mixture of 95% ethanol and MMA/BMA mix (9:1) with 1% (weight/volume) benzoin methyl ether (catalyst). A total of 100 mL is needed.
    - Solution A: 54 mL MMA plus 6 mL BMA (60 mL total), then add 0.6 g benzoin methyl ether powder.
    - Solution B: 95% ethanol.
    - Final solution C: 1:1 mixture of A and B, 50 mL 95% ethanol plus 50 mL solution A.
    - MMA plus 1% benzoin methyl ether solution: 400 mL MMA plus benzoin methyl ether powder (4 g).
    - Samples are infiltrated first with a 1:1 mixture solution for 1 hour.
    - Samples are transferred to a mixture solution of MMA and catalyst for 1 hour. Replace the solution of MMA and catalyst and keep the samples in the solution for another 1 hour.
    - Samples will stay in the MMA and catalyst solution to polymerize. Polymerization is achieved overnight under ultraviolet illumination.
5. Embedding must be done in closed BEEM® Embedding Capsules, gelatin capsules, or other closed containers (BEEM is a registered trademark of Better Equipment for Electron Microscopy, Inc).

Large quantities of catalyzed resin polymerize more rapidly, generating heat that can damage the tissue structure. Keep quantities in final casts as small as possible. The methacrylates in this kit contain the inhibitor needed only to prevent polymerization on storage. The inhibitor does not need to be removed beforehand if the proper ratio of catalyst is used.

Handling Precautions

Caution: Impervious gloves and good laboratory handling procedures are to be employed when working with MMA and BMA Kit ingredients. Care should be taken to avoid skin contact and inhalation of vapors. In case of skin contact, and after removing gloves, immediately and thoroughly wash with soap and water. Work should be conducted in a well-ventilated area. Use of disposable utensils and tools is recommended. The full chemical, physical, and toxicological properties of the products mentioned here are not known. Some people report skin sensitivities to methacrylates. If there is any noticeable irritation, use of the MMA and BMA Kit should be discontinued. All components may cause irritation.

Storage Precautions

The MMA can be safely stored at room temperature. BMA should be stored at 4°C.

Benzoyl peroxide (catalyst) is organic peroxide and should be kept cool and tightly sealed to avoid drying. Avoid grinding or contact with flammable or reducing agents. The catalyst decomposes as it ages; therefore, aged catalyst may require a greater amount, a longer time, or more heat to achieve the same results as fresh catalyst.

For benzoin methyl ether, avoid skin contact at all times.

Disposal of the Material

All the resins and solutions must be properly disposed in a hazardous waste container in accordance with local, state, and federal regulations provided by your institution.

REFERENCES

1. Glauert, A., *Practical Methods in Electron Microscopy,* Vol. 3, Part 1, Elsevier, New York, 1975, p. 154.
2. McMullan, D.M., Hanley, F.L., and Riemer, R.K., A method for selectively limiting lumen diameter in corrosion casting, *Microvasc Res,* 67(3), 215, 2004.
3. Pease, D., *Histological Techniques for Electron Microscopy,* 2nd ed., Academic Press, New York, 1964.
4. Sangiorgi, S., Congiu, T., Manelli, A., Dell'Eva, R., and Noonan, D.M., The three-dimensional microvascular architecture of the human Kaposi sarcoma implanted in nude mice: A SEM corrosion casting study, *Microvasc Res,* 72(3), 128, 2006.
5. Walocha, J.A., Litwin, J.A., and Miodonski, A.J., Vascular system of intramural leiomyomata revealed by corrosion casting and scanning electron microscopy, *Hum Reprod,* 18, 1088, 2003.
6. Kerber, C.W., Heilman, C.B., and Zanetti, P.H., Transparent elastic arterial models. I: A brief technical note, *Biorheology,* 26(6), 1041, 1989.
7. Feiner, L., Webber, A.L., Brown, C.B. et al., Targeted disruption of semaphorin 3C leads to persistent truncus arteriosus and aortic arch interruption, *Development,* 128(16), 3061, 2001.
8. Monnereau, L., Carretero, A., Berges, S. et al., Morphometric study of the aortic arch and its major branches in rat fetuses on the 21st day of gestation, *Anat Embryol,* V209(5), 357, 2005.
9. Ravnic, D.J., Jiang, X., Wolloscheck, T. et al., Vessel painting of the microcirculation using fluorescent lipophilic tracers, *Microvasc Res,* 70(1–2), 90, 2005.

22 Magnetic Resonance Imaging in Stroke Study

Yanlin Wang-Fischer and Souvik Sen

CONTENTS

INTRODUCTION

In 2006, Dr. Sen wrote a very detailed description about magnetic resonance imaging (MRI) in human patients, here we present most of this article with permission from eMedicine.com, 2006.

BACKGROUND

Magnetic resonance imaging is a newly updated and promising tool. MRI is being increasingly used in the diagnosis and management of acute ischemic stroke. This new technique can examine brain longitudinally noninvasively and is the most powerful tool available to assess the number, the spatial distribution, and the size of brain lesions. It has been demonstrated that MRI is a useful and important tool for both clinical diagnostics for patients and research for animal study.[1–26] Using consecutive brain images, investigators were able to quantify edema as an increase in pixel intensity in relation to a defined baseline threshold for discovery of a new drug or for basic research.[1]

The aim of this chapter is to provide simple and up-to-date information about the use of MRI in ischemic stroke patients and stroke research on animals. MRI is a fast-growing technology that is sensitive and relatively specific in detecting changes that occur after strokes. Although it has some limitations, such as high cost, long scanning duration, and decreased sensitivity in the detection of subarachnoid hemorrhages, these constitute exciting challenges in the future of this technology. Recent advances in MRI, including higher strength of magnetic field (1.5 to 3.0 T field strength) yielding better resolution of images, newer sequences of images, and the advent of the open MRI for patients who are claustrophobic or overweight, have led to widespread use of this technology in diagnosis and management of acute stroke. Mounting a small bed into the system allows this technique to be used for laboratory animal studies of stroke.

MECHANISM AND PATHOPHYSIOLOGY OF MAGNETIC RESONANCE IMAGING (MRI)

Nuclear magnetic resonance (NMR) refers to the ability of certain nuclei to emit useful signals when these nuclei are subjected to a strong static magnetic field and then excited by another strong but varying magnetic field. This signal is then recorded during a session and decoded into valuable information. Organic material is made up of a wide variety of molecules that comprise a large number of hydrogen and carbon atoms. There are also intermediate numbers of other atoms like oxygen, nitrogen, phosphorous, iron, and sulfur as well as numerous trace elements such as selenium, chromium, and others. To be useful in NMR, nuclei must be magnetic, that is, have a nuclear magnetic moment. Of all of the atoms within the body, the ^{1}H atom is of the most interest. It generates the largest signal and is therefore the most valuable for *in vivo* NMR experiments.

In the body, some nuclei become excited when positioned in a strong magnetic field; they absorb the radio-frequency (RF) energy of the magnetic field and then release it until they relax completely. The energy is released from the excited tissue over a short period of time according to two relaxation constants known as T1 and T2, and the emitted energy signals are converted into images. The contrasts in the images result from different intensities of these emitted signals, which in turn result from different concentrations of the nuclei in different tissues in the body. Hydrogen (that is, protons) is the most common magnetic resonance (MR)-observable nucleus in the human or animal body and has the advantage of being present in many different tissues in different concentrations. Other organic particles have been tried but demonstrated less spatial resolution than hydrogen. Other biochemical compounds, lactate and N-acetyl aspartate, are under trial to increase understanding of the significance of the different concentrations of these compounds in different pathologic conditions (that is, magnetic resonance spectroscopy [MRS]).

There are three principal stages in an NMR experiment:

1. *Preparation* of the nuclear magnetic system.
2. *Excitation* (or perturbation) of the nuclear magnetic system.
3. *Acquisition* of the signal from the excited nuclear magnetic system before it relaxes back to its equilibrium state.

Preparation of the System

Within a body, the magnetic moments of all of the atoms are random. This leads a body to have a total magnetic moment of zero. However, when placed with a uniform, static magnetic field, the moments will redistribute themselves parallel and antiparallel to the direction of the field. The result of this is to generate a net magnetization within the body. The nuclei resonate in unison and are able to generate a much stronger signal. This process of magnetic moments is called *polarization* and prepares the body for the phase of the NMR experiment.

Excitation of the System

A second magnetic field is applied to the system. This field is rotating in a plane perpendicular to the first, static magnetic field. The rotation is at RF, and it is tuned carefully so that the nuclei will resonate with the field. This resonance causes the moments of the nuclei to move from their polarized positions. After the second field stops or changes (it pulses as well as rotates), the nuclei will return to their polarized state. A signal is emitted by the nuclei during the change.

Acquisition of the Signal

The signal that is generated by the nuclei is picked up by sensitive voltage detection equipment. There are also phase angle detectors that receive information about positioning and timing of the signal. All of this information is processed to generate useful information about the body.

PATHOGENESIS OF IMAGING FINDINGS

Cytotoxic Edema

Regardless of the causes, neuronal ischemia rapidly depletes intracellular adenosine triphosphate (ATP), which results in failure of the membrane-bound ATP-dependent ionic channels responsible for both neuron resting membrane potentials as well as generation of action potentials. This metabolic disorder leads to accumulation of intracellular ions (including calcium ions), creating an intracellular gradient responsible for intracellular accumulation of water, that is, *cytotoxic edema*.

Vasogenic Edema

Cerebral endothelial cells are more resistant to ischemia than neurons and neuroglial cells. About 3 to 4 hours after the onset of ischemia, the integrity of the blood–brain barrier becomes compromised, and plasma proteins are able to pass into the extracellular space. The intravascular water follows when reperfusion occurs, known as *vasogenic edema*; this process begins 3 hours after the onset of stroke and reaches a maximum 2 to 4 days after the onset of stroke (refer to Figure 19.19 in Chapter 19, brain edema after stroke in rats). Reperfusion can also be accompanied by hemorrhagic transformation of the infarct, which is usually related to the volume and site of the infarct and is more common in large cortical infarcts. Changes in MR images due to ischemic stroke follow the vascular territory of the occluded blood vessel, which is characteristic of cerebrovascular disease and helps in differentiating it from other disease entities.

IMAGING TECHNIQUES AND THEIR RELATION TO THE PATHOGENESIS

Commonly used MRI techniques are the following: diffusion-weighted imaging (DWI); perfusion-weighted imaging (PWI); diffusion-perfusion mismatch; echo-planar imaging (EPI); T_1-relaxation time imaging (T1); T_2-relaxation time imaging (T2); magnetic resonance spectroscopy (MRS); magnetic resonance angiography (MRA); and spin density-weighted imaging.

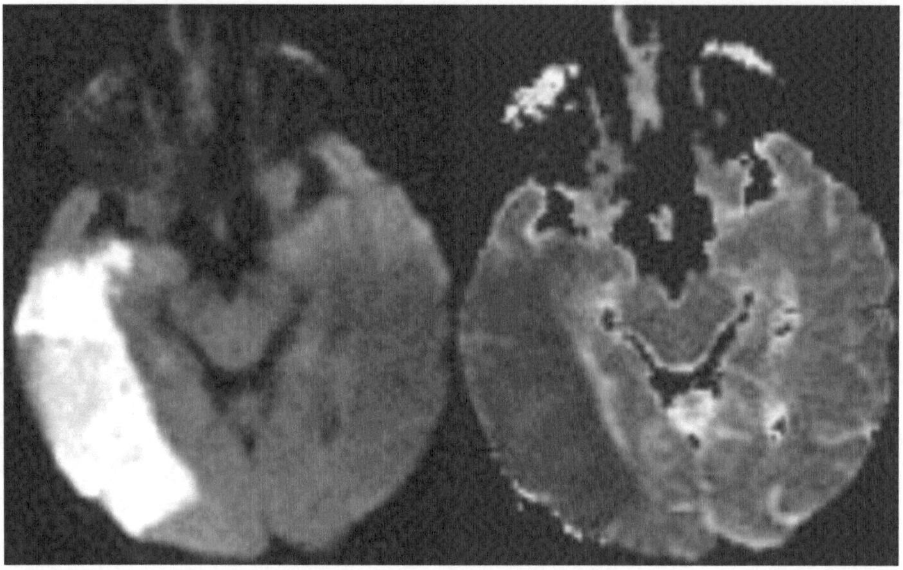

FIGURE 22.1 Magnetic resonance imaging (MRI) in acute stroke. Left: Diffusion-weighted MRI in acute ischemic stroke performed 35 minutes after symptom onset. Right: Apparent diffusion coefficient (ADC) map obtained from the same patient at the same time.

Diffusion-Weighted Imaging (DWI)

- In DWI, the images reflect microscopic random motion of water molecules. A value called the apparent diffusion coefficient (ADC) is measured and captured by this type of imaging. The water molecules move in the direction of the magnetic field gradient; they accumulate a phase shift in their transverse magnetization relative to that of a stationary one, and this phase shift is directly related to the signal attenuation of the image.
- Numerous studies have shown that ADCs in ischemic areas are lower by 50% or more than those of normal brain areas, and they appear as bright areas (that is, hyperintensities) on the DWI (see Figure 22.1). Studies have demonstrated that changes in the ADC occur as early as 10 minutes following onset of ischemia.
- Cytotoxic edema appears following sodium/potassium pump failure, which results from energy metabolism failure due to ischemic insult; this occurs within minutes of the onset of ischemia and produces an increase in brain tissue water of up to 3% to 5%. Reduction in intracellular and extracellular water molecule movement is the presumed explanation for the drop in ADC values.
- The diffusion of water molecules is guarded by biologic barriers in the brain tissue (for example, cell membranes and cellular organelles). The behavior of water molecules is not symmetric and may show uneven distribution of the ADC when measured in one direction; this uneven distribution may give a false impression of a lesion. ADC values are measured in several directions (three, six, or more), and ADC maps are created to produce a direction-insensitive measurement of the diffusion. When ADC is measured in six or more directions, the diffusion motion of all the water molecules (ADC tensor matrix) can be calculated to create what is called *full-diffusion tensor mapping*, which can also be used to visualize white matter tracts.
- Reduction in the ADC also occurs in other conditions such as global ischemia, hypoglycemia, and status epilepticus; it should always be evaluated in relation to the clinical condition of the patient.

- Human studies demonstrated that damage in the areas showing decreased ADC levels is very rarely reversible (in contrast to that in animal models), although a few studies have indicated that intraarterial thrombolysis may occasionally result in disappearance of the diffusion defect. The technique most commonly used to acquire the DWI is an ultrafast one, echo-planar imaging (EPI); this technique decreases scanning time significantly and eliminates movement artifacts.
- The acute drop in ADC is gradually normalized to baseline at 5 to 10 days after ischemia (pseudonormalization); it even exceeds normal levels as time passes, helping in some cases to differentiate among acute, subacute, and chronic lesions.
- DWI is very sensitive and relatively specific in detecting acute ischemic stroke.[12,13,18] DWI findings have shown high levels of diagnostic accuracy; however, recent studies demonstrated that small brainstem lacunar infarctions may escape detection. Normal DWI in patients with symptoms should trigger further investigation for a nonischemic cause of the symptoms. DWI has been shown to reveal diffusion abnormalities in almost 50% of patients with clinically defined transient ischemic attacks (TIAs); it tends to be of higher yield at increasing time intervals from the onset of stroke symptoms.

Perfusion-Weighted Imaging (PWI)

- In PWI, hemodynamically weighted MR sequences are based on passage of MR contrast through brain tissue. With this technique, information about the perfusion status of the brain is available.[11,21] The most commonly used technique is bolus-contrast tracking (other techniques include blood oxygen level and arterial spin tagging). The imaging is based on the monitoring of a nondiffusible contrast material (gadolinium) passing through brain tissue.
- The signal intensity declines as contrast material passes through the infarct area and returns to normal as it exits this area. A curve is derived from these tracing data (that is, signal washout curve), which represents and estimates the cerebral blood volume (CBV).
- An arterial input function can be derived by measuring an artery in lower brain slices or by measuring gadolinium concentration that is proportional to the changes in T2 when gadolinium is used at low doses (<3 mg/kg). The relative cerebral blood flow (CBF) can be calculated by using the CBV and the arterial input function. Then, from the central volume theory, the relative mean transit time (MTT) can be mapped.
- DWI and PWI together have been shown to be superior to conventional MRI both in early phases and up to 48 hours after the onset of stroke.[21] Using both DWI and PWI is very important because together they provide information about location and extent of infarction within minutes of onset; when performed in series, they can provide information about the pattern of evolution of the ischemic lesion. This information may be of great importance in choosing the appropriate treatment modality as well as in predicting outcome and prognosis.
- The lesion usually enlarges on serial DWIs over a period of several days. It has been suggested that this enlargement can be halted if reperfusion (resolution of original PWI lesion) occurs early enough. Lesions that are not large on initial PWI do not show this enlargement.
- The diffusion-perfusion mismatch (see Figure 22.2), i.e., the difference in size between lesions captured by DWI and PWI, usually represents the ischemic penumbra (see Figure 22.3), which is the region of incomplete ischemia that lies next to the core of the infarction. The ischemic penumbra is regarded as an area that is viable but under ischemic threat; it can be saved if appropriate intervention is promptly instituted. The viability of this region could extend up to 48 hours after the onset of stroke. Determining the volume of the ischemic penumbra may be very useful in discovery of new drugs for treatment of stroke and identifying patients who would benefit from thrombolytic therapy and perhaps even conventional treatments such as carotid endarterectomy or blood pressure elevation.

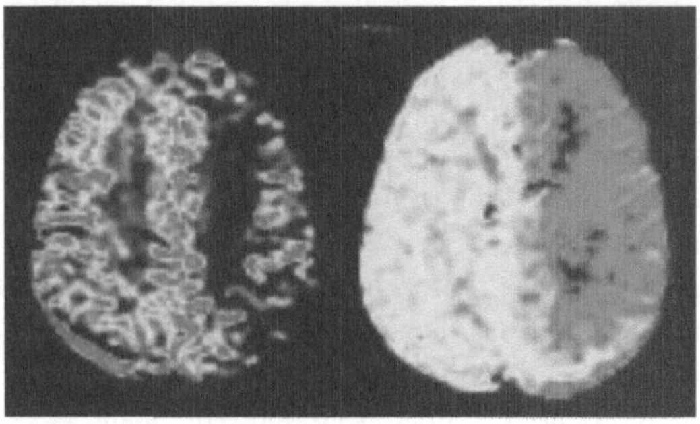

FIGURE 22.2 Magnetic resonance imaging (MRI) in acute stroke. Left: Perfusion-weighted MRI of a patient who presented 1 hour after onset of stroke symptoms. Right: Mean transfer time (MTT) map of the same patient. **(See color insert following page 146.)**

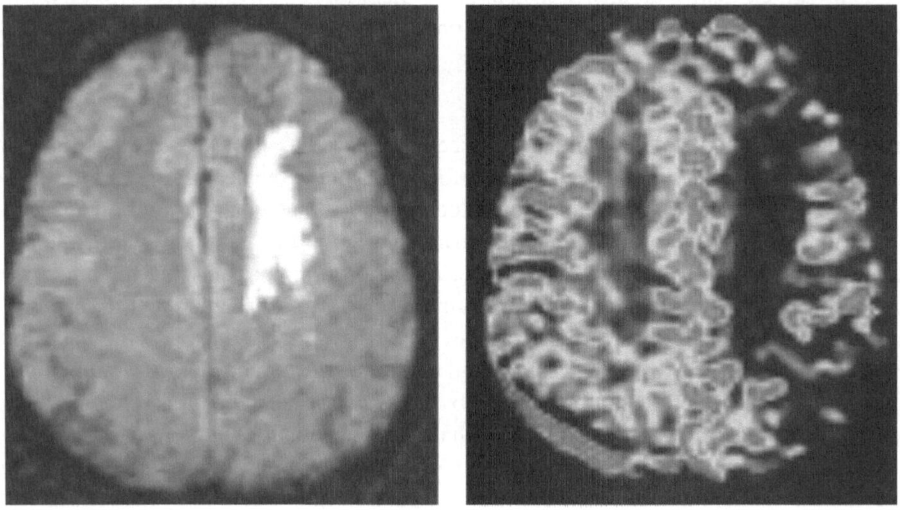

FIGURE 22.3 Magnetic resonance imaging in acute stroke. Diffusion-perfusion mismatch in acute ischemic stroke. The perfusion abnormality (right) is larger than the diffusion abnormality (left), indicating the ischemic penumbra, which is at risk of infarction.

It could also aid in evaluating the risk/benefit ratio of using such treatments in stroke patients.

- One limitation of these techniques is in detection of acute intracerebral hemorrhages; early studies demonstrated that susceptibility imaging could be sensitive in the detection of acute intracerebral hemorrhage. Gradient-recalled echo (GRE) imaging sequences demonstrated the most favorable sensitivity in detecting susceptibility dephasing associated with chronic intracerebral hemorrhages.
- MRI still has some limitations in its application, namely, in patients with metal implants and acutely ill patients requiring close monitoring.

Diffusion-Perfusion Mismatch: Utility in Detecting Ischemic Penumbra

The concept of the ischemic *penumbra* is broadly defined as an area of severely hypoperfused but potentially restorable tissue around the irreversibly infarcted core. Cerebral imaging could potentially be used to rapidly diagnose this partially ischemic region, which might be amenable to therapy. Rapid reperfusion of the ischemic penumbra is the basis for acute stroke treatment, including thrombolytic therapy. Using echoplanar MRI, noninvasive imaging of the penumbra can be performed with the potential for selection of therapy, based on rapid delineation of cerebral pathophysiology. This could facilitate individualized selection of reperfusion therapy beyond the arbitrary time windows confirmed in clinical trials.

The combination of DWI, PWI, and magnetic resonance angiography (MRA) provides a rapid assessment of the ischemic infarct core (DWI), the underlying arterial pathology (MRA), and the tissue at risk (PWI > DWI mismatch). The concept of mismatch, with perfusion larger than diffusion lesion volumes, was postulated as an operational definition of the penumbra by Schlaug and colleagues in the late 1990s.[22]

Kidwell[23] recently suggested two modifications to the MRI-based (PWI > DWI mismatch) penumbral concept. It was shown that the hyperintense DWI lesion does not always progress to frank infarction, and that variable degrees of DWI lesion reversibility occurred with intraarterial thrombolysis. The second modification to the PWI > DWI mismatch concept was that the hypoperfused area on PWI in acute ischemia includes regions of only mildly reduced perfusion, which typically survive, in contrast with more severely ischemic regions, which usually proceed to infarct. These observations led to a modified definition of the penumbra, which is now thought to be more accurately represented by the PWI > DWI mismatch area minus the oligemic rim, but including some of the DWI core with less severely depressed ADC values. Regardless of the precise MRI definition, the penumbra is regarded as potentially salvageable tissue in acute ischemic stroke and has therefore been the focus of research into acute stroke therapies. The finding of potentially salvageable tissue many hours after stroke onset correlates with evidence obtained from ^{15}O and ^{18}F-fluoromisonidazole positron emission tomographic imaging.

Several acute ischemic stroke studies[23–25] are using PWI-DWI mismatch to identify a subgroup of patients who may benefit from reperfusion via recanalization.

Echo-Planar Imaging

- Echo-planar imaging is a new technique that can be used to visualize physiologic parameters in addition to measuring diffusion coefficients of the ischemic brain. Changes in brain oxygenation can be monitored by using gradient echo and EPI, in which deoxygenated blood acts as a susceptibility contrast agent. Gradient echo imaging has the highest sensitivity in detecting early hemorrhagic changes.
- EPI can be used in conjunction with bolus injection of intravenous paramagnetic agents to assess cerebral perfusion and functional changes in CBV.
- In this technique, hypoperfused areas appear as hyperintense signals after injection of the contrast material. This technique is considered a way of reducing the traditionally long scanning time of MRI.

T_1-Relaxation Time Imaging

In T_1-relaxation time imaging (T1), cerebrospinal fluid (CSF) has low signal intensity in relation to brain tissue, and this has been used as an animal study marker.

T_2-Relaxation Time Imaging

The T_2-relaxation time imaging (T2) in which CSF has high signal intensity in relation to brain tissue has been a popular tool for rat stroke study.[1,17] T_2-relaxation time imaging is uniquely suited to the assessment of the development, progression, and regression of brain edema and intracerebral hemorrhage *in vivo*. It can define the number spatial distribution and quantitate the size of brain lesions better than any other imaging modality in rodents.[1–3,5] This technique offers a novel possibility to do repeated measurements and thus monitor the development of cerebral lesions as well as the effect of therapeutic measures at different stages of developing cerebral injury. In the animal study,[1] repeated cerebral MRI measurements revealed two important points. First, approximately 70% of rats developed changes in T_2 MRI before they showed neurological symptoms. Either the affected brain areas were not involved in motor function or they were not large enough to cause disorders in behavior. Thus, T_2 MRI is a powerful tool in the temporal definition of the onset of cerebral lesions.

Magnetic Resonance Spectroscopy (MRS)

- MRS is one of the recent advances in MR technology. The MRS information is gathered in a similar fashion as for MRI. However, rather than providing a spatial map or image, MRS results in a frequency spectrum containing discrete signals from a selected resonating nucleus.[2] The nuclei most commonly studied by *in vivo* MRS include ^{1}H, ^{13}C, ^{19}F, ^{23}Na, and ^{31}P. Due to the high natural abundance of ^{1}H and its ubiquity in all *in vivo* metabolites, it has been the focus of many MRS studies. The magnitude of a metabolite peak in the frequency spectrum is proportional to its concentration in the volume of tissue being examined. Measuring changes in these peak areas, which result from either normal function or disease or from the effects of drugs, allows us to monitor intracellular changes that are not always accompanied by structural changes. MRS evaluates metabolic activity and concentration of certain metabolites in specified areas of the brain. Proton and phosphorus spectroscopic studies have been performed.
- In proton spectroscopy, depression of N-acetyl aspartate, which is considered to be a marker of neurons, is the most consistent finding in acute stroke. This depression may occur within hours after the onset of stroke and continues through the subacute and chronic phases of the stroke, presumably because of loss of neurons.
- Increase in levels of lactate is another important finding and has been attributed to anaerobic metabolism in ischemic tissue. Initial studies of other metabolites, such as choline and creatine, demonstrated decreases in their levels in acute stroke.
- Phosphorus spectroscopy provides information about energy metabolism and pH, depletion of ATP, decrease of tissue pH, and increase of the ratio of inorganic phosphate to phosphocreatine, which has been reported in both human and animal studies.
- Long acquisition times, weak signal, and low spatial resolution of this technique have limited enthusiasm for its use in the clinical management of cerebral ischemia; however, some studies have suggested that MRS results can have prognostic value in stroke.

Magnetic Resonance Angiography (MRA)

- MRA is very sensitive to flow, and based on the difference in signal between moving blood and stationary brain tissue,[9,15] angiographic-like images of the cervicocranial vasculature are produced.
- MRA images are a useful tool in identifying dissections, in that both the true and false lumen of the involved artery can be observed on the source images. Following is a brief description of the two basic techniques.
  - The three-dimensional (3D) time-of-flight (TOF) technique is based on flow-related enhancement; it is the preferred MRA technique. However, it has some limitations, especially flow signal dropout secondary to turbulent flow in the tortuous and stenotic vascular segments, which makes interpretation of stenosis in these areas difficult. These are common predilection sites for atherosclerosis. Also, in slow-flow regions, the spin saturation of the scan causes overestimation of stenosis. In contrast-enhanced studies, it provides more information than standard angiography, especially in detecting critical stenosis of extracranial vessels, but it is less reliable in intracranial critical

stenosis. Always keep in mind that MRA is a flow-dependent technology; absence of flow signal does not mean literally a complete occlusion but rather that flow is below a critical value.

- Two-dimensional (2D) TOF MRA also depends on the relative contrast between flowing blood and stationary tissue; it provides better images than 3D TOF in slow-flow regions. 2D TOF images correlate well with carotid angiography images in depicting cervical bifurcation disease. Its disadvantages, however, are the significant artifacts (for example, stepladder) that often occur, which may obscure vessel details, and the longer scanning time.

- The modified TOF MRA technique, which uses multiple overlapping thin-slab acquisitions (MOTSA), combines the advantages of 2D and 3D TOF techniques. It is very helpful in demonstrating severe stenosis, although the degree of stenosis might be slightly overestimated.
- Two-dimensional phase-contrast (PC) MRA is a technique that is helpful specifically in differentiating slow and absent flow from normal flow; it captures only truly patent vessels. Other imaging sequences (for example, spin-echo sequence or gradient-echo sequence) should be used with PC-MRA to avoid missing lesions such as paravascular hematomas, which are not captured by PC-MRA. PC-MRA also has the disadvantage of signal loss due to turbulent flow in tortuous vessels.

Spin Density-Weighted Imaging

In spin density-weighted imaging, CSF has a density similar to brain tissue.

TYPES OF INFARCTION

Thromboembolic Infarction

Thromboembolic infarction is the most common form of infarction. Typically, it is observed on MRI as a wedge-shaped infarct in the particular vascular distribution. Recent data support the hypothesis that a single infarct in a vascular territory is more likely to be thrombotic than multiple infarcts, which are more likely to be embolic. An animal model of stroke induced by blood clot injection or suture insertion has this form of infarction.

Watershed Infarction

Watershed infarction occurs at the distal margins of specific arterial territories. The distal animal model of stroke has a similar infarction. It can occur both superficially and deep in the brain parenchyma. Common etiologies for this lesion include hypotension, cardiac and respiratory arrest, and proximal arterial stenosis or occlusion. MRI findings follow the pattern of incomplete thromboembolic ischemic infarction in T1 and T2 morphologic and signal changes, with early parenchymal enhancement suggesting early reperfusion. Recent studies showed that this type of infarction could be more readily detected by using DWI.

Lacunar Infarction

Lacunar infarctions are small, deep, cerebral infarctions believed to be caused by intrinsic small-vessel disease secondary to lipohyalinosis and fibrinoid necrosis; they are most frequently observed in patients with hypertension or diabetes mellitus. Common sites for these lesions include basal ganglia, internal capsule, thalamus, brainstem, and cerebellum. MRI findings in these lesions follow the same pattern observed in thromboembolic infarction.

Venous Thrombosis and Infarction

Occlusion of cerebral veins and venous sinuses is usually caused by systemic conditions, such as pregnancy, collagen vascular diseases, inflammatory bowel diseases, and hypercoagulable states as well as local conditions such as infection, neoplasia, and trauma. Occlusion of the venous structure causes outflow obstruction and vascular congestion; these result in parenchymal infarctions and hemorrhages. The characteristics of this kind of infarct are as follows:

1. Patients usually present in the late acute phase or in the subacute phase, which makes the diagnosis difficult because diagnosis at these stages depends on imaging studies.
2. MRI findings in these lesions include loss of venous flow void signal, absence of normal venous enhancement, and visualization of isointense-to-hyperintense signals within the venous channels on both T1 and T2 images. These variable patterns of enhancements are due to mixed blood products, which are present in the lesion.
3. These patterns are usually bilateral, do not respect arterial vascular territories, and have associated hemorrhage.
4. Three-dimensional phase-contrast magnetic resonance venography (MRV) is the preferred technique in the evaluation of venous thrombosis.

MRI Findings in Patients during Different Periods of the Disease

Findings in Stroke: Hyperacute Phase (0 to 24 hours)

- DWI is able to detect ischemic changes within minutes of onset (see Figure 22.1). Reduced proton motion is detected as a decreased ADC.
- Early in the process of cerebral ischemia, PWI, using first-pass contrast bolus injection or spin tagging the protons in the water in blood, reveals reductions of CBF and CBV and an increased MTT of blood through the brain (see Figure 22.2).
- Matched diffusion- and perfusion-weighted abnormalities correlate with the region of infarction and are indicative of permanent neuronal death. Mismatched diffusion and perfusion abnormalities with the perfusion abnormality larger than the diffusion abnormality may be indicative of a region of reversible ischemic penumbra (see Figure 22.3).
- A few hours after stroke onset, a loss of arterial void signal is sometimes observed (30% to 40% of patients); it is best observed on T2-WI (weighted imaging).
- At 2 to 4 hours, T1-WI shows subtle effacement of the sulci due to cytotoxic edema.
- At 8 hours, T2-WI shows hyperintense signal due to both cytotoxic and vasogenic edema.
- At 16 to 24 hours, T1-WI shows hypointense signal due to both cytotoxic and vasogenic edema.
- Contrast-enhanced images show arterial enhancement followed by parenchymal enhancement. The arterial enhancement can be very early (in more than 50% of patients) and is due to slow blood flow; it typically disappears after 1 week.
- Parenchymal enhancement differs in complete and incomplete infarctions. In complete infarction, it starts 5 to 7 days after the stroke and persists for several months. In incomplete infarctions, it can be observed within 2 to 4 hours and usually is more intense than in complete infarction.
- Although conventional MRI sequences most often do not show evidence of stroke in the acute phase, conventional MRI may show signs of intravascular thrombus such as absence of flow void on T2-WI, and hypointense vascular sign on GRE sequence.
- MRI findings in acute ischemic changes in humans are summarized in Table 22.1.

TABLE 22.1
Magnetic Resonance Imaging (MRI) Finding in Acute Ischemic Changes in Humans

Time	MRI Finding	Etiology
2 to 3 minutes	DWI: Reduced ADC	Decreased motion of protons
2 to 3 minutes	PWI: Reduced CBF, CBV, MTT	Decreased CBF
0 to 2 hours	T2-WI: Absent flow void signal	Slow flow or occlusion
0 to 2 hours	T1-WI: Arterial enhancement	Slow flow
2 to 4 hours	T1-WI: Subtle sulcal effacement	Cytotoxic edema
2 to 4 hours	T1-WI: Parenchymal enhancement	Incomplete infarction
8 hours	T2-WI: Hyperintense signal	Vasogenic and cytotoxic edema
16 to 24 hours	T1-WI: Hypointense signal	Vasogenic and cytotoxic edema
5 to 7 days	Parenchymal enhancement	Complete infarction

Notes: ADC, apparent diffusion coefficient; CBF, cerebral blood flow; CBV, cerebral blood volume; MTT, mean transit time; T1, T_1-relaxation time imaging; T2, T_2-relaxation time imaging; WI, weighted imaging.

Findings in Stroke: Acute Phase (1 to 7 Days)

- In this part of the acute phase (1 to 7 days), edema increases (edema maximizes at 48 to 72 hours), and MRI signals become more prominent and well demarcated. The ischemic area continues to appear as an area of hypointensity on T1-WI and as a hyperintense area on T2-WI. Also, the mass effect can be appreciated in this phase.
- In contrast-enhanced images, the arterial enhancement usually persists throughout the acute phase, while the parenchymal enhancement is usually appreciated at the end of this phase in complete infarction. In incomplete infarction, the parenchymal enhancement is usually earlier.
- During this period, reperfusion occurs, and both petechial and frank hemorrhage can be observed, typically 24 to 48 hours after the onset of the stroke. Usually, petechial hemorrhages cause the "fogging" phenomenon due to hemoglobin degradation products, which masks the infarction on both T1-WI and T2-WI.

Findings in Stroke: Subacute Phase (7 to 21 Days)

- In the subacute phase at 7 to 21 days, the edema resolves, and the mass effect becomes less appreciated; however, the infarcted areas still appear as a hypointensity on T1-WI and as a hyperintensity on T2-WI.
- In contrast-enhanced images, the arterial enhancement is usually resolved by this time, and the parenchymal enhancement typically persists throughout this phase. The cortical parenchymal enhancement is usually in a gyriform pattern, while the subcortical enhancement is usually a homogeneous central pattern.

Findings in Stroke: Chronic Phase (>21 Days)

- In the chronic phase, the edema completely resolves, and the infarcted area still appears as a hypointensity on T1-WI and as a hyperintensity on T2-WI. Because of tissue loss in the infarcted area by this time, *ex vacuo* ventricular enlargement and widening of the cortical gyri and fissures take place.
- In contrast-enhanced images, parenchymal enhancement typically persists throughout this phase, also; it usually disappears by 3 to 4 months.

Findings in Transient Ischemic Attacks

- A third to a half of the patients presenting with a TIA have lesions on DWI. A significant proportion of these patients may not reveal a corresponding lesion on T2-WI. PWI may be more sensitive but has not been adequately tested in patients with TIA. DWI-positive TIA lesions do necessarily show as infarction on follow-up MRI.
- Although TIAs have been traditionally defined as transient (<24 hours) neurological deficit of vascular origin, the advent of MRI has led to reconsideration of the definition. Whether DWI-positive TIAs are to be regarded as stroke or TIA is unclear.

Findings in Hemorrhagic Stroke

- GRE and EPI sequences have the ability to detect microbleeds that are clinically silent and not visualized by computed tomographic (CT) scanning or routine MR sequences. These microbleeds are visualized in a fifth to a quarter of patients with ischemic stroke and 5% of elderly asymptomatic individuals. The microbleeds depict hemosiderin deposit and have been histopathologically correlated with prior extravasations of blood. These microbleeds may represent bleeding-prone angiopathy and a higher rate of hemorrhagic transformation from anticoagulation, antithrombotic, and thrombolytic therapy.
- GRE, EPI, and DWI (B0) are sensitive to detecting intraparenchymal hemorrhage (primary intracerebral hemorrhage and hemorrhagic transformation) in the hyperacute stages (first few hours), whereas the conventional T1-WI and T2-WI are sensitive in detecting subacute and chronic bleeding. FLAIR sequences may have a role in detecting extraaxial collections of blood (subdural hemorrhages). Having stated this, the current guidelines do not advocate the use of MR in place of CT scanning to screen patients for thrombolysis.

MRI APPLICATIONS IN STUDY OF RODENTS

The MRI system for rodents is a little different from human use due to small size of the animals. Figure 22.4 shows a sample of the system from Bruker MRI and Concorde microPET®. An animal bed is mounted in a Concorde microPET MRI system for small animal imaging. Since the animal has similar pathogenesis and pathophysiology as the human except for a small body, the parameters used

FIGURE 22.4 A magnetic resonance imaging (MRI) system from Bruker MRI and Concorde microPET. An animal bed is mounted in a Concorde microPET for small animal imaging.

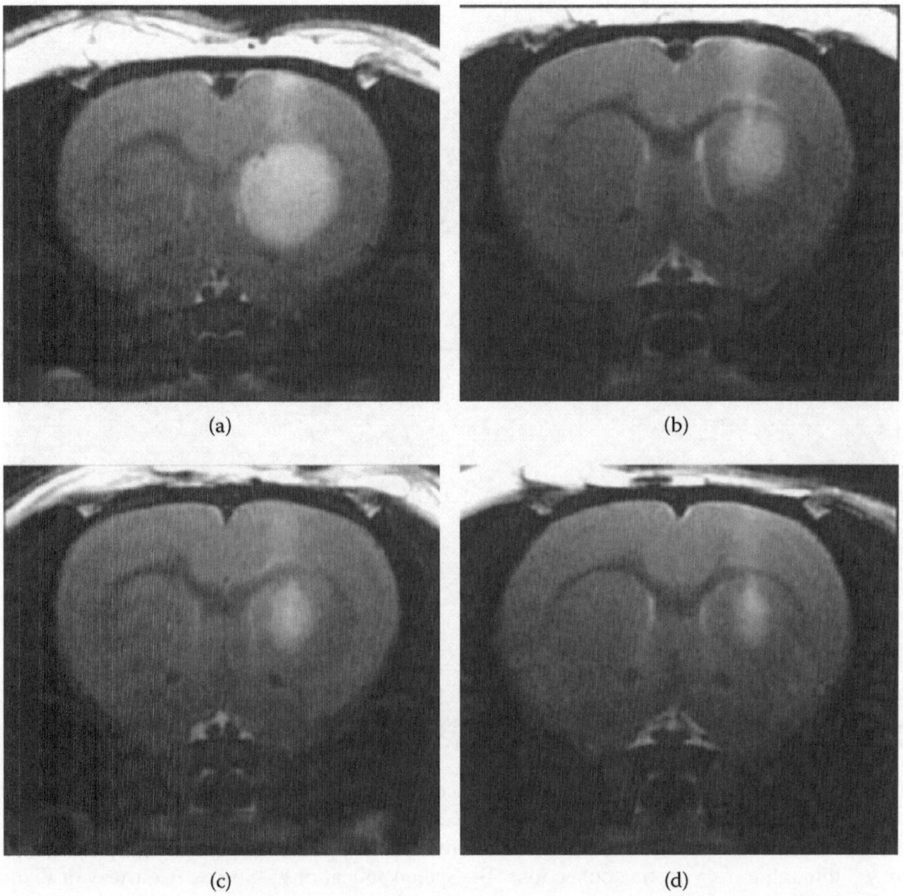

FIGURE 22.5 The size of a lesion in the rat brain is reduced after treatment with a new cerebroprotective drug. Lesion appears bright in the T2W-RARE images. (a) Control; (b) 3-mg dose; (c) 10-mg dose; (d) 30-mg dose. BioSpec Applications—Stroke. (Courtesy of D. Stiller, Boehringer Ingelheim Pharma, Germany.)

for humans can be used in rats. Here, we adopted some MRI imaging pictures from BioSpec® (with their permission) that represent the techniques for stroke study of rats. Detailed information, instrumentation, and software for MRI of rats can be found at BioSpec (www.bruker-biospin.de/NMR).

The following images provide further information:

1. T2-WI (Figure 22.5)
2. Diffusion-weighted MRI (Figure 22.6)
3. Rat brain angiography (Figure 22.7)
4. Multiparameter investigation of stroke (Figure 22.8)
5. Cerebroprotective drug trials (Figure 22.9)
6. Quantitative MR mapping of stroke (Figure 22.10)

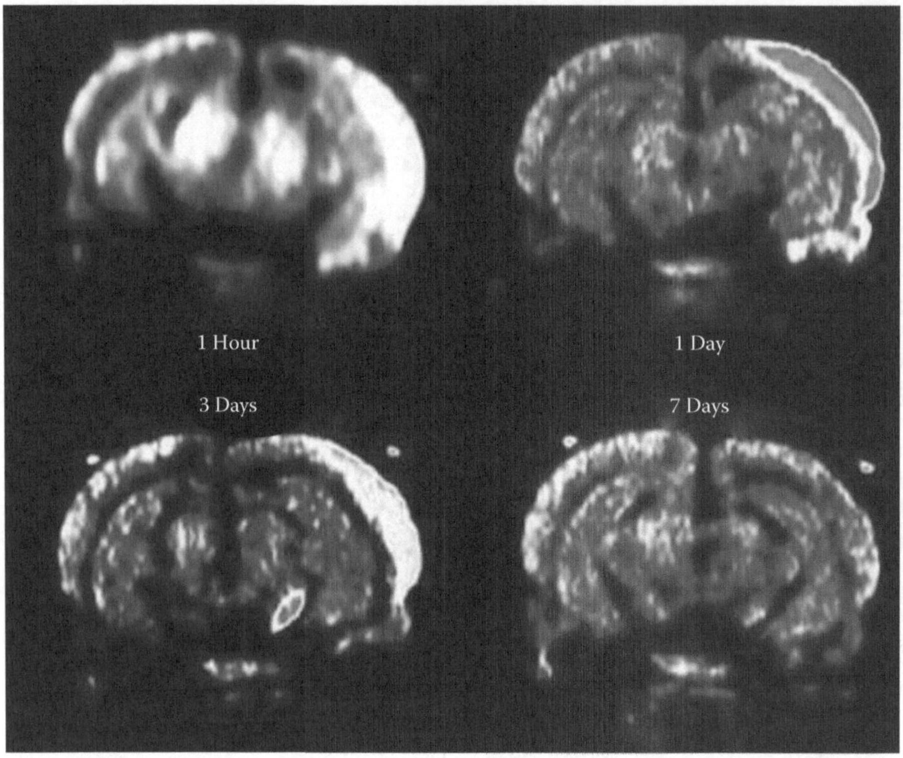

FIGURE 22.6 Diffusion-weighted magnetic resonance imaging (MRI) was used to study cerebroprotective agents following middle cerebral artery occlusion (MCAO) in the rat. The MR images were taken 1 hour, 1 day, 3 days, and 7 days after occlusion of the middle cerebral artery. They show neuronal loss in the substantia nigra at day 3 through transsynaptic connections. BioSpec Applications—Stroke. (Courtesy of P. Juretschke et al., Hoechst Marion Rousel, Core Research, Frankfurt, FRG.)

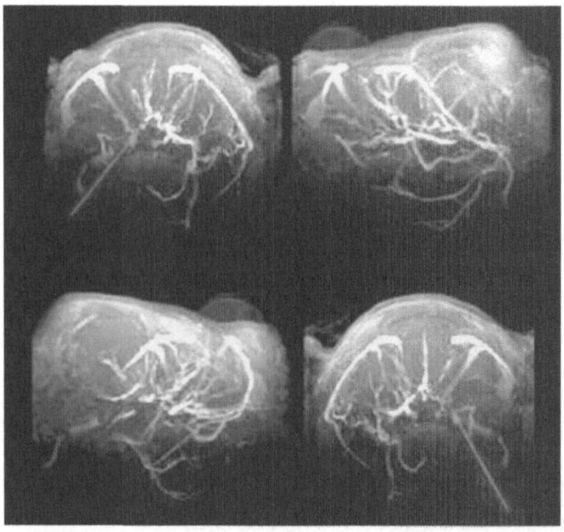

FIGURE 22.7 Four views of an MIP (maximum intensity projection) reconstruction of magnetic resonance imaging (MRI) data obtained with flow-compensated 3D sequence. Drastic reduction of flow is visible at the site of the suture-induced middle cerebral artery occlusion (MCAO) insult (red arrows). These images were obtained without contrast agent. BioSpec Applications—Stroke. (Courtesy of M. Neumaier, U. Pschorn et al., Boehringer Ingelheim Pharma KG, Department of CNS Research, Ingelheim, FRG.) **(See color insert.)**

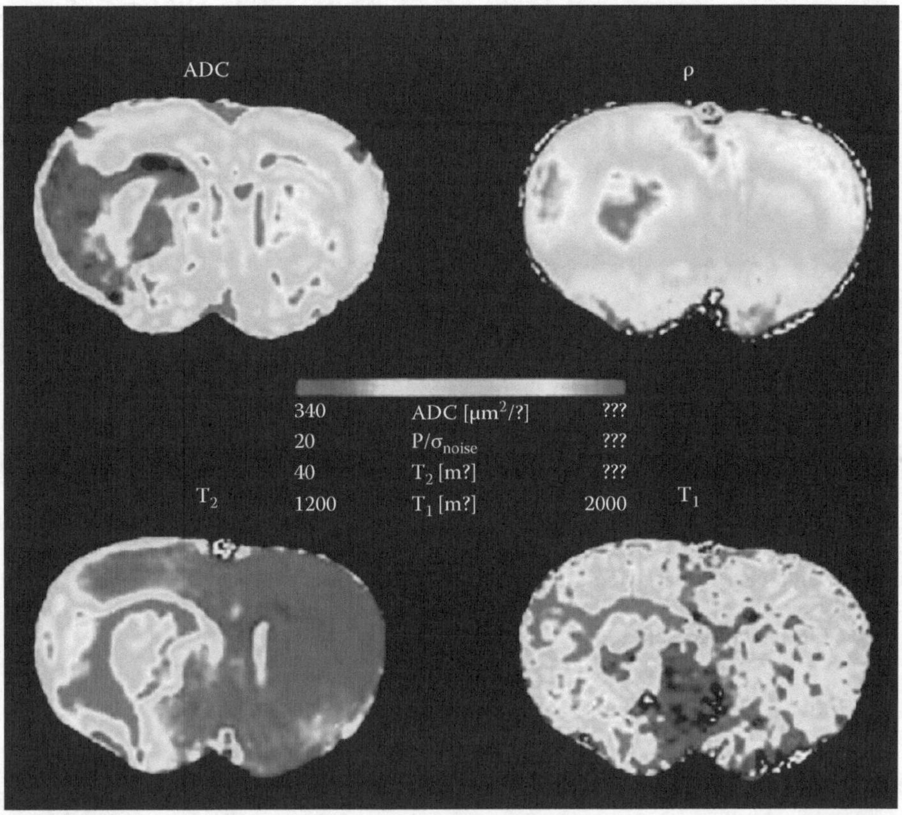

FIGURE 22.8 The damage resulting from a middle cerebral artery occlusion (MCAO) in rat brain can be investigated using several different magnetic resonance imaging (MRI) parameters. Each provides another insight into the changes in the physical environment of the ischemic region. The color-coded parameter maps presented here display the apparent diffusion coefficient (ADC), proton density (P), T2 and T1 relaxation times, all of which show a marked contrast between ischemic (left) and normal brain (right) as well as between the different parameter maps. BioSpec Applications—Stroke. (Courtesy of M. Eis, U. Pschorn, et al., Boehringer Ingelheim Pharma KG, Department of CNS Research, Ingelheim, FRG.) **(See color insert.)**

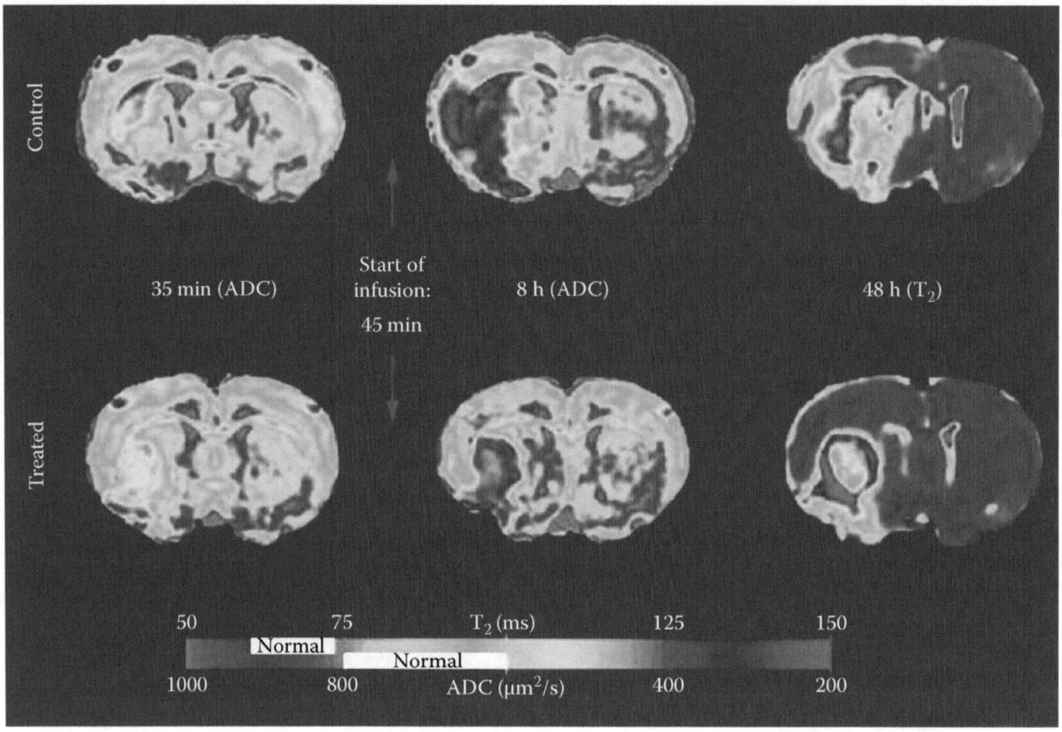

FIGURE 22.9 Rat brains at three time points after suture middle cerebral artery occlusion (MCAO) insult with (bottom row) and without (top row) cerebroprotective drug treatment. The compound reduces the growth of both lesion volume and its severity (note the very different lesion size at the outset). BioSpec Applications—Stroke. (Courtesy of M. Eis, M. Neumaier, and U. Pschorn, Boehringer Ingelheim Pharma KG, Department of CNS Research, Ingelheim, FRG.) **(See color insert.)**

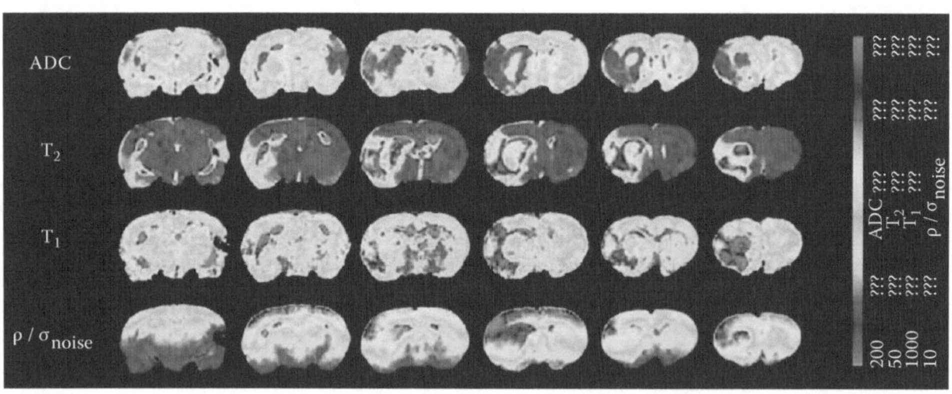

FIGURE 22.10 The extent of ischemic damage was investigated 48 hours after suture middle cerebral artery occlusion (MCAO) in the left hemisphere of a rat using calculated ADC (apparent diffusion coefficient), T2, T1, and proton density maps. The insulted area is divided into a core and an ischemic penumbra, the volumes of which can be assessed by contiguous multislice, full-brain coverage. BioSpec Applications—Stroke. (Courtesy of M. Eis, M. Neumaier, and U. Pschorn, Boehringer Ingelheim Pharma KG, Department of CNS Research, Ingelheim, FRG.) **(See color insert.)**

REFERENCES

1. Blezer, E.L.A. et al., Proteinuria precedes cerebral edema in stroke-prone rats. A magnetic resonance imaging study, *Stroke,* 29, 167, 1998.
2. Brant-Wadzki, M., Weinstein, P., Bartkowski, H., and Moseley, M., MR imaging and spectroscopy in clinical and experimental cerebral ischemia: A review, *AJNR Am J Neuroradiol,* 148, 579, 1987.
3. Donnan, G.A., Baird, A.E., and Christopher, R.L., Diagnosis and imaging of stroke, *J Hypertens,* 12(Suppl 10), S15, 1994.
4. Bradley, W.G. et al., Comparison of CT and MRI in 400 patients with suspected disease of the brain and cervical spinal cord, *Radiology,* 152, 695, 1979.
5. Kertesz, A., Blank, S.E., and Nicolson, L., The sensitivity and specificity of MRI in stroke, *Neurology,* 37, 1580, 1987.
6. McNamara, M.T. et al., Acute experimental cerebral ischemia: MR enhancement using Gd-DTPA, *Radiology,* 158, 701, 1986.
7. Chien, D. et al., MR diffusion imaging of cerebral infarction in humans, *AJNR Am J Neuroradiol,* 13, 1097, 1992.
8. Baird, A.E. and Warach, S., Magnetic resonance imaging of acute stroke, *J Cereb Blood Flow Metab,* 18(6), 583, 1998.
9. Blatter, D.D. et al., Cerebral MR angiography with multiple overlapping thin slab acquisition. II: Early clinical experience, *Radiology*, 183(2), 379, 1992.
10. Chalela, J.A. et al., The hypointense MCA sign, *Neurology,* 28, 58(10), 1470, 2002.
11. Duong, T.Q. and Fisher, M., Applications of diffusion/perfusion magnetic resonance imaging in experimental and clinical aspects of stroke, *Curr Atheroscler Rep,* 6(4), 267, 2004.
12. Fisher, M. and Albers, G.W., Applications of diffusion-perfusion magnetic resonance imaging in acute ischemic stroke, *Neurology*, 10, 52(9), 1750, 1999.
13. Kidwell, C.S. et al., Diffusion MRI in patients with transient ischemic attacks, *Stroke,* 30(6), 1174, 1999.
14. Kidwell, C.S. et al., Magnetic resonance imaging detection of micro-bleeds before thrombolysis: an emerging application, *Stroke,* 33(1), 95, 2002.
15. Medlock, M.D. et al., Children with cerebral venous thrombosis diagnosed with magnetic resonance imaging and magnetic resonance angiography, *Neurosurgery*, 31(5), 870, discussion 876, 1992.
16. Mok, V. et al., Neuroimaging determinants of cognitive performances in stroke associated with small vessel disease, *J Neuroimaging*, 15(2), 129, 2005.
17. Moseley, M.E. et al., Early detection of regional cerebral ischemia in cats: comparison of diffusion- and T2-weighted MRI and spectroscopy, *Magn Reson Med*, 14(2), 330, 1990.
18. Moseley, M.E. et al., Diffusion-weighted MR imaging of acute stroke: Correlation with T2-weighted and magnetic susceptibility-enhanced MR imaging in cats, *AJNR Am J Neuroradiol*, 11(3), 423, 1990.
19. Rincon, F., Anticoagulation and thrombolysis for acute ischemic stroke and the role of diagnostic magnetic resonance imaging, *Arch Neurol,* 61(5), 801, author reply 802, 2004.
20. Runge, V.M., Kirsch, J.E., Wells, J.W., and Woolfolk, C.E., Assessment of cerebral perfusion by first-pass, dynamic, contrast-enhanced, steady-state free-precession MR imaging: an animal study, *AJR Am J Roentgenol*, 160(3), 593, 1993.
21. Schlaug, G. et al., The ischemic penumbra: Operationally defined by diffusion and perfusion MRI, *Neurology,* 22, 53(7), 1528, 1999.
22. Schlaug, G. et al., The ischemic penumbra: Operationally defined by diffusion and perfusion MRI, *Neurology,* 53(7), 1528, 1999.
23. Kidwell, C.S., Evolving paradigms in neuroimaging of the ischemic penumbra, *Stroke*, 35, 2662, 2004.
24. Hillis, A.E., Perfusion-weighted MRI as a marker of response to treatment in acute and subacute stroke, *Neuroradiology,* 46, 31, 2004.
25. Thurnher, M.M., and Castillo, M., Imaging in acute stroke, *Eur Radiol*, 15(3), 408, 2005.
26. Yuh, W.T. et al., MR imaging of cerebral ischemia: Findings in the first 24 hours, *AJNR Am J Neuroradiol,* 12(4), 621, 1991.

23 Administration of Substances and Sampling

Yanlin Wang-Fischer and John McCool

CONTENTS

ADMINISTRATION OF COMPOUNDS OR DRUGS

General Aspects of Administration of Compounds or Drugs

During our *in vivo* rodent experiments, a number of questions have arisen concerning the solutions for injection: Will the pH of the solution affect the animal? What volume is the maximum for a mouse or a rat? Which solvent can be used? Which route should be used for administration and so on? Waynforth and Flecknell (1992)[1] and Yalkowsky (1981)[2] previously addressed topics concerning these and similar questions. To facilitate practical application of their suggestions to laboratory animal research, this chapter summarizes some of their suggestions along with our *in vivo* laboratory animal experiences.

Solution pH for Injection

Rats usually have a wide range of pH tolerance for all routes of administration; a working range is pH 4.5 to 8.0. Solutions of greater acidity can be tolerated orally as long as they do not exceed the equivalent of 0.1*N* HCl, but alkaline solutions are not accepted well by the stomach. The widest tolerance to pH is in the order intravenous > intramuscular > subcutaneous due to the buffering capacity of blood.[1]

Solvents for Injection

1. Normal saline and distilled water: The most common solvents are physiological saline (0.9% sodium chloride) or distilled water. Saline is preferable as it is isotonic with body fluids. Distilled water causes pain by subcutaneous injection and produces some hemolysis by intravenous injection.
2. Phosphate-buffered saline (PBS), balanced salt solution (Hank's BSS), and tissue culture media. These solvents have been used in our animal studies successfully.
3. Sodium acetate (from 20 to 100 m*M*) can be used for intranasal delivery.
4. Some substances may require a more complex solvent to render them suitable for injection. The following materials combined with distilled water or saline can be administered by any of the injection routes:
   a. 60% (volume/volume) propane-1,2-diol (propylene glycol). Propylene glycol is an organic compound (a diol alcohol), usually a tasteless, odorless, and colorless clear oily liquid that is hygroscopic and miscible with water, acetone, and chloroform.
   b. 0.5% (weight/volume) carboxymethyl cellulose (CMC) for insoluble compounds.
   c. 10% (volume/volume) ethyl alcohol.
   d. 10% Tween-80 [polyoxythylene (20) sorbitan monooleate]: Tween-80 (polysorbate 80), a hydrophilic nonionic surfactant commonly used as an ingredient in dosing vehicles for preclinical *in vivo* studies, functions as an emulsifier.
   e. 50% (volume/volume) dimethylformamide (DMF; *N,N*-dimethylformamide) is a clear liquid miscible with water and the majority of organic solvents. It is a common solvent for *in vivo* animal studies.
   f. 50% (volume/volume) dimethylsulfoxide (DMSO) is suitable for less-soluble compounds. DMSO is a highly polar organic liquid used widely as a chemical solvent. Due to its ability to penetrate biological membranes, it is often used as a vehicle for animal studies. It is also used to protect tissue during cryopreservation for histological

staining (see Chapter 20). DMSO shows a range of pharmacological activity, including analgesia and antiinflammatory action.

g. Glycerin (2% to 3%) is a good solvent for intranasal administration.

The concentrations listed here are the maximum that are practicable. Use of concentrations at the lower end of the range is recommended. Although solvents such as ethanol, acetone, benzene, carbon tetrachloride, and DMF can be used in undiluted form, these solvents are extremely toxic and have to be used in minute quantities. Also, their toxicity is dependent on the route of administration. Neat acetone or benzene is often used to apply materials to the skin without apparent harm. Neat solvents should only be used as a last choice.

5. Vegetable oils (e.g., olive oil, peanut oil) are suitable for lipid-soluble substances. Oils cannot be injected intravenously. If it is necessary to inject lipoidal substances intravenously, a 15% (volume/volume) oil-water emulsion is recommended.

Materials can be injected as a suspension. However, dosage may not be precise due to the tendency for the suspended particles to sediment. The suspended particles should be finely divided if the intravenous route is to be used. By adding a drop or two of Tween-80, flocculent suspensions can be more evenly prepared. Following intravenous injection, the particles will be filtered out in the capillary beds of the extremities and the lung, modifying the distribution of injected material and sometimes causing pulmonary distress to the animals. It is recommended that solutions be filtered before intravenous injection.

Injection Volumes

There are numerous guidelines available to assist investigators in selecting appropriate injection volumes including (1) institutional animal care and use committees (IACUCs) in their choice and application of survival rodent dosing and bleeding techniques in the United States; (2) EFPIA (European Federation of Pharmaceutical Industries Associations); and (3) ECVAM (European Center for the Validation of Alternative Methods); EFPIA and EVCAM have provided a *Good Practice Guide to the Administration of Substances and Removal of Blood* (2000) (available at www.eslav.org/efpia.htm). Recommended volumes for injection are summarized in Table 23.1.

TABLE 23.1
Suggested Maximum Volume for Injection into an Adult (>200 g) Rat

Route	Maximum Volume/Site	Comments
Subcutaneous	5 to 10 mL	Neck, back, abdomen. Use smaller than a 21-gauge needle. Large volumes absorbed relatively slowly.
Intramuscular	0.2 to 0.3 mL	Quadriceps, gluteals, triceps. Use smaller than a 22-gauge needle. Large volumes disrupt muscle fibers, not retained within the muscle, and cause pain.
Intraperitoneal	Up to 10 mL	Lower right quadrant of abdomen; use smaller than a 22-gauge needle. Only material in isotonic fluid should be given in large volume. Much smaller volumes should be used for irritant material.
Intravenous	0.5 to 3 mL slowly	Lateral tail or saphenous veins; use a 25-gauge or smaller needle. Inject over 1 to 2 minutes.
Intranasal	25 to 50 µL	25 to 30 µL for one naris; use a small pipette tip; if the amount is over than 30 µL, alternately use two nares 2 minutes apart. pH 4.0 to 8.0.
Intradermal	0.05 to 0.1 mL	Use smaller than a 25-gauge needle.
Tracheal	40 µL	General anesthesia required.
Gavage	5 mL	Use a balled tip needle.

Sources: From References 1 and 9.

These volumes should be reduced if the material is likely to irritate the tissues. The maximum quantity of injection solution also depends on the route of administration, pH of the solvents, and rate of the injection. In some instances (for example, subcutaneous or intramuscular injections), a greater volume can be given by multisite injection. However, for intramuscular injection, no more than 0.2 mL should be administered to each injection site in an adult rat. Very rapid injection (less than 30 seconds), even of the animal's own blood, can produce acute cardiovascular failure and can be lethal.

Absorption of the Injected Substances

Absorption from all sites of administration is dependent on drug solubility and local conditions.

1. Absorption of the solution form is quicker than from the solid form (suspension form).
2. Absorption in the stomach may be slower at a low pH due to precipitation of the drug substance in the gastric fluid.
3. High-concentration dose formulations are absorbed more rapidly than low-concentration formulations.
4. Absorption is virtually instant after intravenous injection.
5. The peritoneal cavity offers a large absorbing surface; the absorption is rapid but about four times slower than intravenous injection.
6. Absorption intramuscularly is quicker than subcutaneously.
7. Absorption by gavage is variable and is dependent on
   a. pH of substance
   b. Solubility
   c. Concentration
   d. Rate of gastric emptying
8. Absorption by epidermis is proportional to the lipid solubility of the drug.

Dose Conversions between Human and Animal

For dose conversions between humans and animals, see Table 23.2. To use this table,

1. If the dose in a dog is 10 mg/kg, the total dose for a 12-kg dog is 12 kg × 10 mg/kg = 120 mg. Referring to Table 23.2, the intersection of a 70-kg human and a 12-kg dog is 3.1. The dose of a human with body weight of 70 kg is 120 mg × 3.1 = 372 mg, or 372/70 = 5.31 mg/kg.
2. If the dose of erythropoietin in a human is 300 IU/kg once daily, the total dose for a 70-kg person is 70 kg × 300 IU/kg = 21,000 IU. Referring to Table 23.2, the cross point of a 70-kg human and a 200-g rat is 0.018. Thus, the dose for a rat with body weight 200 g is 21,000 IU × 0.018 = 378 IU, or 378 × 1000/200 = 1890 IU/kg. For the mouse, the intersection between a 70-kg human and a 20-g mouse is 0.0026; the dose for a mouse with body weight 20 g is 21,000 IU × 0.0026 = 54.6 IU, or 54.6 × 1000/20 = 2730 IU/kg.

The conversion between human and animal is based on the body surface area of each species. A more general conversion can be applied to determine appropriate dosing for rodents from recommended human doses as follows:

The dose for a rat is 6 to 10 times that of a human by kilogram body weight.
The dose for a mouse is 7 to 11 times that of a human by kilogram body weight.

TABLE 23.2
Dose Conversion between Human and Animal*

Unknown / Known	20-g Mouse	200-g Rat	400-g Guinea Pig	1.5-Kg Rabbit	2-kg Cat	4-kg Monkey	12-kg Dog	70-kg Human
20-g Mouse	1	7	12.25	27.8	29	64.1	124.2	387.9
200-g Rat	0.14	1	1.74	3.9	4.2	9.2	17.8	56
400-g Guinea Pig	0.08	0.57	1	2.25	2.4	5.2	10.2	31.5
1.5-kg Rabbit	0.04	0.25	0.44	1	1.08	2.4	4.5	14.2
2-kg Cat	0.03	0.23	0.41	0.92	1	2.2	4.1	13
4-kg Monkey	0.016	0.11	0.19	0.42	0.45	1	1.9	6.1
12-kg Dog	0.008	0.06	0.1	0.22	0.24	0.52	1	3.1
70-kg Human	0.0026	0.018	0.031	0.07	0.076	0.16	0.32	1

* Proportion of body surface of human being to laboratory animals; these data interpreted from a Chinese textbook for graduate students in animal study: *Common Use Biological Data in Experimental Animals*, Hunan Medical University (in Chinese, 1983).

ROUTES AND METHODS OF ADMINISTRATION OF SUBSTANCES

Compounds to be administered to rats can be given through the gastrointestinal tract (orally, for example, in water or feed or by gavage tube inserted into the stomach) or parenteral injection (injected systemically through a variety of routes). The average daily consumption of feed and water for an adult rat is 15 to 25 g and 30 to 50 mL, respectively. The following volumes can be injected into an adult rat safely: 2 to 5 mL subcutaneously, 0.1 to 0.2 mL intramuscularly (0.1 mL per site), 1.5 to 2.5 mL intravenously, 3 to 5 mL into the stomach, 3 to 5 mL intraperitoneally, and 25 to 50 μL or 100 μL/kg intranasally. Usually, 25 to 30 μL per naris is recommended in a rat because rats tend to swallow the solution when more than 30 μL per naris is administered. Alternately, for rats, injection into two nares 2 minutes apart is suggested if the amount is over 30 μL (refer to Table 23.1).

Gastrointestinal Tract

Oral or Per Os

Substances may be admitted orally (per os, p.o.) by addition to the food or drinking water, by use of a capsule or pill, or by instillation into the mouth using a mechanical device, such as a syringe. Capsules or coated pills are rarely used in rabbits or rodents. When used, capsules or pills are placed in the mouth near the back of the tongue, and the animal is induced to swallow by stroking the throat.

Gavage

Only experienced staff may perform the gavage[3,4] procedure.

Equipment

Gavage needles of the correct size for the animal or a flexible, blunt-tip catheter may be used. Metal gavage tubes are available in a variety of lengths and diameters to accommodate the size of the animal to be dosed. Table 23.3 summarizes our recommended gavage needles based on our laboratory experiences using several different types and sizes. Needles can be found at www.vwr.com or other animal supply companies.

TABLE 23.3
Recommended Sizes of Balled Gavage Needles

Species	Body Weight (g)	Gauge	Length (cm)	Ball Diameter (mm)	Shape
Mice	14	24	2.5	1.25	Straight
	15 to 20	22	2.5 to 4	1.25	Straight
	20 to 25	20	2.5 to 5	2.25	Straight/curved
	25 to 35	18	2.5 to 7	2.25	Straight/curved
Rats	50 to 120	20 to 18	2.5 to 4	2.25	Curved
	120 to 200	18 to 16	5 to 7.5	2.25	Curved
	200 to 300	16	7.5 to 10	3	Curved
	300 to 350	14 to 13	7.5	4	Curved

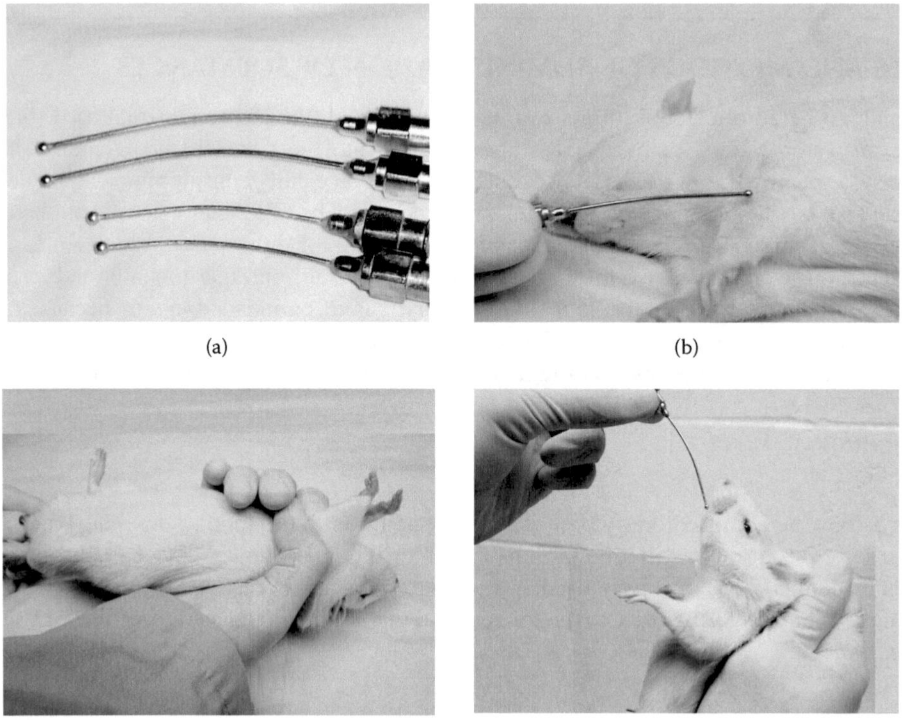

(a) (b) (c) (d)

FIGURE 23.1 Gavage needles (a), measurement of needle length (b), restraint rat by hand (c), and insertion of the needle into the esophagus (d). (Courtesy of Johns Hopkins University, www.jhu.edu/animalcare/rat.htm.)

Procedure

1. Specialized ball-ended feeding needle or flexible catheters should be used (Figure 23.1a). The ball at the end of the gavage needle protects the oropharyngeal tissues and makes inadvertent endotracheal passage less likely.

2. Estimate the distance that the needle needs to be inserted into the rat (Figure 23.1b). It is important that the gavage needle be both the correct length and diameter. Measure the distance between the nose and the last rib (stomach); the needle should be no longer than this. Excessively long or improperly handled stomach tubes may penetrate and rupture the pharyngeal mucosa or the stomach wall. If difficulty advancing the needle is encountered, the bulb may be too large.
3. Use a single-handed restraint technique to hold the rat (Figure 23.1c).
4. Extend the head and neck. Keep the nose, head, and spine aligned so that the esophagus is straight. *Do not* allow the rodent to tip back as you perform the gavage.
5. Use a free hand to guide the gavage needle into the animal's mouth. Position the needle toward the center and run the bulb along the roof of the mouth, then up and over the base of the tongue and gently down into the esophagus. Positioning the tube to the right or left in the mouth can facilitate passage into the esophagus. The rat will usually reflexively swallow as the feeding tube approaches the pharynx, facilitating entry into the esophagus (Figure 23.1d).
6. Advance the gavage tube gently without resistance. If resistance is encountered, the animal struggles, or the animal reacts violently (coughing, gasping), cease advancing the gavage needle and slowly remove it. Accidental insertion of the gavage tube into the trachea or damage to the esophagus may otherwise occur. If you accidentally place the needle in the trachea or lungs, you may drown the animal.
7. Once the desired position is attained, inject the material and withdraw the gavage needle. Gavage volumes are about 10 mL per kg of body weight (approximately 2.5 mL may be administered to a 250-g rat, or a 20-g mouse may have 0.2 mL administered). Monitor the animal after the procedure to ensure that there are no adverse effects. Any signs of distress, such as gasping and frothing at the mouth or nose, may indicate injury or inhalation of the foreign materials.

Suggestions

The gavage technique is best done in unanesthetized animals with an intact swallow reflex. However, light anesthesia may be required in some larger animals. This technique is similar for the rat, mouse, guinea pig, and rabbit.

PARENTERAL

Parenteral routes of administration include injections into various compartments of the body. Sites used for collection of blood from veins may also be used for intravenous administration. Intraperitoneal administration is one of the most frequently used parenteral routes in rodents. In stroke studies, cistern magna, brain ventricles, and intranasal are also common sites for giving substances. Materials given intramuscularly must be given in small volumes. Absorption via this route, however, is more rapid than via subcutaneous administration. Regardless of the route to be used, it is essential that the animal be securely restrained to avoid injury to personnel and to animals. The investigator should know the physiological properties of the substance for injection. Considerable tissue damage and discomfort can be caused by irritating vehicles or drugs. Maximum volumes of administration have been discussed (refer to Table 23.1).

Intravenous

Tail Vein Injection

Anatomy of the Tail Vessels

There are two lateral veins, one ventral artery, and a dorsal vein in the tail.[5] Some may have two veins and two arteries located laterally and centrally, respectively (see Figure 23.2).

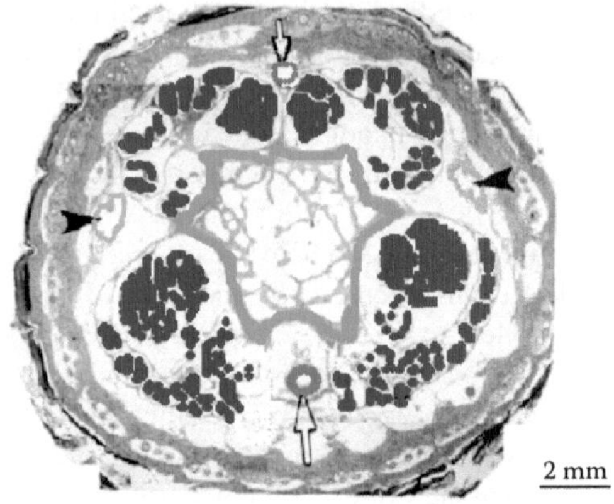

FIGURE 23.2 Dorsal structure of vessels at the middle section of the rat tail. Arrowheads indicate the veins, and arrows indicate the dorsal vein or arteries.

Equipment

Depending on the size of the rat, 30- to 25-gauge needles 0.5 to 1 inch long can be used. A rat restrainer and warming lamp or warm water is also important. The techniques described for rats also apply for mice. The animal can be restrained in a cylindrical restrainer (Figure 23.3a), in a cage lid (Figure 23.3b), or in a rodent triangular restraint bag (Figure 23.3c). Light anesthesia with 1% to 2% isoflurane is helpful for restraint. Prolonged intravenous administration/sampling may be accomplished by jugular vein or tail vein catheterization. This requires a surgical approach.

Intravenous injections are usually made into the lateral tail vein. The tail should be warmed prior to performing the procedure either by massaging it with warm, wet gauze or by placing the animal under a heating lamp. The tail vein is easier to see in nonpigmented rats. A fine-gauge needle (25 or 27 gauge) should be used for this procedure. Personal preference is a 27-gauge butterfly needle for this procedure.

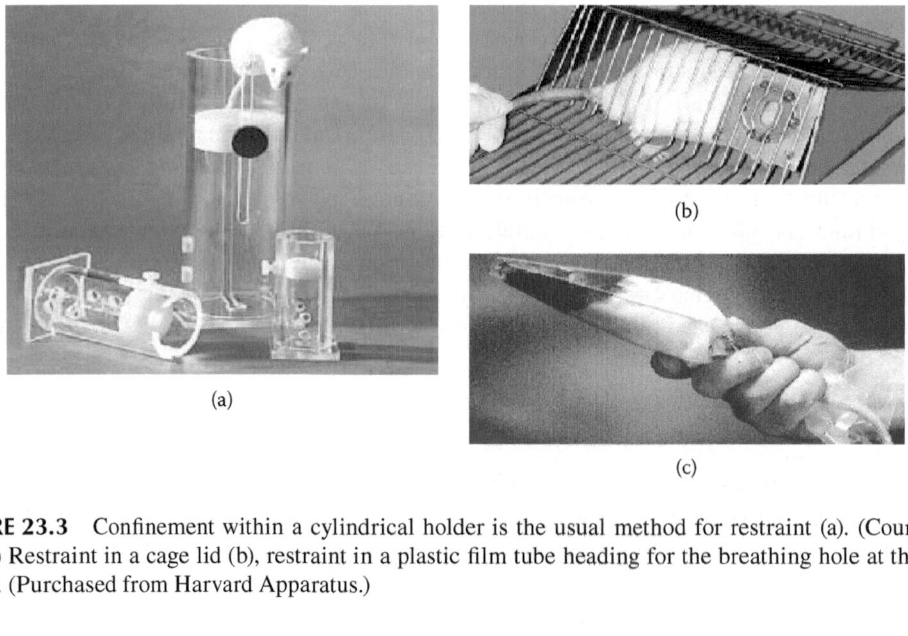

FIGURE 23.3 Confinement within a cylindrical holder is the usual method for restraint (a). (Courtesy of VWR.) Restraint in a cage lid (b), restraint in a plastic film tube heading for the breathing hole at the small end (c). (Purchased from Harvard Apparatus.)

Procedure for Tail Vein Injection

1. Weigh the animal prior to the injection. The injection volume will be determined by the body weight of an animal. A rule of thumb is that the injection volume should be no more than 2 mL in each bolus injection.
2. Blood flow to the tail vein can be improved to facilitate the procedure by initially placing the rat in a cage, chamber, or bowl with a heating lamp above it (about 50 cm) or by massaging the tails with warm, wet gauze. Care should be taken not to overexpose the animals to the heat source, both with regard to time and intensity, as dermal burns may occur. More than one animal can be warmed at a time.
3. A rat is transferred into a holding device, which restrains the rat while allowing access to the tail vein (see Figure 23.3).
4. The lateral tail vein is identified on either side, the solution is injected using the smallest syringe possible (with a 25- or 27-gauge butterfly needle for rats). The needle is inserted into the vein at about a 15° to 30° angle to the tail skin. Start at the tip of the tail and move closer to the tail base if you need to stick the animal more than once.
5. The operator should be able to see the needle inside the vein or at least see the blood flow back into the syringe hub. The dose formulations should be visualized as it is injected into the vein.
6. After removing the needle, apply gentle pressure to the injection site with gauze for 1 to 3 minutes or until bleeding stops. The rats can immediately be returned to their cages after injection.

Cannulation of Tail Vein for Long-Term Intravenous Infusion

Equipment

Polyethylene PE-10 tubing (about 30 cm long, VWR, Cat. No. 63019-003) connected to PE-50 polyethylene tubing (15 cm long, VWR, Cat. No. 63019-047) by 3M superstrong adhesive is recommended for this procedure. PE-50 tubing is connected to a 23-gauge needle and a syringe. The PE-10 tubing is advanced through the blunt end of a 20-gauge needle, leaving the sharp end exposed for insertion into the animal's vein (Figure 23.4).

Procedure

1. A rat is warmed in a warm box with a heating lamp for 15 minutes.
2. Restrain the rat so that the tail is protruding out of the restraint apparatus. The rat can be under slight anesthesia with isoflurane to help restrain it.
3. Mark the veins and arteries with a small ink mark at the base of the tail.
4. A rubber band is place at the base of the tail as a tourniquet. Place the rat in lateral recumbence to expose the lateral vein.
5. The 20-gauge needle tip (with PE-10 tubing inside the needle) is inserted into the skin and vein at an angle of about 15° to 30° to the tail skin at the terminal third of the tail.
6. Successful entry in the vein is verified by a observing the flow of blood into the tubing.
7. Holding the needle in place, advance the PE-10 tubing into the vein until resistance is met (reach the tail base). Blood should flow into the PE-10 tubing during insertion of the

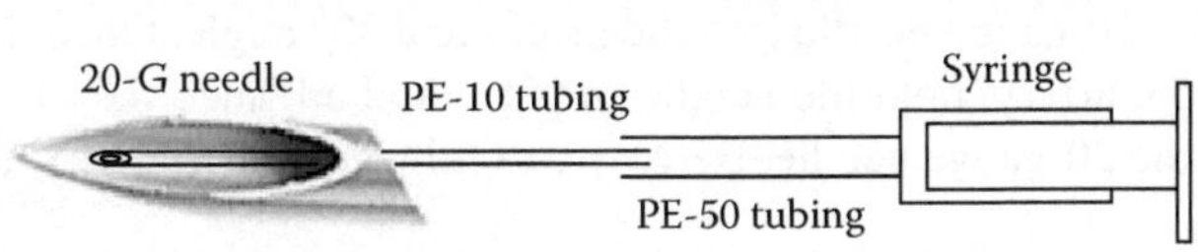

FIGURE 23.4 Structure of the tail vein cannulation set. PE-10 tubing is inserted into a 20-gauge needle; the other end is connected to PE-50 tubing, which is attached to a syringe.

FIGURE 23.5 The tail of the rat is placed in a metal spring to protect it from biting.

tubing. After successful insertion, loosen and remove the rubber band. Pull out the 20-gauge needle, leaving only the PE-10 tubing in the vein. The other end of the PE-10 tubing is connected to PE-50 tubing attached to a syringe.

8. Wrap the tail and the PE-10 tubing with tape to keep the tubing in the vein. The tail is placed in a metal spring to protect it from being bitten (Figure 23.5). If blood flow into the tubing is not observed during the insertion, the tubing is likely not in the vein. The procedure should be repeated at a location 0.5 to 1 cm cranial to the first attempt.

Tail vein cannulation can be used to collect blood samples by placing the PE tubing into a collection tube. The cannula can be maintained for future infusion; PE-10 and PE-50 tubing must be filled with heparinized saline (9 U/mL), and the open end of PE-50 tubing should be closed by flame heat. Rats are returned to a special cage with their tails kept outside through a hole in the back of the cage. The cannulation can be kept for 2 to 3 days.

Cannulation of the Tail Artery

Equipment

Equipment for cannulation of the tail artery is the same as for tail vein cannulation.

Procedure

1. A rat is placed in a face-down prone position or face-up supine position to expose tail arteries.
2. Either about 0.1 mL lidocaine is injected intradermally into the incision area for local anesthesia or the rat is anesthetized with 2.5% isoflurane.
3. A 1-cm-long and 0.5- to 1-mm-deep incision is made at the terminal third of the tail. Care must be taken not to cut too deep.
4. The artery is on the surface of the tailbone just under the skin between tail tendons.
5. Carefully dissect the artery and place a curved microdissecting forceps under the artery.
6. Two silk threads are placed under the artery.
7. Insert the tip of a 20-gauge needle into the artery at a 30° angle. Once the needle has been advanced into the artery, hold the needle in place and advance the PE-10 tubing into the artery through the 20-gauge needle. Be very careful when pushing the tubing so the artery is not ruptured.
8. Successful entry in the artery is verified by observing the flow of blood into the tubing. After observing blood flow into the PE-10 tubing, pull the needle out of the artery and continue to push the PE tubing into the artery to the base of tail (resistance will be felt).

9. Tighten the silk threads on the artery over the tubing and seal the wound with Tissumend (tissue glue) to stop bleeding.
10. Inject 1 mL heparinized normal saline (9 U/mL) into the tubing and seal the end of the PE tubing by melting it near a flame or make a node at the end.
11. Wrap the tail and the PE-10 tubing with tape to keep the tubing in the artery.
12. The rat is placed back into its home cage with the tail in a metal spring that is pulled out through the hole at the back of the cage to protect from biting. The catheter can be kept for 2 to 3 days.

Jugular Vein Cannulation

The right jugular vein is generally chosen for the cannulation. External jugular veins are very shallow and lie between skin connective tissue and sternocephalic muscle.

Equipment

- Silicone catheter, 0.025-inch inside diameter (VWR, Cat. No. 60985-702), surgical instruments (see Chapter 10)
- 60% polyvinypyrrolidone (PVP) (Cat. No. PVP-10, Sigma, molecular weight 10,000) prepared by mixture of 5 g PVP, 2.75 mL normal saline, and 0.5 mL 10,000 IU heparin to yield a final volume of about 8.3 mL 60% PVP
- Heparin (J.A. Webster, 10,000 U/mL, 4 mL/bottle, Cat. No. 560570)

Anesthesia

Rats can be anesthetized with a mixture of ketamine (25 mg/kg) and xylazine (2.5 mg/kg) by intraperitoneal injection

Surgical Procedure

1. Place eye cream on both eyes of a rat to protect the eyes from drying. Shave the ventral and back neck areas such that the prepared areas are two to three times larger than the surgical incision; local skin is aseptically cleaned by 0.5% Betadine and 75% alcohol or 2% chlorhexidine.
2. Place the rat in a face-up supine position with the head, four legs, and tail fixed on a surgical board with adhesive tape.
3. A 2-cm midline incision is made at the ventral neck.
4. The tissues are gently blunt dissected to expose at least a 1-cm length of the jugular vein.
5. A pair of slightly curved microforceps is placed under the vein; two silk threads are placed under the vein.
6. The cephalic end of the jugular vein is ligated to occlude the blood flow; the thread is fixed on the board to give the vein slight tension.
7. The thread at the cardiac end is tied loosely on the vein; the forceps is removed.
8. Being careful not to cut through the whole vein, a tiny diagonal hole toward the heart is made on the vein by a pair of microspring scissors.
9. Hold the incision open with a pair of microdissecting tweezers, insert a silicone catheter into the tiny hole, and gently advance it into the vein. The length of the catheter in the vein is dependent on animal body weight (100- to 200-g rat inserted 2.7 cm, 250- to 350-g rat inserted 3.5 cm). The other end of catheter is connected to a syringe filled with heparinized saline (9 U/mL) (Figure 23.6a).
10. The cranial side of the ligature is then tied on the vein and catheter. Verify that blood can flow into the catheter after the ligature. Flush the vein and catheter with heparinized saline.

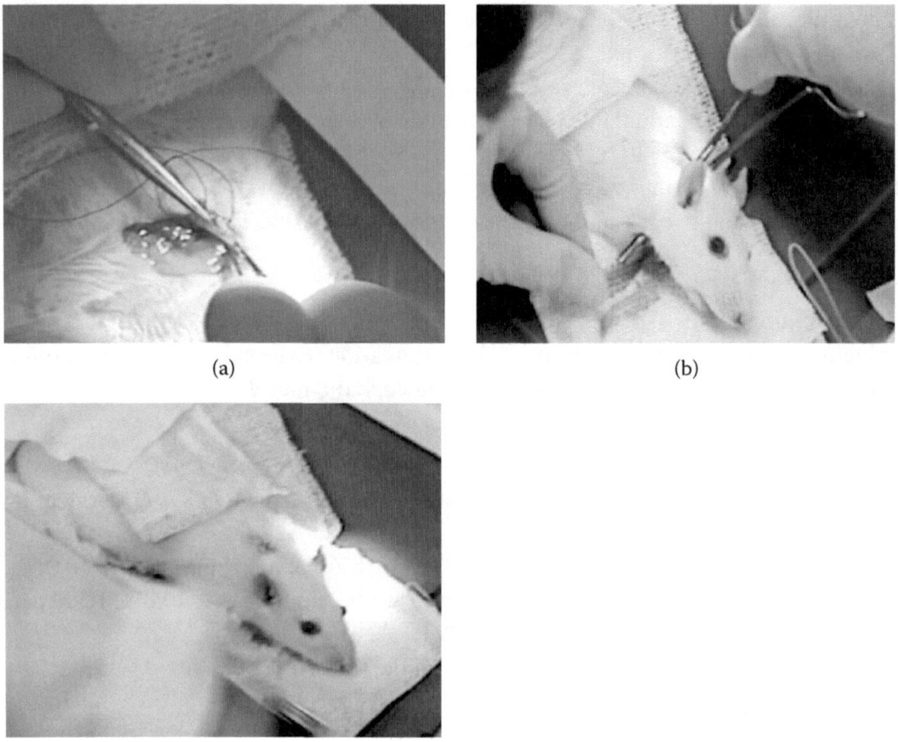

(a) (b) (c)

FIGURE 23.6 Procedure for right jugular vein cannulation: A silicon catheter was inserted into the jugular vein (a); the end of catheter was subcutaneously drawn through to the ventral incision (b) and remains free in the back of the neck (c).

11. A suspension of 60% PVP-heparin-saline (500 to 625 IU/mL heparin) is injected into the catheter; make a tiny air bubble between heparinized saline and PVP layers so that you can see the PVP enter the catheter close to the incision. Stop the injection when the air bubble reaches the ligature so the PVP stays in the catheter and does not enter into the blood flow.
12. The free end of the catheter is sealed.
13. The rat is turned to a prone position with face down.
14. A midline incision (1 cm long) is made on the back of the neck.
15. A curved forceps is inserted subcutaneously toward the side of ventral neck incision, and the free end of catheter is grasped and pulled across the left side of the neck back out through the ventral neck incision (Figure 23.6b).
16. Administer 0.25% bupivicaine topically to the local incision sites to relieve postoperative pain.
17. The incisions are closed by suture or clip with the catheter free outside (Figure 23.6c).
18. Furazolidone powder or another antibiotic is sprayed on the incisions to protect from infection.

Bolus Injection

In most studies using the intravenous route, the test substance is given over a short period, generally less than 1 minute. Such relatively rapid injections require the test substance to be compatible with blood and not too viscous. When large volumes are required to be given, the injection material should be warmed to body temperature. The rate of injection is an important factor in intravenous administration, and it is suggested that, for rodents, the rate should not exceed 3 mL/minute.

TABLE 23.4
Circulating Blood Volume in Laboratory Animals

Species	Total Blood Volume (mL/kg) (6% Body Weight)	Maximum Blood Volume for Survival (15% Blood Volume)	Examples
Mouse	63 to 80	12 mL/kg	0.3 mL in a 25-g mouse
Rat	58 to 70	9 mL/kg	2.25 mL in a 250-g rat
Rabbit	44 to 70	9 mL/kg	2.7 mL in a 300-g rabbit

Slow Intravenous Injection

Because of the expected clinical application of the compound or because of limiting factors such as solubility or irritancy, it may be necessary to consider administering substances by slow intravenous injection. Typically, different techniques would be applied for slow injection to minimize the possibility of extravascular injection of material. For slow intravenous injection over the course of 5 to 10 minutes, a butterfly needle can be used, or better still an intravenous cannula may be taped in place in a superficial vein (short term) or surgically placed some time prior to use (longer term or for multiple injections).

Continuous Infusion

For reasons of solubility or clinical indication, it may be necessary to consider continuous infusion, but careful consideration is needed if infusions are prolonged. The volume and rate of administration will depend on the substance being given and take account of fluid therapy practice. As a guide, the volume administered on a single occasion will be less than 10% of the circulating blood volume over 2 hours. Information on circulating blood volumes is available in Table 23.4.

Intraperitoneal

Equipment

Needles with 23 to 25 gauge and 5/8 to 1 inch are recommended. Restraint is best accomplished with light anesthesia or by having a second person hold the rat in a head-down, stretched-out position to avoid injury to internal organs or major blood vessels.

Procedure

1. Intraperitoneal injections are usually made in the lower right quadrant of the abdomen.
2. A rat is restrained with its head tilted lower than the body.
3. After swabbing the lower right quadrant with alcohol, a fine-gauge needle is introduced slowly through the skin, subcutaneous tissue, and abdominal wall.
4. Withdraw the syringe plunger to ensure that the needle is not in the bladder or intestines.
5. If nothing is withdrawn, inject the material slowly. If the needle accidentally entered the bladder or intestines, withdraw and discard the needle and syringe and repeat the procedure using uncontaminated materials.

Intramuscular

Equipment

A 25- to 26-gauge, 1/2- to 5/8-inch needle with a 1-mL syringe is recommended. For rat restraint, see Figure 23.1c.

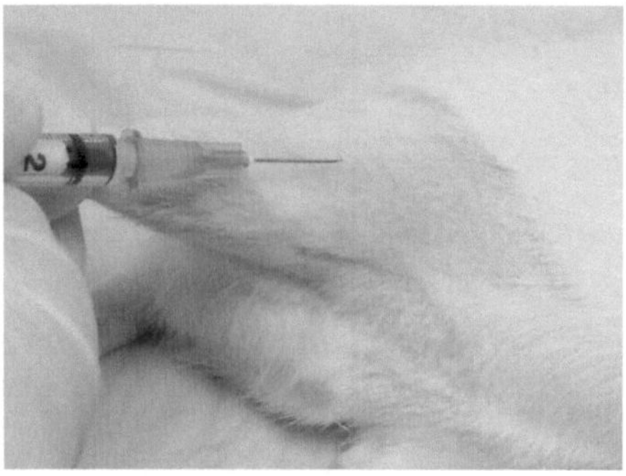

(a)

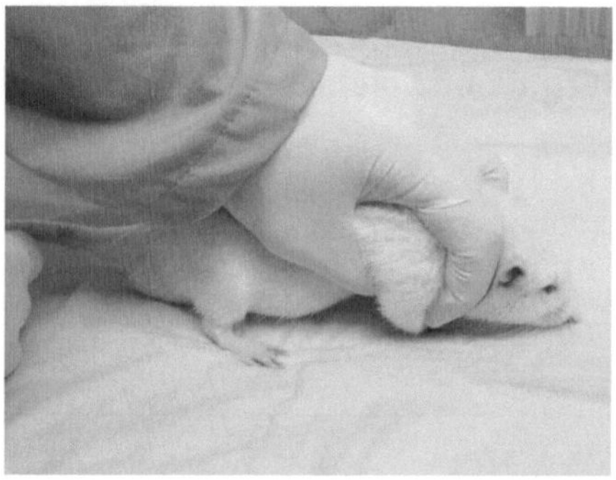

(b)

FIGURE 23.7 Intramuscular injection at the anterior thigh muscle (a); restraint of rat for subcutaneous injection (b). (Courtesy of Johns Hopkins University, www.jhu.edu/animalcare/rat.htm.)

Procedure

1. The back and hind leg muscles are used. Intramuscular injections are usually not recommended in rats because of the small muscle mass.
2. A fine-gauge needle should be used to make injections in the anterior thigh muscle (Figure 23.7a).

Subcutaneous

Equipment

A 23-gauge, 1-inch needle is recommended. For rat restraint, see Figure 23.7b.

Procedure

1. Subcutaneous injections are usually made into the loose skin over the neck or flank. Be sure to use adequate restraint. The rat skin is thick and difficult to penetrate. Care should be taken to avoid accidental human injections.
2. Tent the skin and insert the needle 5 to 10 mm through the skin before making the injection. Verify that the needle has not exited through the "back end" of the injection site. Lack of resistance to the injection is indicative that you are in the right location.
3. Check for leak back, especially if a larger volume is injected.

Intradermal

Equipment

A 25- to 27-gauge, 1-inch needle is recommended. The rat is restrained as discussed.

Procedure

1. Intradermal injection is typically used for assessment of immune, inflammatory, or sensitization response. The location is dependent on the study purpose.
2. Material may be formulated with an adjuvant. Volumes of 0.05 to 0.1 mL can be used depending on the thickness of skin.
3. Cleanse area, hold skin taut, and inject fluid (bevel side up). A small bubble just beneath the skin will be evident (see Figure 23.8).

Intranasal Administration

Background

The blood–brain barrier (BBB) presents a major challenge in developing a new drug for neurological diseases or injuries because it prevents a number of potential therapeutic agents from reaching the brain. Intracerebroventricular (ICV) administration has been used for animal studies; however, it may not be practical in humans as it requires surgery with associated risks of infection and other

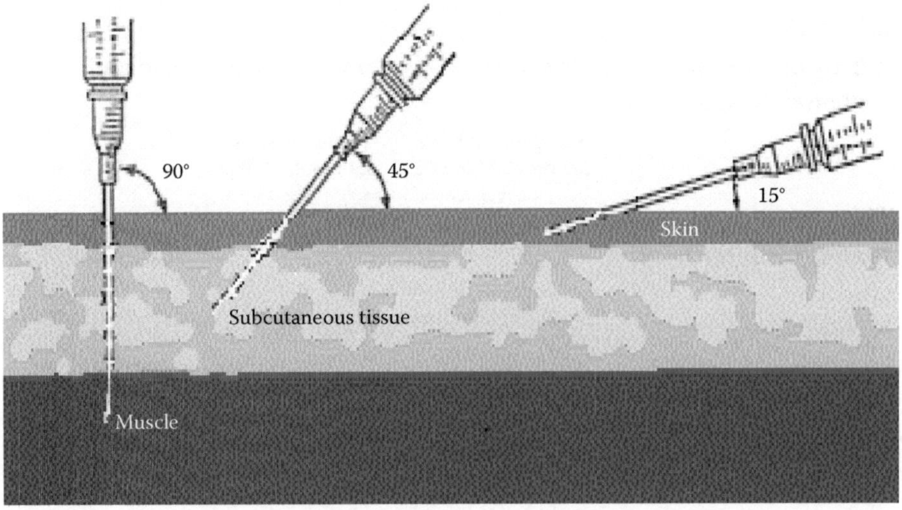

FIGURE 23.8 A schema shows the intradermal, subcutaneous, and intramuscular injections.

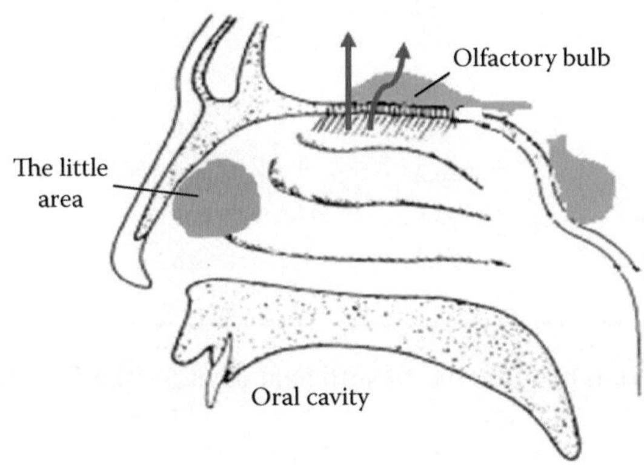

FIGURE 23.9 Olfactory neural pathway providing both intraneuronal and extraneuronal pathways into the brain. The extraneuronal pathway delivers drug directly to the brain parenchymal tissue, to the cerebrospinal fluid (CSF), or to both within minutes. Drugs also can be delivered into blood after intranasal dosing through the nasal mucosa.

complications. Intranasal (IN) administration is a noninvasive method of bypassing the BBB and delivering drugs to the brain directly from the nasal cavity along pathways that seem to be associated with the peripheral olfactory and trigeminal systems[6,7] through both intraneuronal and extraneuronal pathways. The extraneuronal pathway delivers drug directly to the brain parenchymal tissue and the cerebrospinal fluid (CSF) within minutes. The intraneuronal pathway involves axonal transport and requires hours to days for drugs to reach different brain segments (see Figure 23.9).[6,7] Also, as the nose is rich in capillaries, this is a viable route for delivery of drugs into the circulatory system.

Intranasal administration seems a very promising way to administer neurological treatments. It has been reported that the olfactory neuroepithelium is the only area of the body in which an extension of the central nervous system (CNS) comes into direct contact with the external environment.[8] Thorne, Frey et al. (2000)[6] reported evidence for delivery of insulin-like growth factor 1 (IGF-1) to the CNS from the nasal cavity along both olfactory and trigeminal pathways. They radiolabeled IGF-1, delivered it intranasally, and subsequently found rapid delivery (within 30 minutes) of the radiolabel to the CNS, with concentrations in the nanomolar range in the olfactory bulb, frontal cortex, hippocampal formation, cerebellum, and brainstem. Nanomolar concentrations of IGF-1 have been shown to provide protection against neuronal injury such as hypoglycemia.[6,9] It is suggested

TABLE 23.5
Comparison of Intranasal Delivery of 125-I-Hypocretin-1 in Awake and Anesthetized Mice

	Anesthetized Mice 5.5 nmol, 15.6 μCi n = 5		Awake Mice 6.9 nmol, 19.8 μCi n = 4		
Tissue Concentration (n*M*)	**Means**	**SE**	**Means**	**SE**	**Ratio of Anesthetized/Awake**
Blood	31.1	4.7	140.2	16.4	0.2
Olfactory epithelium	16051	3618	647.5	320.6	24.8
Olfactory bulbs	324	83	24.9	11.5	13.0
Anterior olfactory nucleus	126	43	16.6	5.8	7.6
Trigeminal nerve	1007	301	91.5	21.8	11.0
Frontal and parietal cortex	55	21.9	9.7	1	5.7
Striatum	13.5	2.8	5.5	0.5	2.5
Septal nucleus	13.5	1.5	7.4	1.2	1.8
Thalamus	22.9	7.5	7.4	0.5	3.1
Hypothalamus	61.1	19.2	13	1.3	4.7
Hippocampus	20.6	4	7.3	0.6	2.8
Brainstem	33.2	11.4	9	0.6	3.7
Cerebellum	26.2	7.4	10.2	0.9	2.6
Upper cervical spinal cord	95.8	21.5	22.7	2	4.2
Lower cervical spinal cord	24.8	6.5	15.6	4.1	1.6

Note: SE, standard error.

that intranasal administration may produce a sufficient level for the CNS to protect against neurological disease or injury.

Procedure for Intranasal Administration of Drugs in Rats

1. A rat is sedated with 3% isoflurane in O_2 for about 1 minute before intranasal dosing or with other anesthetics such as pentobarbital, dependent on study purpose. If the study intent prohibits the use of anesthetics, intranasal administration can be performed on unanesthetized rats, but the variation will be greater due to the activity of swallowing and sneezing by rats following intranasal administration. The exposure levels in the brain and blood are about four to five times larger in rats under sedation during the intranasal dosing than rats dosed awake (our unpublished data). Also see Dr. Frey's data in mice (Table 23.5).
2. A rat is placed on its back, and a total of 25 μL drug solution or 100 μL/kg can be given as nose drops by a pipette tip. Volumes greater than 30 μL (100 μL/kg in a big rat) may be administered by alternating drops every 2 minutes between the left and right nares (see Figure 23.10).
3. The mouth is held closed during the administration to ensure that the drops are naturally inhaled into the upper nasal cavity.
4. The pH of solution can be 4.0 to 8.0.

Intranasal Administration in Mice

The method of intranasal administration of drug formulations to rats can also be applied to mice. Mice can be sedated with isoflurane (the same as rats) or restrained with a hand (no anesthesia); in this case, some drug may be swallowed during the administration. About 10 to 15 μL of solution per mouse or 500 μL/kg can be administrated into nares with a small pipette tip.

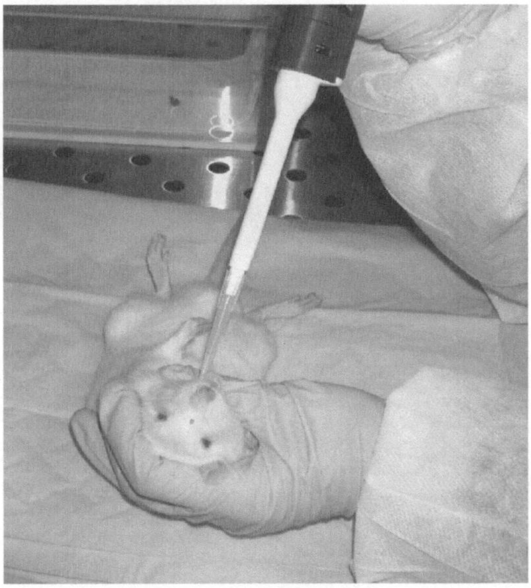

FIGURE 23.10 Position of the rat for intranasal administration of drug.

Dr. Frey's group[10] compared intranasal administration of drug on anesthetized mice and unanesthetized mice. They found that intranasal administration of drug on anesthetized mice had better drug delivery to the brain. We abstracted some of their data and summarize it in Table 23.5 (courtesy of Dr. Frey's group).

Problems and Limitation of Intranasal Delivery

1. The position of rats during intranasal administration is very important. Figure 23.11 shows the relationship of nose anatomy in different positions of rats. A rat on its back facilitates intranasal delivery of drug into brain.
2. Swallow and sneeze activity during IN dosing: Rats have spontaneous swallow and sneeze activity in response to intranasal stimulant while they are awake. Based on our experience with intranasal administration to rats, about 85% of drug could be swallowed following application to an unanesthetized rat. Animals under anesthesia lack this swallow response. There are different anesthetics available for this purpose, such as pentobarbital, ketamine, and isoflurane; researchers can choose the one to fit their study purpose. We use isoflurane due to its quick induction of sleep (in 1 to 2 minutes) and quick waking up (30 seconds to 1 minute), which fits our study purpose. Isoflurane has very low solubility in blood and body tissues, and the lungs eliminate most of the inspired isoflurane (refer to Chapter 8). Our studies have shown that isoflurane has no effect on infarct size, food intake, foot fault test, body temperature, and body weight (our unpublished data).
3. Only certain drugs can be administered intranasally.
4. If applied to unanesthetized animals, care should be taken to prevent mechanical damage to their nose cavity from struggling.
5. A quick way should be available to assess nasal tolerability since some compounds may damage the nasal mucosa.

Intracisternal

For intracisternal administration, refer to Chapter 19 on spinal fluid collection.

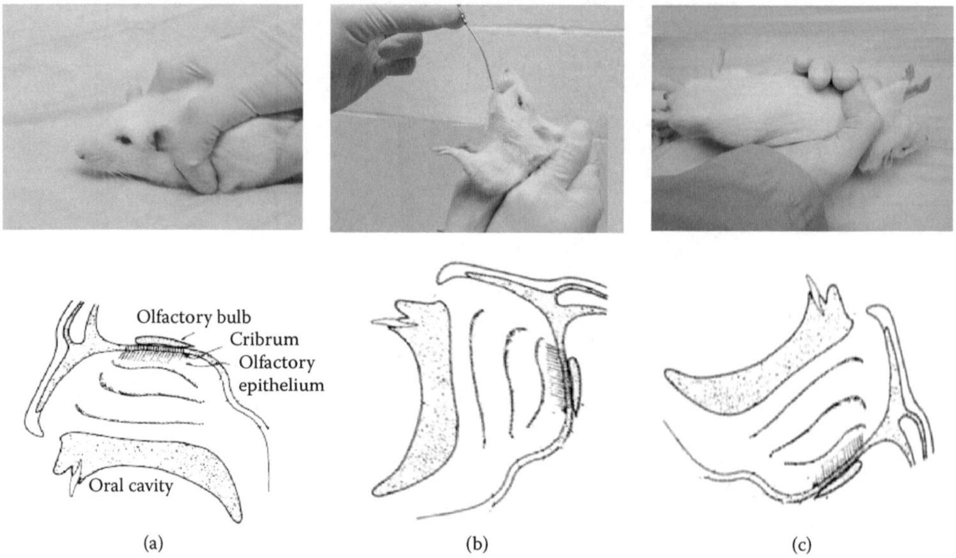

FIGURE 23.11 Relationship of nasal anatomy in different positions of the rat. A rat in its natural position (a) or a rat in its straight down position (b) does not facilitate intranasal delivery of drug from nose to brain. A rat laying on its back facilitates intranasal delivery of drug into brain (c).

Concept of Microperfusion of the Brain

The concept of microperfusion in the brain is to deliver substances through the external carotid artery to the internal carotid artery directly to the infarct areas in the brain. The advantages of this technique are as follows:

1. The first metabolic degeneration of compound by the liver, as would be the case with general intravenous injection of drug, would be avoided.
2. If a compound has a very high binding power to serum albumin, using microperfusion administration can lessen serum protein binding due to the short duration of time spent in the blood prior to reaching the target.
3. It avoids systemic toxicity from the drug or compound.

A drawback to microinfusion of the brain is that it is not very suitable for clinical patients, but it is a good way for studying compound efficacy research in animals.

Minipump Implantation

Minipump implantation utilizes an ALZET, or equivalent, osmotic pump with catheter implanted into a vein (jugular or femoral) for long-term infusion. Detailed information about the pump is available at www.alzet.com. This section, adapted from their information with their permission plus our own experience in use of the pumps, is intended to help investigators understand this technique.

Introduction

ALZET osmotic pumps are miniature pumps that continuously deliver test agents at controlled rates in mice, rats, and other laboratory animals. When implanted subcutaneously or intraperitoneally, these pumps serve as a constant source for prolonged drug delivery. Continuous delivery allows the effects of test agents to develop fully and reproducibly, especially when an agent has a short half-life. In addition to systemic administration, targeted delivery can be conducted by directing drug solutions to an area remote from the site of implantation. This is accomplished by attaching a catheter to

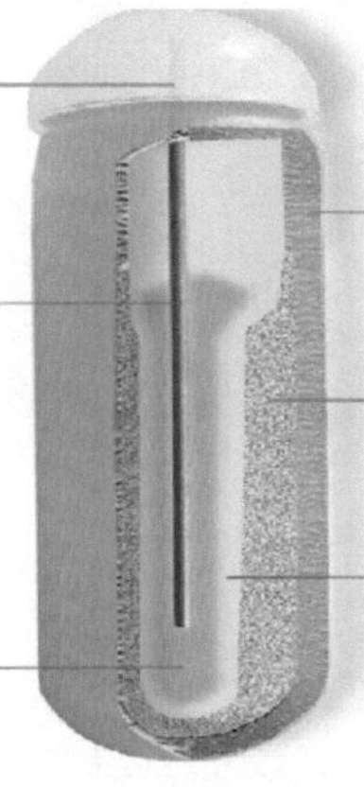

FIGURE 23.12 Structure of an osmotic minipump showing how the pump works. (Courtesy of Durect, ALZET Osmotic Pumps, www.alznet.com, 2006.)

the pump. For example, jugular vein infusion, brain infusion, or microperfusion can be accomplished with a minipump. The information discussed next was adapted from ALZET with permission.

Principle of Operation

ALZET osmotic pumps are composed of three concentric layers with the following characteristics (Figure 23.12):

1. Drug reservoir
2. Osmotic sleeve
3. Rate-controlling capability
4. Semipermeable membrane

"An additional component, called the flow moderator, is a 21 gauge stainless steel tube with a plastic end-cap. This is inserted into the body of the osmotic pump after filling. The drug reservoir, the innermost compartment of the pump, is a cylindrical cavity molded from a synthetic elastomer. The reservoir wall is chemically inert to most aqueous drug formulations, and dilute acids, bases and alcohols (for more information on compatibility, see 'Vehicles and Agents Delivered with ALZET Pumps' at http://www.alzet.com). It is also impermeable, blocking any exchange of material between the drug reservoir and the surrounding osmotic sleeve.

"Outside the reservoir is the osmotic sleeve, a cylinder containing a high concentration of sodium chloride. It is the difference in osmotic pressure between this compartment and the implantation site that drives the delivery of the test solution. The osmotic agent maintains constant osmotic activity during the lifetime of the pump, resulting in a consistent rate of delivery.

"Water crosses the semi-permeable membrane and adds bulk to the salt sleeve, which causes compression of the flexible reservoir. This results in the delivery of test solution through the flow moderator at the same rate that water enters the pump. The rate at which water enters the osmotic sleeve is regulated by the water permeability of the semipermeable membrane, its dimensions, and the osmotic pressure difference across the membrane. Consequently, the delivery profile of the pump is independent of the drug formulation dispensed. Drugs of various molecular configurations, including ionized drugs and macromolecules, can be dispensed continuously in a variety of vehicles at constant rates."

Rate and Duration

While the volume delivery rate of the pump is fixed, different mass delivery rates of test agents are achieved by varying the concentration of agent in the solution or suspension used to fill the

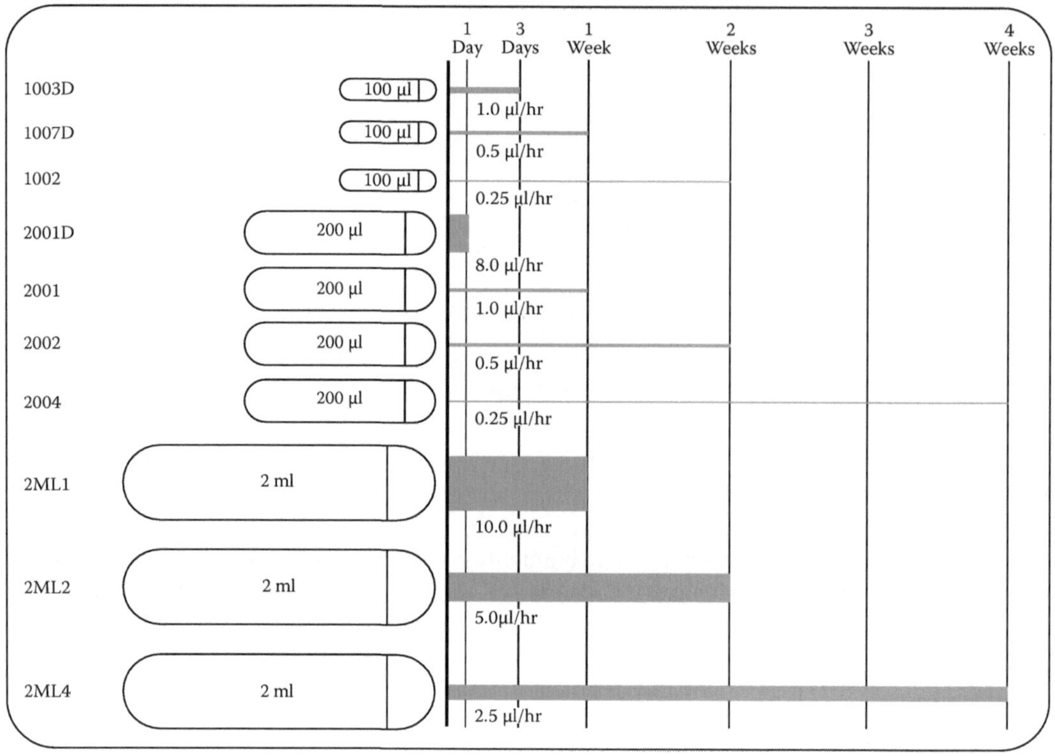

FIGURE 23.13 ALZET osmotic pump models. (Courtesy of Durect, ALZET Osmotic Pumps, www.alznet.com, 2006.)

pump. Figure 23.13 shows different size pump models, and Figure 23.14 shows the actual size of the pumps.

Pump Performance

"Each lot of pumps is tested in DURECT laboratories to determine the exact pumping rate and reservoir volume and to ensure accurate compound delivery. DURECT estimates the *in vivo* pumping rate of ALZET osmotic pumps by measuring their pumping rate *in vitro* in 0.9% saline at 37°C (±0.5°C). This *in vitro* testing method gives a good measure of reproducibility over time, both within pumps and between pumps, and allows an estimation of the pumping rate to be expected in homeothermic animals for which 0.9% saline is isotonic. For example, in rats and mice the mean pumping rates of subcutaneously or intraperitoneally implanted osmotic pumps are within 5% of the *in vitro* rate. The *in vivo* and *in vitro* pumping rates of osmotic pumps are specified for 37°C operating conditions. Both temperature and osmolality affect the rate at which water crosses the semipermeable membrane and enters the osmotic sleeve."

The functional verification, selecting vehicle, filling, troubleshooting, priming, and more, are described in the manufacturer's instructions (see Figure 23.15).

Brain Injection on Rats

Background

Neuron cells are grouped in the brain by function. Inserting a probe into certain brain locations (including brain ventricular space) allows scientists to research neurological problems in different location of the brain.

Equipment

Equipment includes an anesthesia device, drug, stereotaxic equipment, and an 11-gauge guide probe.

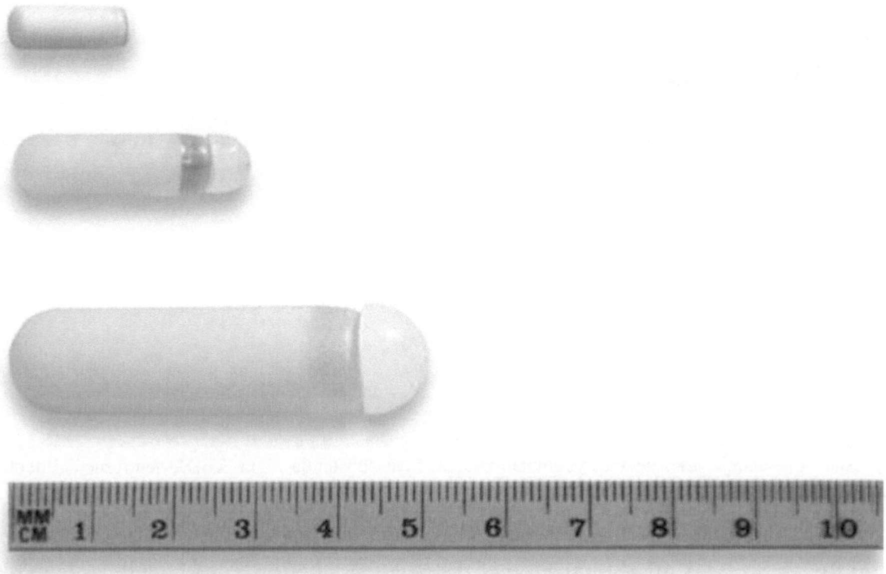

FIGURE 23.14 ALZET pumps are available in three sizes. The figure shows the actual size. (Courtesy of Durect, ALZET Osmotic Pumps, www.alznet.com, 2006.)

Procedure

1. Rats are anesthetized with a 65-mg/kg i.p. pentobarbital.
2. Eye cream is placed on both eyes to protect the eyes from drying out. The top of the head is shaved. Local skin is cleaned by 0.5% Betadine and 75% alcohol or 2% chlorhexidine.
3. The rat is placed in a face-down prone position. The head is held in place by placing two ear bars from the stereotaxic equipment into the right and left ear. The teeth hook is placed between the upper and lower teeth to help fix the head in a flat position.

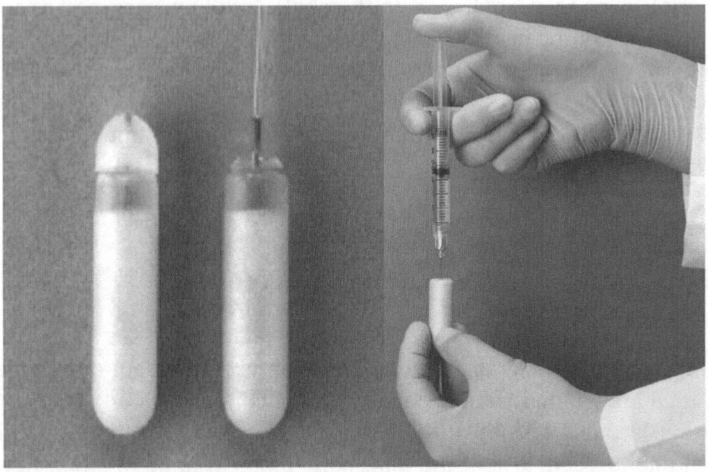

FIGURE 23.15 Filling procedure for a 200-μL pump before and after the attachment of catheters. The white cap is removed. The delivery port (filling tube) is exposed. Fill the pump using a syringe and a blunt-tipped filling tube. Holding the pump in an upright position, fill the reservoir, allowing air to escape around the filling tube. (Courtesy of Durect, ALZET Osmotic Pumps, www.alznet.com, 2006.)

TABLE 23.6
Brain Injection Coordinates

Region	A–P (Anterior–Posterior)	M–L (Middle Line–Lateral)	D–V (Dorsal–Venture, Deep)	Angle
IC	–6.6/–8.8	±1.5	–3	30°
Frontal	+2.7	±2.6	–2	
Cerebellum	–11.8	±2.5	–4.2	
Hippocampus	–3.6	±4.4	–2.4	30°
Ventricle	–0.9	±1.4	–3	
VMH	–2.6	±3.8	–8.3	20°
SNr	–5.3	±2.2	–7.2	
LH	–2.6	±4	–8	13°

Notes: IC, islands of calleja; LH, lateral hypothalamic area; SNr, substantia nigra; VMH, ventromedial nucleus of the hypothalamus.

4. A midline incision is made on the top of the head. Use cotton-tipped applicators to remove the thin layer of skull muscle.
5. The bragma and lambda are exposed. They are the bone marks for orientation. Mark the bragma with a marker.
6. If bleeding occurs, use epinephrine (1:1000, 0.1 mL) locally to stop bleeding or cauterize local small vessels with a cauterizer.
7. Set up principal points using the bragma and lambda as bone marks.
8. The 11-gauge guide probe is used to make a hole on the skull. The guide probe is fixed on the probe holder, which has three directional ruler grids to measure the distances and allow stereoorientation.
9. Injection to different locations of the brain requires different measurement. For brain injection coordinates, see Table 23.6.
10. After the drilling procedure, the probe is inserted and fixed on the skull with 382 instant adhesive gel and 712 accelerator spray to expedite drying.
11. Dental material mixture is applied over the skull area, covering half the height of the probe, to help fix the probe in place (dental material: hygienic cold-cure denture resin, type II class I powder, pink color, mixture with liquid; the mixture takes about 5 to 10 minutes to form a solid).
12. The animal is returned to its cage. After 5 to 7 days recovery time, the rat is ready for *in vivo* experimentation.

BLOOD COLLECTION

Circulating Blood Volumes and Maximum Blood Volume for Survival Collection in Lab Animals

The calculation of volumes for blood sampling relies on accurate data on circulating blood volumes. Generally, total body blood volume equals 6% of the animal's body weight (60 mL/kg). For example, a 300-g rat has about 18 mL total blood volume. Blood withdrawal (like volume fluid replacement) should not exceed 10% of the total body blood volume in any 2-week period. Up to 20% of the total body blood volume may be withdrawn over any 2-week period if it is accompanied with double the volume fluid replacement (preferably lactated Ringer's) given subcutaneously, intraperitoneally, or intravenously at the time of withdrawal. Table 23.4 gives the circulating blood volumes of the species commonly used in safety evaluation studies.[11] Hematocrit must be monitored and fluid replacement considered for protocols that require blood collection in larger volumes or at more frequent intervals.

Orbital Bleed

The infraorbital sinus is a system of dilated venous channels at the back of the orbit. Blood can be collected from this area in *anesthetized* rats using a hematocrit microtube. Always use protective eye ointment after collection attempts.

Equipment

Equipment includes hematocrit microtubes (preferably heparinized), eye ointment, tubes for blood, and instruments for anesthesia.

Procedure

1. Anesthetize a rat.
2. Apply pressure to the external jugular vein caudal to the mandible with a thumb and gently elevate the upper eyelid with the index finger of the same hand.
3. Insert the hematocrit tube into the conjunctiva of the middorsal globe.
4. Gently direct the hematocrit tube in a caudal and medial direction until blood is obtained.
5. Once the desired amount of blood is obtained, discontinue the external jugular pressure and remove the hematocrit tube. (About 0.5 mL blood can be safely collected at each time point.)
6. Gently place pressure on the eye globe to provide hemostasis.

Advantage

An orbital bleed is a fast way to collect blood.

Disadvantages

1. The procedure requires the use of anesthesia.
2. It may cause ocular trauma.
3. It cannot obtain multiple samples over a short period of time.
4. Special training is required for proficient use of this technique.
5. This technique may cause long-term pain and distress.

Potential Considerations

1. Consider analgesia for rodents in pain.
2. Limit the number of retroorbital bleedings per eye.

Tail Bleed

Lateral Tail Vein Bleed

Equipment

A 25- to 27-gauge needle and microcapillary tube (heparinized) and a rat restrainer are needed for the procedure.

Procedures

1. Restrain a rat.
2. Clean the area over the tail vein 3 cm from the base of the tail; use an acceptable antiseptic scrub.
3. Apply heat (warm water, heat lamp, etc.) to the tail to facilitate vasodilation of the tail vein; take precautions to prevent thermal burns.
4. Insert a syringe equipped with a 25-gauge needle at a 45° angle toward the vein.
5. Apply slight negative pressure during needle insertion (this will help determine if the needle has entered the vein).
6. Use gentle negative pressure to collect blood. *Note*: Too much pressure will collapse the vein.
7. Apply direct pressure to the incision for 1 to 3 minutes to facilitate hemostasis.
8. Alternatively, a 25-gauge needle (without syringe) can be inserted into the tail vein; blood accumulating in the needle hub is collected into a hematocrit tube.

Advantages

1. Serial blood collections may be obtained.
2. If done correctly, this technique is not likely to cause long-term pain and distress.

Disadvantages

1. Anesthesia should be considered for this technique.
2. Blood collection is slow in mice and may take 1 to 2 minutes.
3. Temporary or permanent damage may occur to the vein during routine blood collection.
4. The procedure requires personnel training for proficient use.

Tail Sectioning (Technique Suitable for Mice and Rats)

Equipment

Equipment includes a microcapillary tube (heparinized), rat restrainer, and a sterile scalpel blade.

Procedure

1. Restrain a rat; this will require a restriction device.
2. Clean the tail with an appropriate antiseptic solution.
3. With a sterile scalpel blade, make a transverse section through the long axis of the tail 2 mm from the tip.
4. Use a hematocrit tube or blood-collecting tube to collect blood dripping from the sectioned tail.
5. Massage the tail by passing the thumb and index finger from the base to the tip of the tail if blood flow is inadequate.
6. Apply direct pressure to the incision for 1 to 3 minutes to facilitate hemostasis; a drop of tissue glue may be required in some rats to close the surgical incision.
7. Repeated blood sampling may be obtained by two methods; method selection is based on the length of time after the initial incision:

   0 to 24 hours: Bleeding may be restarted by removing the clot; *always try this method first.*

   After 24 hours: Follow the protocol; however, make the new cut through the tail 1 mm from the tip.

Advantages

1. The procedure is rapid, simple, and easy.
2. Multiple blood samples may be obtained from the initial tail excision by removing the scab.

Disadvantages

1. The method requires equipment for rodent restraint.
2. Multiple tail excisions may be required for repeated blood collection.
3. The risk of causing pain and distress increases with repeated tail cuts, especially if the tail vertebrae are damaged.

Potential Considerations

1. Consider analgesia for rodents in pain.
2. To prevent pain and discomfort, *limit the number of tail incisions.*

Lateral Saphenous Vein (Technique Suitable for Mice and Rats)

Equipment

A microcapillary tube, rat restrainer, 25-gauge needle, and a sterile scalpel blade are needed.

Procedure

1. Restrain a rat; this may require two people or a restriction device for a rat.
2. Shave the hair in the lateral saphenous vein area with a scalpel blade (the vein is located caudal and lateral to the fibula and tibia).
3. Clean the shaved area.
4. Extend the hind leg applying gentle downward pressure above the knee joint; this will help improve venous filling.
5. Puncture the saphenous vein with a 25-gauge needle.
6. Collect the blood accumulating over the incision using a hematocrit tube (heparinized).
7. Apply direct pressure to the incision for 1 to 3 minutes to facilitate hemostasis.
8. Repeated blood samples may be obtained by removing the scab.

Figure 23.16 shows this procedure on a mouse.

Advantages

1. Serial blood collections may be obtained.
2. This technique has the least potential to cause long-term pain and distress.

Disadvantages

1. Blood collection is slow, and it may take 1 to 2 minutes to collect a sample.
2. The procedure requires specialized equipment or assistants.
3. The method requires personnel training for proficient use.

Potential Considerations

1. Provide analgesia for rodents in pain.
2. Sedation may be required for rodents that are difficult to handle.

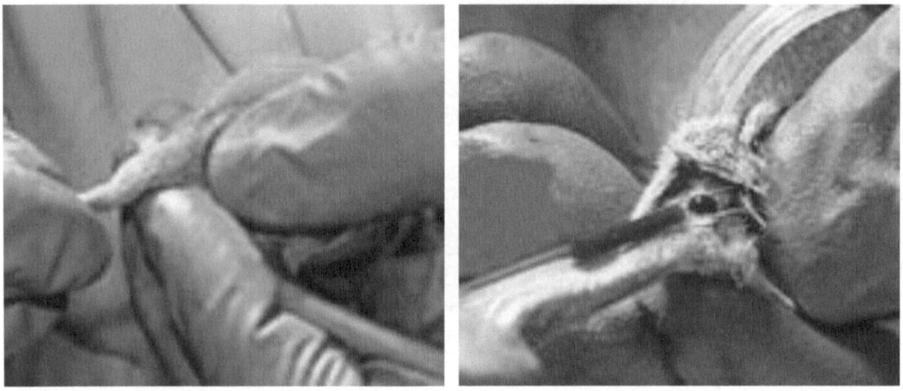

FIGURE 23.16 Collection of blood from the lateral saphenous vein in a mouse. (Courtesy of Johns Hopkins University, www.jhu.edu/animalcare/rat.htm.)

Intracardial Puncture Blood Collection

Equipment

A 22-gauge needle attached to a 5- to 10-mL syringe, tubes for blood, anesthesia device, or drugs are needed. Cardiac puncture is the preferred technique for terminal collection of large blood volumes.

Procedure

1. General anesthesia is administered.
2. The animal is placed on a solid surface with its ventrum exposed.
3. The xiphoid process is palpated at the caudal aspect of the animal's sternum. A notch is present on both sides of this process.
4. A 22-gauge needle attached to a 5- to 10-mL syringe is inserted approximately 30° to 45° from the horizontal axis of the sternum, just behind the xiphoid cartilage and slightly left of the middle and directed toward the heart as determined by palpating for the apex beat (Figure 23.17).

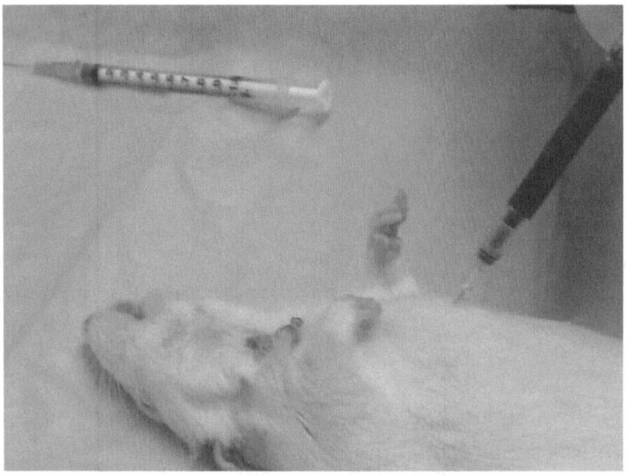

FIGURE 23.17 Procedure for intracardiac puncture to collect blood. The needle is inserted about 30° to 45° from the horizontal axis of the sternum, just behind the xiphoid cartilage and slightly left of the middle, and directed toward the heart as determined by palpating for the apex beat. (Courtesy of Johns Hopkins University, www.jhu.edu/animalcare/rat.htm.)

5. Negative pressure should be applied by placing slight backward pull on the plunger once it has been inserted beneath the skin.
6. Reflux of blood is apparent once the needle has penetrated the heart.

Advantages

Up to 10 mL of blood can be obtained from the heart of a deeply anesthetized rat.

Disadvantages

This is a terminal procedure and cannot be repeated for multiple bleedings.

Considerations

This procedure is performed as a terminal event only, and general anesthesia is required. The animal must be sacrificed at the completion of the procedure before it awakens from anesthesia.

Cannulation of Vessels for Chronic Blood Collection

Common Carotid Artery (CCA) Cannulation

The left carotid artery is generally chosen for CCA cannulation because a catheter on this side will pass into the aortic arch and lie on the descending aorta; a catheter in right carotid artery is easily passed through the ascending aorta into left ventricular and causes arrhythmias.

Equipment

For surgical instrument, see Chapter 10; use a PE-50 catheter, 60% PVP (see the section on jugular vein cannulation).

Anesthesia

Rats are anesthetized with a mixture of 25 mg/kg ketamine and 2.5 mg/kg xylazine given intraperitoneally or 2% to 3% isoflurane in oxygen.

Surgical Procedure

1. Apply eye cream on both eyes to protect the eyes from drying. The ventral and back neck areas are shaved; the shaved area should be two to three times larger than the surgical incision. Local skin is cleaned with 0.5% Betadine and 75% alcohol or 2% chlorhexidine.
2. A rat is placed in a face-up supine position. A 10-cc syringe or similar size tube is placed under the neck to help expose the CCA. The tail and four legs of the rat are fixed on a surgical board with adhesive tape.
3. A 2-cm midline incision is made on the ventral neck.
4. A blunt dissection is made between the left sternohyoid, sternomastoid, and omohyoid muscles to bare the left CCA; care should be taken as the vagus nerve is very close to the CCA.
5. A pair of curved artery forceps is placed under the CCA to help expose the CCA. Place three threads under the CCA. One ligature in made near the cephalic end of the artery so that the end of the thread is fixed on the board to give the artery slight tension. The other two ligatures are positioned loosely on the cardiac end of the CCA.
6. A small vessel clamp is clipped on the cardiac end of the CCA (near the heart side) to stop blood flow. The forceps is then removed.
7. Taking care not to cut through the CCA, a tiny diagonal hole toward the heart is made on the left CCA by a pair of microspring scissors.

8. Hold the incision open with microdissecting tweezers, insert the PE-50 catheter into the hole, and advance it into the CCA. The catheter is connected to a syringe filled with heparinized normal saline (9 U/mL).
9. The two ligatures are lightly tightened around the CCA and the catheter, the catheter near the syringe is clamped by a blood vessel rongeur, the vessel clamp is then released, and the catheter is advanced toward the heart. The distance is determined as described previously by body weight (inserted distance 2 cm for 300-g rat, 1.5 cm for 200-g rat, and 1 cm for 100-g rat). The tip of the catheter lies in the aortic arch. Most often, it lies downstream in the descending portion.
10. The CCA and catheter are then tightened. A drop of Super Glue is placed over the ligature to ensure that the catheter stays in the artery. Make sure blood is flowing after releasing the forceps. Flush the CCA and catheter with heparinized saline.
11. A suspension of 60% PVP-heparin-saline (500 to 625 IU/mL heparin) is injected into the catheter; make a tiny air bubble between heparin-saline and PVP layers so that you can see the PVP mixture flow through the catheter into the CCA; stop the injection while the bubble passes the Super Glue node so that PVP does not enter the circulation.
12. The free end of the catheter is sealed by flame heating.
13. The rat is turned to a face-down prone position.
14. A 1-cm midline incision is made on the back of the neck.
15. A curved forceps is inserted subcutaneously toward the side of the ventral neck incision, and the free end of the catheter is grasped and pulled back through the ventral incision (see Figure 23.6b and 23.6c).
16. Bupivicaine is topically administered to the local incision sites to relieve postoperative pain.
17. The incisions are closed by suture or clips with the catheter free outside.
18. Furazolidone (4%) powder or another antibiotic is applied on the incisions to protect from infection.

REFERENCES

1. Wayneforth, H. and Flecknell, P.A., *Experimental and Surgical Technique in the Rat*, 2nd ed., Academic Press, London, 1992.
2. Yalkowsky, S.H., ed., *Techniques of Solubilization of Drugs*, Marcel Dekker, New York, 1981.
3. Fox, J.G., Anderson, L.C., Lowe, M., and Quimby, F.W., eds., *Laboratory Animal Medicine*, 2nd ed., Academic Press, New York, 2002.
4. Poole, T., ed., *The UFAW Handbook on the Care and Management of Laboratory Animals*, 7th ed., Blackwell Science, Oxford, U.K., 1999.
5. Staszyk, C., Bohnet, W., Gasse, H., and Hackbarth, H., Blood vessels of the rat tail: A histological reexamination with respect to blood vessel puncture methods, *Lab Anim*, 37, 121, 2003.
6. Thorne, R.G., Pronk, G., and Frey, W.H. II, Delivery of insulin-like growth factor-I to the brain and spinal cord along olfactory and trigeminal pathways following intranasal administration: a noninvasive method for bypassing the blood-brain barrier, *Soc Neurosci Abstr*, 26, 1365, 2000.
7. Chen, X.Q., Fawcett, J.R., Rahman, Y.E., Ala, T.A., and Frey, W.H. II, Delivery of nerve growth factor to the brain via the olfactory pathway, *J Alzheimers Dis*, 1(1), 35, 1998.
8. Hilger, A., Applied anatomy and physiology of the nose. In: *Boies's Fundamentals of Otolaryngology*, Adams, G.L., Boies, L.R., and Hilger, P.A., eds., 6th ed., Saunders, Philadelphia, 1989, p. 177.
9. Liu, X.F., Fawcett, J.R., Thorne, R.G., and Frey, W.H. II, Non-invasive intranasal insulin-like growth factor-I reduces infarct volume and improves neurological function in rats following middle cerebral artery occlusion, *Neurosci Lett*, 308(2), 91, 2001.

10. Hanson, L.R., Martinez, P.M., Taheri, S., Kamsheh, L., Mignot, E., and Frey, W.H. II, Intranasal administration of hypocretin 1 (orexin A) bypasses the blood-brain barrier and targets the brain: A new strategy for the treatment of narcolepsy, *Drug Deliv Technol,* 2006. Available at: www.drugdeliverytech.com/cgi-bin/articles.cgi?idArticle=240.
11. EFPIA (European Federation of Pharmaceutical Industries Associations) and ECVAM (European Center for the Validation of Alternative Methods), *A Good Practice Guide to the Administration of Substances and Removal of Blood, Including Routes and Volumes*, Draft Document, 2000. Available at: www.eslav.org/eslav.

24 Study Design in Animal Models of Stroke

Yanlin Wang-Fischer and Lee Koetzner

CONTENTS

INTRODUCTION

Stroke research is worth nothing if it is not reproducible, and research is not reproducible if basic principles of sound experimental design are not followed. Concern about experimental design in clinical research has been prevalent for some time,[1] and the effects of poor clinical trial design are widely appreciated.[2] However, the effects of poor preclinical stroke study design on reproducibility have only recently been acknowledged.[3] This chapter is a guide to the basic issues of experimental design.

EXPERIMENTAL FACTORS IN STROKE STUDIES

Experiments—in stroke research or in other fields—are attempts to measure the effects due to the hypothesis, while controlling or eliminating the effects due to extraneous factors.[3] Identifying the factors in an experimental design will help the experimenter better understand the results.

TREATMENT FACTORS

The simplest experimental designs focus on only one treatment factor; these are called *single-factor designs*. For example, a single-factor study might look at groups treated with bone morphogenetic protein 7 (BMP-7) or a control treatment. Since BMP-7 has effects on bone, kidneys, heart, lungs, and brain, the experimenter would focus on BMP-7 actions in the brain. Experiments can have multiple-factor designs: A two-factor design might look at BMP-7 combined with hypothermia (with vehicle controls for BMP-7 and normothermic controls for hypothermia).

Treatment factors should be standardized. This will limit the possibility of results being influenced by extraneous factors, for example, differences between suppliers in compound purity or differences between surgeons in skill level. Another factor that can play a role in rodent experiments is the time of day: Animal core temperature has a circadian rhythm, which can create a difference between morning and evening experiments.[4]

SUBJECTS (ANIMALS)

The choice of animals is very important for the study. Animals—even inbred rodents—differ not only by strain, age, sex, and weight but also by vendor, breeding site, estrus cycle, diet, bedding, and handling. Experimenters who neglect these factors put their data at risk. For example, the size of a rat influences anatomical landmarks; as a result, it is suggested that body weight should vary by no more than 20 g between groups.

EXPERIMENTAL EFFECTS

An experimenter's choice of endpoints determines the results of the study. These endpoints must be relevant, sensitive, and accurate. *Relevance*, the relationship between the measured parameter and pathology or function, can be guided by a literature review. The endpoint should be sensitive to the experimental factors in the study. For example, if brain edema is used as an endpoint for a stroke model, the magnitude of the edema resulting from stroke has to be big enough to be sensitive to the experimental factors. *Accuracy* is the degree to which observed values approximate actual variables. This will directly affect the reproducibility of the findings.

EXPERIMENTAL DESIGN

Experimental design is the process of organizing the experiment properly to ensure that the right type of data is available to answer the questions of interest as clearly and efficiently as possible. The specific questions that the experiment is intended to answer must be clearly identified before carrying out the experiment. We should also attempt to identify known or expected sources of variability in the experimental units since one of the main aims of an experiment is to reduce the effect of these sources of variability on the answers to questions of interest. That is, we design the experiment to improve the precision of our answers. In some ways, the design of a study is more important than the analysis. A poorly analyzed study can be reanalyzed, but a badly designed study can never be retrieved. Consideration of design is also important because the design of a study will govern how the data are to be analyzed. Experimental design involves randomization of animals or samples, replication, and control of bias by blinded design.

Randomization

A randomized experiment is a form of study used in the testing of the efficacy of medicines, devices, or medical procedures. It is widely considered the most reliable form of scientific evidence because it controls most of the biases that can easily compromise the validity of medical research.[2]

Randomization Procedures

There are several issues to consider in generating the randomization sequences.[5] Since most statistical tests are most powerful when the groups compared have equal sizes, the randomization procedure should generate groups of equal size. Depending on the structure of the randomization procedure, investigators may be able to infer the next group assignment by guessing which of the groups has been assigned the least up to that point. This interferes with blinding and can lead to biased measurements. Finally, if important covariates that are related to the outcome are ignored in the statistical analysis, estimates arising from that analysis may be biased. The potential magnitude of that bias, if any, will depend on the randomization procedure.

Complete Randomization

In the commonly used and intuitive complete randomization[6] procedure, each animal is randomly assigned to one group. It is simple and offers a good deal of robustness against both selection and accidental biases. Its main drawback is the possibility of imbalances between the groups. Imbalance is mainly a concern for small sample sizes, like those used for *in vivo* stroke studies ($n < 20$).

Permuted Block Randomization

Permuted block randomization[6] is often used for *in vivo* studies. In this form of restricted randomization, blocks of animals are created such that balance is enforced within each block. For example, if *A* is the experimental group and *B* is the control group, then a block of four animals may be ordered as *AABB*, *ABAB*, *BAAB*, *ABBA*, *BABA*, and *BBAA*, with each order occurring with a probability of 1/6. Note that there are equal numbers of animals assigned to the experiment and the control group in each block.

Permuted block randomization has several advantages. In addition to promoting group balance at the end of the trial, it also promotes balance in the sense that treatment groups are evenly distributed throughout the observation period. This is particularly important when an experiment enrolls subjects over a long period of time or when transient factors could influence a subset of the results. Unfortunately, by enforcing within-block balance, permuted block randomization can bias observations. Toward the end of each block, the investigators may be able to identify the group with the least assignment up to that point; predicting future group assignment becomes progressively easier. The likelihood of this sort of bias can be reduced by blinding and by larger numbers of groups.

Replication

Replication is repeating an entire study so that the variability associated with the study can be estimated. Replications and repeated measurements are different. Replications should not be confused with *repeated measurements*, which refer to taking several measurements of a phenomenon within a study. *Reproducibility* is one of the main principles of the scientific method, and refers to the ability of a test or experiment to be accurately replicated.

The results of an experiment performed by a particular researcher or group of researchers are generally evaluated by other independent researchers by reproducing the original experiment. They repeat the same experiment themselves, based on the original experimental description, and determine whether their experiment gives results similar to those reported by the original group. Experiments that cannot be reliably reproduced are generally not considered to provide useful scientific evidence. Investigators whose results prove to be highly reproducible are typically given more credence by other scientists than those investigators whose results are less reproducible.

Sample Size

Reproducibility is also related to the sample size in each study. If the groups in a study are too small, the study will not be able to answer the question posed and is a waste of time and money. However, studies should not be too large because animals and resources will be wasted if fewer animals would have sufficed. Estimating a proper sample size to ensure a significant result is an important part of study design. The size of the groups should be sufficient to allow meaningful scientific interpretation of the data generated. Thus, the number of animals or isolated preparations should be adequate to demonstrate or rule out the presence of biologically significant effects of the test substance. This should take into consideration both the size and the variability of the biological effect. Appropriate negative and positive control groups should be included in the experimental design. In well-characterized *in vivo* test systems, some controls may not be necessary; however, the exclusion of controls from studies must be justified.

The appropriate sample size depends on four critical quantities: the type I and type II error rates α and β, the variability of the data σ^2, and the effect size d. In a test, the effect size is the amount by which we would expect the two treatments to differ or is the difference that would be clinically worthwhile. Usually, α and β are fixed at 5% and 20% (or 10%), respectively. A simple formula for a two-group parallel test with a continuous outcome is that the required sample size per group is given by $n = 16\sigma^2/d^2$ for two-sided α of 5% and β of 20%. For example, in a test of reduced infarct size in stroke rats, if an effect for infarct size is 40 mm^3 (that is, the infarct difference between treated and untreated is 40 mm^3) and the between-subject standard deviation is 35 mm^3, we would require $n = 16 \times 35^2/40^2 = 16 \times 1225/1600 = 12$ rats per group in the study. The required sample size increases in proportion to the square of the standard deviation of the data (the variance) and decreases in proportion to the square of the effect size. Doubling the effect size reduces the required sample size by four; it is much easier to detect large effects. In practice, determining the optimum sample size usually requires the experimenter to consider other criteria, such as animal resources, budget, and staff time; the formula is used to back-calculate the detectable effect size. The optimum sample size also can be inferred by reference to the published literature.

Blinding

Single Blind

In a single-blind test, the researcher knows the identity of the treatment, but the patient does not. Because the patient does not know which treatment is being administered, there should be no placebo effect. In practice, since the researcher knows, it is possible for them to treat the patient differently or to subconsciously hint to the patient important treatment-related details, thus influencing the outcome of the study. Many investigators assume that animal studies are single blinded due to the animal's unawareness.

Double Blind

The double-blind method is an important part of the scientific method and is used to prevent research outcomes from being influenced by observer bias. Blinded research is an important tool in many fields of research, from medicine to psychology and the social sciences and forensics. Double-blind experiments are held to achieve a higher standard of scientific rigor.

In a double-blind experiment, neither the individuals nor the researchers know which treatments belong to the control group and which belong to the experimental group. Only after all the data are recorded (and in some cases, analyzed) do the researchers learn which individuals are which. For instance, in a stroke study, one researcher allocates a series of codes to different treatments (or vehicle). The second researcher is told the codes but not their allocation. Since the second researcher does not know, bias cannot influence the study. In this system, there often is a more realistic distribution of body weight of animals and level of brain injury in animals. Therefore, double-blind (or randomized) tests are preferred as they tend to give the most accurate results.[3]

Double-blind methods can be applied to any experimental situation for which there is the possibility that the results will be affected by conscious or unconscious bias, either in treatment or measurement, on the part of the experimenter. Double-blinding is relatively easy to achieve in drug studies by formulating the investigational drug and the control (either a placebo or an established drug) to have identical appearance (color, taste, etc.). Patients or animals are randomly assigned to the control or experimental group and given codes by a study coordinator, who also encodes the drugs. Neither the patients or animals nor the researchers monitoring the outcome know which patient or animal is receiving which treatment until the study is over and the code is broken.

Effective blinding can be difficult to achieve if the treatment is notably effective or the treatment is very distinctive in taste or has side effects that allow the researcher or the subject to guess group assignment.[7] It is also difficult to use the double-blind method to compare surgical and nonsurgical interventions. A good protocol will foresee these potential problems to ensure blinding is as effective as possible.

Triple Blind

Some randomized controlled tests are considered triple blinded, although the meaning of this may vary according to the exact study design. The most common meaning is that the subject (animals), researcher, and person administering the treatment are blinded to what is being given. Alternately, it may mean that the researcher and statistician are blinded. These additional precautions are often in place in studies described as double blind, and thus the term *triple blinded* is infrequently used. Sometimes, triple blind is used to mean that multiple investigators are all blinded to the protocol (such as the person giving the treatment and a pathologist who interprets the results). However, it indicates an additional layer of security to prevent undue influence of study results by anyone directly involved with the study.

TYPES OF CONTROL GROUPS

In Vitro Studies: Concept of Controlled Experiments

A controlled experiment compares the results obtained from an experimental sample against a control sample, which is identical to the experimental sample except for the one factor with the effect that is being tested. It is good practice to have several replicate samples for the test being performed and to have both a positive control and a negative control. The results from replicate samples can be averaged, or if one of the replicates is obviously inconsistent with the results from the other samples, it can be discarded as the result of an experimental error (a procedural mistake, equipment failure, recording error, or other problem). Most often, tests are done in duplicate or triplicate.

A *positive control* is a treatment or procedure that is very similar to the actual experimental treatment but is known from previous experience to give a positive result. A *negative control* is known to give a negative result. The positive control confirms that the basic conditions of the experiment were able to produce a positive result, even if none of the actual experimental samples produce a positive result. The negative control demonstrates the baseline result obtained when a test does not produce a measurable positive result; often, the value of the negative control is treated as a "background" value to be subtracted from the test sample results. Sometimes, the positive control takes the form of a standard curve. A historical control is the data from previous studies or references. Since historical controls are conducted in different experiments, and therefore may involve different experimental conditions, they are a weak form of control.

An example that is often used in teaching laboratories is a protein assay. Students might be given a fluid sample containing an unknown (to the student) amount of protein. It is their job to correctly perform a controlled experiment in which they determine the concentration of protein in this sample. The teaching lab would be equipped with a protein standard solution with a known protein concentration. Students could make several positive control samples containing various dilutions of the protein standard (to make a standard concentration-response curve). Negative control samples would contain all of the reagents for the protein assay but no protein. In this example, all samples are performed in duplicate. The assay is a colorimetric assay in which a spectrophotometer measures the amount of protein in samples by detecting a colored complex formed by the interaction of protein molecules and the added dye. In this illustration, the results for the diluted test samples can be compared to the results of the standard curve to determine an estimate of the amount of protein in the unknown sample.

In Vivo Studies

In vivo studies have the same logical structure as *in vitro* studies and can make use of all the same experimental controls.[8] However, the implementation of these controls is different.

Placebo Control

A *placebo* is an inactive treatment or procedure—a negative control. Placebo controls were reviewed in 1955 by Henry K. Beecher[9] and are also known as the subject-expectancy effect, that is, the phenomenon that a patient's symptoms can be alleviated by an otherwise-ineffective treatment since the individual expects or believes that it will work. Some people consider this to be a remarkable aspect of human physiology; others consider it to be an illusion arising from the way medical experiments are conducted. The phenomenon is not fully understood by science.[10] Although most discussions of placebo effects involve human subjects, there have been reports of placebo-like effects in animals.[11,12]

Placebo-Controlled Studies

Beecher[9] reported that about a quarter of patients who were administered a placebo (for example, against back pain) reported relief or diminution of pain. Remarkably, not only did the patients report improvement, but also the improvements themselves were often objectively measurable, and the same improvements were typically not observed in patients who did not receive the placebo. Because of this effect, government regulatory agencies approve new drugs only after tests establish not only that patients respond to them, but also that their effect is greater than that of a placebo (by way of affecting more patients, by affecting responders more strongly, or both). Such a test is called a *placebo-controlled study*. Because a doctor's belief in the value of a treatment can affect his or her behavior, and thus what his or her patient believes, such trials are usually conducted in double-blind fashion; that is, not only are the patients unaware when they are receiving a placebo, also the doctors are made unaware. It has even been shown that mock surgery can have placebo effects,[13] although few surgery efficacy studies include these controls. To merit approval of a drug, the group receiving the experimental treatment must experience a greater benefit than the placebo group. However, many studies show some benefit in the placebo group. For example, Khan et al. published a meta-analysis of studies of investigational antidepressants and found a 30% reduction in suicide and attempted suicide in the placebo groups and a 40% reduction in the treated groups.[14] Many physicians consider nontreatment groups unethical when there is an approved, efficacious treatment; therefore, many studies do not include an untreated group.[1] As a result, determining the actual size of the placebo effect, compared to totally untreated patients, is difficult.

How the Placebo Effect Works

There are three main hypotheses for how the placebo effect works: the subject expectancy effect, conditioning, and motivation.

The *subject-expectancy effect* attributes the placebo effect to conscious or unconscious manipulation by patients in reporting improvement. Hróbjartsson, Norup, and Götzsche[15–17] argued in their articles, "Most patients are polite and prone to please the investigators by reporting improvement, even when no improvement was felt."[15] Subjective bias can also be unconscious, with the patient believing he or she is improving even if this is not the case.

Classical conditioning is a type of associative learning by which the subject learns to associate stimuli with a particular response. In this case, the stimuli are the treatment (perceived as medicine), and the response is the relief of symptoms. It is difficult to tell the difference between conditioning and the expectancy effect when the outcome is subjective and reported by the patient. However, conditioning can result in measurable biological changes similar to the changes seen with the real treatment or drug. Studies showing that placebo treatments result in changes in brain function similar to the real drug may be examples of conditioning resulting in objectively measurable results.[18–20]

Motivational explanations of the placebo effect have typically considered the placebo effect to be an outcome of one's desire to feel better, reduce anxiety, or cooperate with an experimenter or health care professional.[21,22] The motivational perspective is supported by recent research showing that nonconscious goals for cooperation can be satisfied by confirming expectations about a treatment.[23]

Choice of a Placebo for Preclinical Stroke Studies

Any solution used for dissolving an experimental compound can be used as a placebo. It is given by the same route in the same volume. For instance, in a study of BMP-7, the protein has to be dissolved in 2% $NaHCO_3$, pH 4.0, for intravenous dosing. Therefore, the vehicle for this study should be the same: 2% $NaHCO_3$ i.v.

Dose-Response Control

Experimental groups receive the drug at different doses (at least three different doses to create a dose-response curve). The doses to be given are based on the drug potency. If the doses are selected correctly, one experiment can include doses with full efficacy, some efficacy, and no efficacy.

Positive Control

A *positive control* is a substance that is known to give a positive result. A good example of a positive control in stroke studies is dextrorphan. Dextrorphan hydrochloride is a glutamate antagonist (a noncompetitive *N*-methyl-d-aspartate [NMDA] channel blocker) that is neuroprotective in experimental models of focal brain ischemia. Dextrorphan, the *O*-demethylated metabolite of the commonly used antitussive dextromethorphan, has been shown to attenuate hypoxic neuronal injury in culture[24,25] and to significantly reduce ischemic neuronal injury in animal stroke models.[26–29] Dextrophan has also been used in human patient trials.[30]

Sham Surgery

In an *in vivo* study, sham control animals undergo procedures similar to the experimental group but without key steps that make the disease model. For example, in the distal stroke model, animals undergo the opening of a brain window and incision on the neck but without occlusion of middle cerebral artery and common carotid artery. The purpose of sham surgery is to exclude placebo effects from the surgical procedure; these effects can be remarkably strong.[13]

Historical Control

Data from previous experiments can be used as a historical control. However, since it is nearly impossible to exactly reproduce the conditions of an *in vivo* experiment, historical controls are weak controls.[4]

PHARMACOKINETIC STUDIES

A simple definition of *pharmacokinetics* (PK) is "how the body works on a drug," determining drug concentrations in the body. *Pharmacodynamics* (PD) can similarly be defined as "how a drug works on the body," resulting in measurable drug effects. Combining these two ideas leads to the concept of dose-concentration-effect, which is called PKPD modeling. PK is the study of the disposition of drugs in the body (that is, their absorption, distribution, metabolism, and elimination, or ADME). It deals with a mathematical description of the rates of drug movement into, within, and out of the body. The body is a very complex system, and a drug undergoes many steps as it is being absorbed, distributed through the body, and metabolized or excreted. Although the details of drug kinetics are complicated, we can often approximate drug kinetics using simple mathematical models.[31,32]

The *elimination half-life* is the time taken for the plasma concentration to fall to half its original value. This kind of PK information is necessary for a new compound efficacy study to choose the first dose and the route of administration for the drug. For example, a drug with an elimination half-life of 30 minutes might work when given by continuous infusion but not by intravenous bolus. The experiments that provide these data can be very simple: Drugs are infused, and then samples are collected at set time points. The modeling of PK parameters from these data is beyond the scope of this chapter; readers are encouraged to consult published references.[31,32]

REFERENCES

1. Kazdin, A.E., ed., *Methodological Issues and Strategies in Clinical Research*, American Psychological Association, Washington, DC, 1992.
2. Ioannidis, J.P.A., Contradicted and initially stronger effects in highly cited clinical research, *JAMA*, 294, 218, 2005.
3. Sena, E. et al., How can we improve the preclinical development of drugs for stroke? *Trends Neurosci*, 30, 433, 2007.
4. Claassen, V., *Neglected Factors in Pharmacology and Neuroscience Research*, Elsevier, Amsterdam, 1994.
5. Lachin, J.M., Matts, J.P., and Wei, L.J., Randomization in clinical trials: Conclusions and recommendations, *Controlled Clin Trials*, 9(4), 365, 1988.
6. Cochran, W.G. and Cox, G.M., *Experimental Designs*, 2nd ed., Wiley, New York, 1957/1992.
7. Munjack, D.J. et al., Actual medication versus therapist guesses: In a blind study, how blind is blind? *J Clin Psychopharmacol*, 9, 148, 1989.
8. Kerlinger, F.N., *Foundations of Behavioral Research*, 3rd ed., Harcourt Brace Jovanovich, Fort Worth, Texas, 1986.
9. Beecher, H.K., The powerful placebo, *JAMA*, 159, 1602, 1955.
10. Barfod, T.S., Placebos in medicine: placebo use is well known, placebo effect is not, *BMJ*, 330, 45, 2005.
11. Cobb, L.A. et al., An evaluation of internal-mammary–artery ligation by a double-blind technique, *N Engl J Med*, 260, 1115, 1959.
12. Goldberg, S.R. and Schuster, C.R., Conditioned suppression by a stimulus associated with nalorphine in morphine-dependent monkeys, *J Exp Anal Behav*, 10, 235, 1967.
13. Herrnstein, R.J., Placebo effect in the rat, *Science*, 138, 677, 1962.
14. Khan, A., Warner, H.A., and Brown, W.A., Symptom reduction and suicide risk in patients treated with placebo in antidepressant clinical trials: An analysis of the Food and Drug Administration database, *Arch Gen Psychiatry*, 57, 311, 2000.
15. Hróbjartsson, A. and Norup, M., The use of placebo interventions in medical practice—A national questionnaire survey of Danish clinicians, *Eval Health Prof*, 26, 153, 2003.

16. Hróbjartsson, A. and Götzsche, P., Is the placebo powerless? An analysis of clinical trials comparing placebo with no treatment, *N Engl J Med,* 344, 1594, 2001.
17. Hróbjartsson, A. and Götzsche, P., Is the placebo powerless? Update of a systematic review with 52 new randomized trials comparing placebo with no treatment, *J Intern Med,* 256, 91, 2004.
18. Leuchter, A.F. et al., Changes in brain function of depressed subjects during treatment with placebo, *Am J Psychiatry,* 159, 122, 2002.
19. Sauro, M.D., Endogenous opiates and the placebo effect: a meta-analytic review, *J Psychosom Res,* 53, 115, 2005.
20. Wager, T.D. et al., Placebo-induced changes in FMRI in the anticipation and experience of pain, *Science,* 303, 1162, 2004.
21. Margo, C.E., The placebo effect, *Surv Ophthalmol,* 44, 31, 1999.
22. Price, D.D. et al., An analysis of factors that contribute to the magnitude of placebo analgesia in an experimental paradigm, *Pain,* 83, 147, 1999.
23. Geers, A.L. et al., Goal activation, expectations, and the placebo effect, *J Pers Soc Psychol,* 89, 143, 2005.
24. Choi, D.W., Peters, S., and Viseskul, V., Dextrorphan and levorphanol selectively block *N*-methyl-d-aspartate receptor-mediated neurotoxicity on cortical neurons, *J Pharmacol Exp Ther,* 242, 713, 1987.
25. Goldberg, M.P., Pham, P.C., and Choi, D.W., Dextrorphan and dextromethorphan attenuate hypoxic injury in neuronal culture, *Neurosci Lett,* 80, 11, 1987.
26. George, C.P. et al., Dextromethorphan reduces neocortical ischemic neuronal damage *in vivo, Brain Res,* 440, 375, 1988.
27. Steinberg, G.K., Saleh, J., and Kunis, D., Delayed treatment with dextromethorphan and dextrorphan reduces cerebral damage after transient focal ischemia, *Neurosci Lett,* 89, 193, 1988.
28. Steinberg, G.K. et al., Protection after transient focal cerebral ischemia by the *N*-methyl-d-aspartate antagonist dextrorphan is dependent upon plasma and brain levels, *J Cereb Blood Flow Metab,* 11, 1015, 1991.
29. Graham, S.H., Chen, J., and Simon, R.P., A dose response study of dextrorphan in permanent focal ischemia, *Neurosci Lett,* 160, 21, 1993.
30. Albers, G.W. et al., Tolerability and pharmacokinetics of the *N*-methyl-d-aspartate antagonist dextrorphan in patients with acute stroke, *Stroke,* 26, 254, 1995.
31. Neubig, R.R., The time course of drug action. In: *Principles of Drug Action: The Basis of Pharmacology,* 3rd ed., Pratt, W.B. and Taylor, P., eds., Churchill Livingstone, New York, 1990, chap. 4.
32. Pratt, W.B., The entry, distribution and elimination of drugs. In: *Principles of Drug Action: The Basis of Pharmacology,* 3rd ed., Pratt, W.B. and Taylor, P., eds., Churchill Livingstone, New York, 1990, chap. 3.

25 Common Biochemical and Physiological Parameters in Rats

Yanlin Wang-Fischer and Lee Koetzner

CONTENTS

INTRODUCTION

The purpose of this chapter is to provide scientists with a range of normal or expected values for selected hematology, serum chemistry, and physiological parameters. These data were obtained from Charles River Laboratories and other resources. Different analytical methods as well as environmental and technique variables can influence the values obtained for most parameters. The values also vary with strain and age. For these reasons, care should be taken in using these data; they are not intended as a substitute for normative data collected within a single institution. Rats included in this publication were mostly single housed with free access to water. The animal rooms were generally maintained at average temperatures of 22°C with an average relative humidity of 30% to 70%. A 12 hour/12 hour light/dark cycle was employed in all studies. Since Sprague-Dawley (SD) rats are used very often in stroke studies, this chapter generally provides data for SD rats.

BODY WEIGHT AND AGE

Body weight can be used to infer age, using a growth curve; we have adapted these data from Charles River Laboratories with their permission (see Table 25.1).

TABLE 25.1
Expected Ages at Target Body Weights

	Age (days)	
Weight (g)	Male	Female
40 to 50	21	21
51 to 75	27	28
76 to 100	31	31
101 to 120	35	36
121 to 140	38	42
141 to 160	42	47
161 to 180	45	56
181 to 200	49	67
201 to 225	53	82
226 to 250	57	92
251 to 275	65	
276 to 300	72	
301 to 350	84	
351 and up	93	

Source: From Reference 1.

HEMATOLOGY

Hematological parameters are summarized in Table 25.2 based on data from References 1 and 2 with permission (see Table 25.2).

BLOOD BIOCHEMISTRY

Biochemical parameters for rat blood are summarized in Table 25.3. Most data in this table are reprinted from References 1 and 2 with permission; a few data are reproduced from different publications (see Table 25.3).[3–5]

PHYSIOLOGICAL PARAMETERS

Table 25.4 shows the physiological parameters.[2]

URINE CHEMISTRY

Table 25.5 shows the urine chemistry (means ± standard error, n = 10).[2,6]

BRAIN WEIGHTS AND MORPHOMETRY

We have adapted data on brain dimensions from those published by Barnett and colleagues[6] at Charles River Laboratories to help researchers understand normal rat brain morphology.

METHODS AND ANIMALS

"Young (11 days old) and adult Sprague Dawley (70–80 days old) rats were selected for brain evaluation and were anesthetized with sodium pentobarbital. Rats were perfused *in situ* with 10% neutral

TABLE 25.2
Hematology

Parameters	Male, Mean ± Standard Deviation	Female, Mean ± Standard Deviation
RBC (×10⁶/μL)	8.00 ± 0.47	8.10 ± 0.43
Hg (g/dL)	16.22 ± 0.81	16.09 ± 0.72
HCT (%)	56.83 ± 3.00	54.28 ± 2.57
MCV (fL)	70.75 ± 2.25	67.67 ± 2.49
MCH (pg)	20.32 ± 0.38	19.90 ± 0.69
MCHC (g/dL)	28.83 ± 0.56	29.59 ± 0.57
CHCM (g/dL)	27.65 ± 0.68	28.42 ± 0.55
Plt (×10³/μL)	1125.1 ± 128.8	1112.4 ± 182.6
WBC (×10³/μL)	10.37 ± 2.01	8.96 ± 1.16
Neut (%)	9.89 ± 3.10	8.93 ± 2.61
Neut (×10³/μL)	1.11 ± 0.20	0.68 ± 0.10
Lymp (%)	84.39 ± 2.80	85.94 ± 3.01
Lymp (×10³/μL)	8.90 ± 1.87	7.95 ± 1.15
Mono (%)	2.58 ± 0.70	2.29 ± 0.60
Mono (×10³/μL)	0.10 ± 0.04	0.08 ± 0.03
Eos (%)	1.20 ± 0.25	1.18 ± 0.25
Eos (×10³/μL)	0.10 ± 0.04	0.10 ± 0.06
Baso (%)	0.48 ± 0.35	0.35 ± 0.25
Baso (×10³/μL)	0.07 ± 0.03	0.08 ± 0.04

Source: From References 1 and 2.

Notes: Baso, basophils; CHCM, cellular hemoglobin concentration mean; Eos, eosinophils; HCT, hematocrit; Hg, hemoglobin; Lymp, lymphocytes; MCH, mean corpuscular hemoglobin; MCHC, mean corpuscular hemoglobin concentration; MCV, mean corpuscular volume; Mono, monocytes; Neut, neutrophils; Plt, platelets; RBC, red blood cell count; WBC, white blood cell count. SD rats age 56 to 70 days (adult); bled from cardiac puncture; analyzed by Bayer Advia 120, nonfasted values.

buffered formalin, the calvaria were then removed and the heads immersed in 10% neutral buffered formalin. After at least 48 hours fixation, the brains from both young rats and adult rats were dissected free, leaving the olfactory bulbs intact and attached to the brain. The brains were then weighed and a Vernier caliper was used to obtain two linear measurements from each intact brain, the anterior–posterior length of the cerebrum extending from the anterior pole to the posterior pole, exclusive of olfactory bulbs and anterior-posterior length of the cerebellum, extending from the anterior edge of the cortex to the posterior pole.

"Within the coronal section passing through the optic chiasm at a magnification of 40× [Figure 25.1]: in this and other sections, the optic chiasm was used as the morphologic landmark at histologic trim. In those sections where the optic nerve and chiasm were lost during processing and sectioning, the correct location was confirmed by the presence of the anterior commissure.

"*Thickness of the parietal cortex*: A measurement of the dorsolateral portion of the cerebral cortex within the coronal section taken through the optic chiasm at a magnification of 40× [Figure 25.1].

"*Height of the cerebellum:* A measurement taken at the level of the deep cerebellar nuclei, including lobes 1–6 and extending from the roof of the fourth ventricle to the dorsal surface (maximum height of the cerebellum), at a magnification of 20× [Figure 25.2].

TABLE 25.3
Blood Biochemistry

Parameters	Male (n = 52) (Mean ± Standard Deviation)	Female (n = 50) (Mean ± Standard Deviation)
ALK (U/L)	329.30 ± 93.82	196.00 ± 38.21
ALT (U/L)	46.90 ± 6.69	44.80 ± 7.67
AST (U/L)	107.30 ± 21.70	137.50 ± 56.64
GGT (U/L)	0.50 ± 0.58	0.40 ± 0.48
LDH (IU/L)[3]	300.00 ± 26.00	N/A
Glucose (mg/dL)	103.00 ± 20.00	102 ± 15.00
Tpr (g/dL)	7.10 ± 0.62	7.40 ± 0.38
Albumin (g/dL)	4.50 ± 0.51	4.80 ± 0.52
Globulin (g/dL)	4.00 ± 0.85	4.00 ± 0.85
A/G	0.80 ± 0.28	0.80 ± 0.28
Tbil (mg/dL)	0.20 ± 0.05	0.10 ± 0.05
BUN (mg/dL)	14.20 ± 2.67	14.90 ± 3.36
Creatinine (mg/dL)	0.40 ± 0.07	0.40 ± 0.07
Chol (mg/dL)	93.90 ± 16.20	89.50 ± 11.82
Triglycerol (mg/dL)	122.2 ± 54.12	87.60 ± 20.85
Glycerol (mg/dL)[4]	0.34 ± 0.01	N/A
FFA (mEq/L)	0.70 ± 0.10	0.70 ± 0.10
Calcium (mg/dL)	13.10 ± 0.50	13.30 ± 0.32
Chloride (mEq/L)	98.90 ± 1.43	98.70 ± 1.49
Phosphorus (mg/dL)	12.70 ± 1.67	12.50 ± 0.90
Potassium (mEq/L)	7.50 ± 0.68	7.50 ± 0.68
Sodium (mEq/L)	148.50 ± 1.39	147.80 ± 2.71
pH	7.35 ± 0.11	7.35 ± 0.11
CO_2 (mM)	24.00 ± 4.00	24.00 ± 4.00
pCO_2 (mm Hg)	42.00 ± 6.00	42.00 ± 3.00
CO_2CP (mEq/L)	23.30 ± 0.40	23.30 ± 0.40
O_2 (mL/dL)	18.60 ± 2.00	18.60 ± 2.00
Insulin (pM)	152 ± 23	152 ± 23
Leptin (ng/mL)[5]	0.51 ± 0.10	N/A
PTH (pg/mL)*	405.1 ± 38	N/A

Source: From References 1 and 2.

Notes: A/G, albumin/globulin ratio; ALK, alkaline phosphatase; ALT, alanine aminotransferase (SGPT); AST, aspartate aminotransferase (SGOT); BUN, blood urea nitrogen; Chol, cholesterol; CO_2CP, carbon dioxide combining power; FFA, free fatty acid; GGT, gamma glutamyl transferase; LDH, lactate dehydrogenase; PTH, parathyroid hormone; Tbil, total bilirubin; Tpr, total protein. SD rats age 56 to 70 days (adult); bled from cardiac puncture; analyzed by Hitachi 717, Olympus AU 640e.

* PTH level in blood serum is from our unpublished data, by enzyme-linked immunosorbent assay (ELISA).

TABLE 25.4
Physiological Parameters

	Male	Female
Body temperature (°C)	37.00 ± 1.10	37.00 ± 1.10
Mean BP (mm Hg, awake)	98.00 ± 20.00	98.00 ± 20.00
Systolic BP (mm Hg)	129.7 ± 2.1	129.7 ± 2.1
Heart rate (time/minutes)	328.00 ± 150.00	328.00 ± 150.00
Respiratory rate (time/minutes)	85.50 ± 20.00	85.50 ± 20.00
Blood volume (mL/kg)	64.10 ± 6.50	64.10 ± 6.50
Urine volume (mL/100 g body weight/day)	3.30 ± 0.20	3.30 ± 0.20
Lifespan (years)	3.00 ± 0.50	3.00 ± 0.50
Adult weight (g)	300 to 500	250 to 300
Birth weight (g)	5 to 6	6-May
Allergens	Dander, urine protein, salivary protein	

Source: From Reference 2.
Note: BP, blood pressure.

TABLE 25.5
Urine Chemistry (Means (Standard Error, n = 10)

Urine protein (mg/24 hours)	39.2 ± 2.9
Albumin excretion (mg/24 hours)	2.9 ± 0.38
Na^+ excretion (mmol/24 hours)	1.01 ± 0.06
Urine creatinine (mmol/24 hours)	1.18 ± 0.05
Glomerular filtration rate (mL/minute/100 g)	0.318 ± 0.02
Specific gravity	1.0105 ± 0.001

Source: From References 2 and 6.

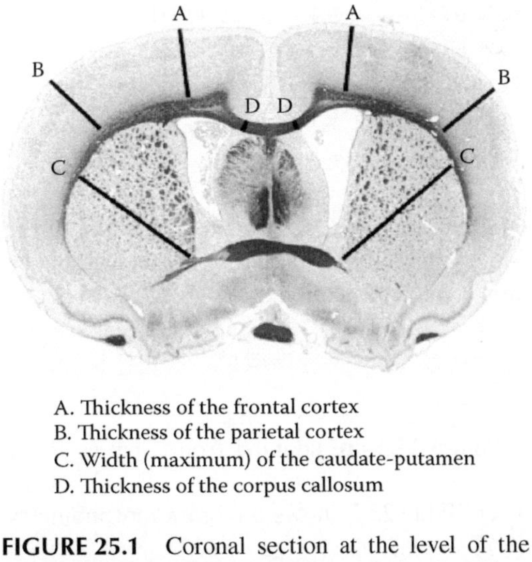

FIGURE 25.1 Coronal section at the level of the optic chiasm (adult).[7] (Data sheet of Charles River Laboratories. With permission.)

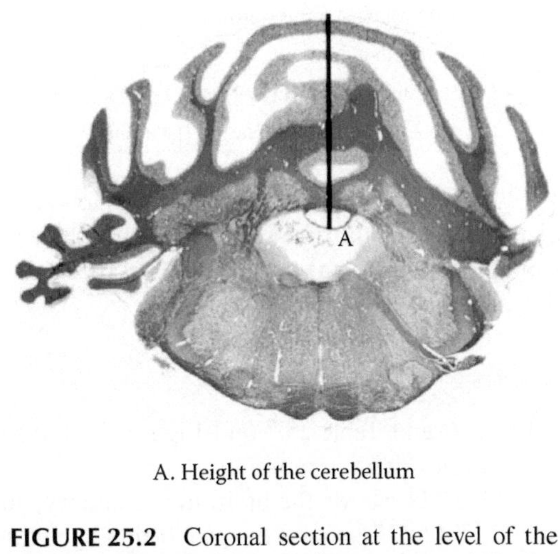

FIGURE 25.2 Coronal section at the level of the deep cerebellar nuclei (adult).[7] (Data sheet of Charles River Laboratories. With permission.)

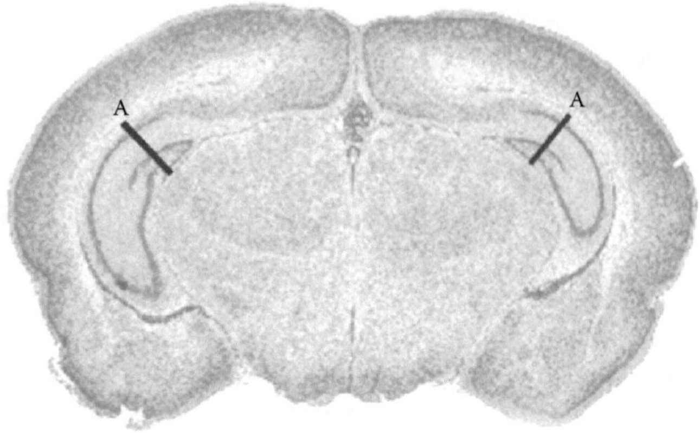

A. Thickness of the hippocampal gyrus

FIGURE 25.3 Coronal section at the level of the hypothalamus (day 11).[7] (Data sheet of Charles River Laboratories. With permission.)

TABLE 25.6
Brain Morphometry (Juvenile)

	Males (Postnatal Day 6)		Females (Postnatal Day 11)	
Parameters	Mean	Range	Mean	Range
Brain weight (g)	1.236	1.132 to 1.32	1.214	1.084 to 1.343
Anterior–posterior cerebrum (mm)	12.2	10.5 to 12.88	12.24	10.8 to 12.98
Anterior–posterior cerebellum (mm)	5.3	3.2 to 6	5.2	3.1 to 6
Frontal cortex (μm)	1428	1264 to 1551	1472	1273 to 1616
Parietal cortex (μm)	1504	1409 to 1629	1515	1410 to 1626
Caudate-putamen (μm)	2349	2052 to 2488	2311	1938 to 2530
Corpus callosum (μm)	292.1	272 to 312	284.3	251 to 331.2
Hippocampal gyrus (μm)	1042	948 to 1136	1005	919 to 1060
Cerebellum (μm)	3350	3005 to 3606	3344	2856 to 3756
External germinal layer (μm)	36.7	30.3 to 40.6	38.9	35.9 to 44.8

Sources: From Reference 7. (Data sheet of Charles River Laboratories. With permission.)

"*Thickness of the hippocampal gyrus:* A measurement on the dorsal to lateral portion of the dentate gyrus within the section taken at the level of the hypothalamus. Measurements were taken from the hippocampus on both sides of the brain section at a magnification of 40× [Figure 25.3] and the median value recorded."

Results

Table 25.6 and Table 25.7 and Figures 25.1 through Figure 25.3 are adapted from Charles River Laboratories (with permission).

Table 25.6 shows the brain morphometry (juvenile).[7] Table 25.7 shows the brain morphometry (adult).[7] Table 25.8 shows the proportion by volume (%).[7]

TABLE 25.7
Brain Morphometry (Adult)

Parameters	Males Mean	Males Range	Females Mean	Females Range
Brain weight (g)	2.282	2.127 to 2.413	2.071	1.933 to 2.151
Anterior–posterior cerebrum (mm)	15.83	14.08 to 16.73	15.32	13.83 to 15.88
Anterior–posterior cerebellum (mm)	7.2	6.3 to 7.6	7	5.8 to 7.7
Frontal cortex (μm)	1792	1660 to 1838	1709	1628 to 1818
Parietal cortex (μm)	1871	1776 to 1956	1764	1656 to 1905
Caudate-putamen (μm)	3244	2920 to 3624	3080	2834 to 3379
Corpus callosum (μm)	272.3	243.2 to 290.4	269.1	246.3 to 291.6
Hippocampal gyrus (μm)	1654	1552 to 1819	1538	1420 to 1602
Cerebellum (μm)	5116	4648 to 5419	4878	4592 to 5028

Source: From Reference 7. (Data sheet of Charles River Laboratories. With permission.)

TABLE 25.8
Proportion by Volume (%)

	Rat	Human
Cerebral cortex	31	77
Basal ganglia	7	4
Diencephalon	6	4
Midbrain	4	1
Hindbrain	7	2
Cerebellum	10	10
Spinal cord	35	2

Source: From Reference 7.

CEREBROSPINAL FLUID

Sharma and colleagues[8] reported cellular and protein parameters in cerebrospinal fluid (CSF) collected from rats thorough a percutaneous CSF collection technique. Male Fischer-344 rats weighing 200 to 300 g were purchased from Harlan Sprague-Dawley.

Table 25.9 shows the erythrocyte and nucleated cell counts and protein concentration in CSF from F-344 rats.[8]

ELECTROCARDIOGRAPH PARAMETERS

Table 25.10 shows electrocardiographic (ECG) parameters in normal SD rats.[2]

TABLE 25.9
Erythrocyte and Nucleated Cell Counts and Protein Concentration in Cerebral Spinal Fluid from F-344 Rats

	Mean ± Standard Deviation	N
Total volume (L)	60.8 ± 3.5	59
Total red blood cell count (103/L)	6.2 ± 1.2	30
Total white blood cell count (per L)	2.3 ± 4.9	30
Neutrophils (per L)	0.2 ± 0.3	14
Macrophages (per L)	1.1 ± 2.2	14
Lymphocytes (per L)	1.9 ± 3.4	14
Total protein (mg/dL)	17.1 ± 2.7	29

Source: From Reference 8.

TABLE 25.10
Electrocardiogram in Normal Sprague-Dawley Rats

Parameters	Interval (seconds), Mean ± Standard Deviation	Voltage (mV, Lead II), Mean ± Standard Deviation
P	˜0.015 ± 0.0037	0.015 ± 0.0037
P-P	0.049 ± 0.007	
QRS	˜0.015 ± 0.0015	˜Q 0.03 ± 0.017
		R 0.775 ± 0.226
		S 0.225 ± 0.147
Q-T	0.0787 ± 0.0137	
S-T	No S-T section	
T	0.0638 ± 0.0134	0.145 ± 0.055
Heart rate	358 ± 47 (beats/minute)	

Source: From Reference 2.

REFERENCES

1. Giknis, M.L.A. and Clifford, C.B., Clinical laboratory parameters for Crl:CD (SD) rats. Available at: www.criver.com, accessed March 2006.
2. You, J.L., *Common Used Physiological and Biochemical Parameters for Experimental Animals*, Hunan Medical University for Graduate Students in Medicine, Changsha, Hunan, 1983.
3. Vuguin, P. et al., Food deprivation limits insulin secretory capacity in postpubertal rat, *Pediatr Res*, 49, 468, 2001.
4. Levy, J.R. et al., Leptin responses to glucose infusions in obesity-prone rats, *Am J Physiol Endocrinol Metab,* 279, E1088, 2000.
5. Dobrian, A.D. et al., Oxidative stress in a rat model of obesity-induced hypertension, *Hypertension,* 37, 554, 2001.
6. Barnett, J.F., Jr., Giknis, M.L.A., and Clifford, C.B., Postnatal growth, development and behavioral/functional evaluation in Crl:CD (SD)IGS BR rats. Available at: www.criver.com, accessed March 2006.
7. Swanson, L.W., Mapping the human brain: past, present, and future, *Trends Neurosci*, 18, 471, 1995.
8. Sharma, A.K. et al., Development of a percutaneous cerebrospinal fluid collection technique in F-344 rats and evaluation of cell counts and total protein concentrations, *Toxicol Pathol,* 34(4), 393, 2006.

Index

A

B

C

D

K

L

M

N

O

P

R

S

T

U

V

W

Z

For Product Safety Concerns and Information please contact our EU representative GPSR@taylorandfrancis.com Taylor & Francis Verlag GmbH, Kaufingerstraße 24, 80331 München, Germany

Batch number: 10722641

Printed by Printforce, the Netherlands